STATA BASE REFERENCE MANUAL
VOLUME 4
S–Z
RELEASE 12

A Stata Press Publication
StataCorp LP
College Station, Texas

The suggested citation for this software is

StataCorp. 2011. *Stata: Release 12*. Statistical Software. College Station, TX: StataCorp LP.

Title

> **sampsi** — Sample size and power for means and proportions

Syntax

> sampsi $\#_1$ $\#_2$ $\left[\ ,\ options\right]$

options	Description
Main	
<u>onesample</u>	one-sample test; default is two-sample
sd1(*#*)	standard deviation of sample 1
sd2(*#*)	standard deviation of sample 2
Options	
<u>a</u>lpha(*#*)	significance level of test; default is alpha(0.05)
<u>p</u>ower(*#*)	power of test; default is power(0.90)
<u>n</u>1(*#*)	size of sample 1
<u>n</u>2(*#*)	size of sample 2
<u>r</u>atio(*#*)	ratio of sample sizes; default is ratio(1)
pre(*#*)	number of baseline measurements; default is pre(0)
post(*#*)	number of follow-up measurements; default is post(1)
<u>nocont</u>inuity	do not use continuity correction for two-sample test on proportions
r0(*#*)	correlation between baseline measurements; default is r0()=r1()
r1(*#*)	correlation between follow-up measurements
r01(*#*)	correlation between baseline and follow-up measurements
<u>onesided</u>	one-sided test; default is two-sided
<u>m</u>ethod(*method*)	analysis method where *method* is post, change, ancova, or all; default is method(all)

Menu

sampsi

Statistics > Power and sample size > Tests of means and proportions

sampsi with repeated measures

Statistics > Power and sample size > Tests of means with repeated measures

Description

sampsi estimates require sample size or power of tests for studies comparing two groups. sampsi can be used when comparing means or proportions for simple studies where only one measurement of the outcome is planned and for comparing mean summary statistics for more complex studies where repeated measurements of the outcome on each experimental unit are planned.

If n1(#) or n2(#) is specified, sampsi computes power; otherwise, it computes sample size. For simple studies, if sd1(#) or sd2(#) is specified, sampsi assumes a comparison of means; otherwise, it assumes a comparison of proportions. For repeated measurements, sd1(#) or sd2(#) must be specified. sampsi is an immediate command; all its arguments are numbers; see [U] **19 Immediate commands**.

Options

_____ ⌐Main⌐ _____

onesample indicates a one-sample test. The default is two-sample.

sd1(#) and sd2(#) are the standard deviations of population 1 and population 2, respectively. One or both must be specified when doing a comparison of means. When the onesample option is used, sd1(#) is the standard deviation of the single sample (it can be abbreviated as sd(#)). If only one of sd1(#) or sd2(#) is specified, sampsi assumes that sd1() = sd2(). If neither sd1(#) nor sd2(#) is specified, sampsi assumes a test of proportions. For repeated measurements, sd1(#) or sd2(#) must be specified.

_____ ⌐Options⌐ _____

alpha(#) is the significance level of the test. The default is alpha(0.05) unless set level has been used to reset the default significance level for confidence intervals. If a set level $\#_{\text{lev}}$ command has been issued, the default value is alpha(1 − $\#_{\text{lev}}$/100). See [R] **level**.

power(#) = $1 - \beta$ is the power of the test. The default is power(0.90).

n1(#) and n2(#) are the sizes of sample 1 and sample 2, respectively. One or both must be specified when computing power. If neither n1(#) nor n2(#) is specified, sampsi computes sample size. When the onesample option is used, n1(#) is the size of the single sample (it can be abbreviated as n(#)). If only one of n1(#) or n2(#) is specified, the unspecified one is computed using the formula ratio = n2()/n1().

ratio(#) is the ratio of sample sizes for two-sample tests: ratio = n2()/n1(). The default is ratio(1).

pre(#) specifies the number of baseline measurements (prerandomization) planned in a repeated-measure study. The default is pre(0).

post(#) specifies the number of follow-up measurements (postrandomization) planned in a repeated-measure study. The default is post(1).

nocontinuity requests power and sample size calculations without the continuity correction for the two-sample test on proportions. If not specified, the continuity correction is used.

r0(#) specifies the correlation between baseline measurements in a repeated-measure study. If r0(#) is not specified, sampsi assumes that r0() = r1().

r1(#) specifies the correlation between follow-up measurements in a repeated-measure study. For a repeated-measure study, either r1(#) or r01(#) must be specified. If r1(#) is not specified, sampsi assumes that r1() = r01().

r01(#) specifies the correlation between baseline and follow-up measurements in a repeated-measure study. For a repeated-measure study, either r01(#) or r1(#) must be specified. If r01(#) is not specified, sampsi assumes that r01() = r1().

onesided indicates a one-sided test. The default is two-sided.

method(post | change | ancova | all) specifies the analysis method to be used with repeated measures. change and ancova can be used only if baseline measurements are planned. The default is method(all), which means to use all three methods. Each method is described in *Methods and formulas*.

Remarks

Remarks are presented under the following headings:

Studies with one measurement of the outcome
Two-sample test of equality of means
One-sample test of mean
Two-sample test of equality of proportions
One-sample test of proportion
Clinical trials with repeated measures

Studies with one measurement of the outcome

For simple studies, where only one measurement of the outcome is planned, sampsi computes sample size or power for four types of tests:

1. Two-sample comparison of mean μ_1 of population 1 with mean μ_2 of population 2. The null hypothesis is $\mu_1 = \mu_2$, and normality is assumed. The postulated values of the means are $\mu_1 = \#_1$ and $\mu_2 = \#_2$, and the postulated standard deviations are sd1(#) and sd2(#).

2. One-sample comparison of the mean μ of a population with a hypothesized value μ_0. The null hypothesis is $\mu = \mu_0$, and normality is assumed. The first argument, $\#_1$, to sampsi is μ_0. The second argument, $\#_2$, is the postulated value of μ, that is, the alternative hypothesis is $\mu = \#_2$. The postulated standard deviation is sd1(#). To get this test, the onesample option must be specified.

3. Two-sample comparison of proportion p_1 with proportion p_2. The null hypothesis is $p_1 = p_2$, and the postulated values are $p_1 = \#_1$ and $p_2 = \#_2$.

4. One-sample comparison of a proportion p with a hypothesized value p_0. The null hypothesis is $p = p_0$, where $p_0 = \#_1$. The alternative hypothesis is $p = \#_2$. To get this test, the onesample option must be specified.

Examples of these follow.

Two-sample test of equality of means

▷ Example 1

We are doing a study of the relationship of oral contraceptives (OC) and blood pressure (BP) level for women ages 35–39 (Rosner 2006, 331–333). From a pilot study, it was determined that the mean and standard deviation of BP of OC users were 132.86 and 15.34, respectively. The mean and standard deviation of OC nonusers in the pilot study were found to be 127.44 and 18.23. Because it is easier to find OC nonusers than users, we decide that n_2, the size of the sample of OC nonusers, should be twice n_1, the size of the sample of OC users; that is, $r = n_2/n_1 = 2$. To compute the sample sizes for $\alpha = 0.05$ (two-sided) and the power of 0.80, we issue the following command:

```
. sampsi 132.86 127.44, p(0.8) r(2) sd1(15.34) sd2(18.23)

Estimated sample size for two-sample comparison of means

Test Ho: m1 = m2, where m1 is the mean in population 1
                   and m2 is the mean in population 2
Assumptions:

           alpha =   0.0500  (two-sided)
           power =   0.8000
              m1 =   132.86
              m2 =   127.44
             sd1 =    15.34
             sd2 =    18.23
           n2/n1 =     2.00

Estimated required sample sizes:

              n1 =       108
              n2 =       216
```

We now find out that we have only enough money to study 100 subjects from each group. We can compute the power for $n_1 = n_2 = 100$ by typing

```
. sampsi 132.86 127.44, n1(100) sd1(15.34) sd2(18.23)

Estimated power for two-sample comparison of means

Test Ho: m1 = m2, where m1 is the mean in population 1
                   and m2 is the mean in population 2
Assumptions:

           alpha =   0.0500  (two-sided)
              m1 =   132.86
              m2 =   127.44
             sd1 =    15.34
             sd2 =    18.23
  sample size n1 =       100
              n2 =       100
           n2/n1 =     1.00

Estimated power:

           power =   0.6236
```

We did not have to specify n2(#) or ratio(#) because ratio(1) is the default.

◁

One-sample test of mean

▷ Example 2

Suppose that we wish to test the effects of a low-fat diet on serum cholesterol levels. We will measure the difference in cholesterol levels for each subject before and after being on the diet. Because there is only one group of subjects, all on the diet, this is a one-sample test, and we must use the onesample option with sampsi.

Our null hypothesis is that the mean of individual differences in cholesterol level will be zero; that is, $\mu = 0 \, \text{mg}/100 \, \text{mL}$. If the effect of the diet is as large as a mean difference of $-10 \, \text{mg}/100 \, \text{mL}$, then we wish to have power of 0.95 for rejecting the null hypothesis. Because we expect a reduction in level, we want to use a one-sided test with $\alpha = 0.025$. From past studies, we estimate that the standard deviation of the difference in cholesterol levels will be about $20 \, \text{mg}/100 \, \text{mL}$. To compute the required sample size, we type

```
. sampsi 0 -10, sd(20) onesam a(0.025) onesided p(0.95)
```

Estimated sample size for one-sample comparison of mean
  to hypothesized value

Test Ho: m =   0, where m is the mean in the population

Assumptions:

```
         alpha =   0.0250  (one-sided)
         power =   0.9500
 alternative m =      -10
            sd =       20
```

Estimated required sample size:

```
           n =       52
```

We decide to conduct the study with $n = 60$ subjects, and we wonder what the power will be at a one-sided significance level of $\alpha = 0.01$:

```
. sampsi 0 -10, sd(20) onesam a(0.01) onesided n(60)
```

Estimated power for one-sample comparison of mean
  to hypothesized value

Test Ho: m =   0, where m is the mean in the population

Assumptions:

```
         alpha =   0.0100  (one-sided)
 alternative m =      -10
            sd =       20
 sample size n =       60
```

Estimated power:

```
         power =   0.9390
```

◁

Two-sample test of equality of proportions

▷ Example 3

We want to conduct a survey on people's opinions of the president's performance. Specifically, we want to determine whether members of the president's party have a different opinion from people with another party affiliation. Using past surveys as a guide, we estimate that only 25% of members of the president's party will say that the president is doing a poor job, whereas 40% of members of other parties will rate the president's performance as poor. We compute the required sample sizes for $\alpha = 0.05$ (two-sided) and the power of 0.90 by typing

```
. sampsi 0.25 0.4
```

Estimated sample size for two-sample comparison of proportions

Test Ho: p1 = p2, where p1 is the proportion in population 1
                  and p2 is the proportion in population 2

Assumptions:

```
         alpha =   0.0500  (two-sided)
         power =   0.9000
            p1 =   0.2500
            p2 =   0.4000
         n2/n1 =   1.00
```

Estimated required sample sizes:

```
          n1 =      216
          n2 =      216
```

To compute the power for a survey with a sample of $n_1 = 300$ members of the president's party and a sample of $n_2 = 150$ members of other parties, we type

```
. sampsi 0.25 0.4, n1(300) r(0.5)

Estimated power for two-sample comparison of proportions

Test Ho: p1 = p2, where p1 is the proportion in population 1
                    and p2 is the proportion in population 2

Assumptions:

                 alpha =    0.0500  (two-sided)
                    p1 =    0.2500
                    p2 =    0.4000
        sample size n1 =      300
                    n2 =      150
                 n2/n1 =      0.50

Estimated power:

                 power =    0.8790
```

◁

One-sample test of proportion

▷ Example 4

Someone claims that females are more likely than males to study French. Our null hypothesis is that the proportion of female French students is 0.5. We wish to compute the sample size that will give us 80% power to reject the null hypothesis if the true proportion of female French students is 0.75:

```
. sampsi 0.5 0.75, power(0.8) onesample

Estimated sample size for one-sample comparison of proportion
  to hypothesized value

Test Ho: p = 0.5000, where p is the proportion in the population

Assumptions:

                 alpha =    0.0500  (two-sided)
                 power =    0.8000
         alternative p =    0.7500

Estimated required sample size:

                     n =       29
```

What is the power if the true proportion of female French students is only 0.6 and the biggest sample of French students we can survey is $n = 200$?

```
. sampsi 0.5 0.6, n(200) onesample

Estimated power for one-sample comparison of proportion
  to hypothesized value

Test Ho: p = 0.5000, where p is the proportion in the population

Assumptions:

                 alpha =    0.0500  (two-sided)
         alternative p =    0.6000
        sample size n =      200

Estimated power:

                 power =    0.8123
```

◁

❑ Technical note

r(warning) is saved only for power calculations for one- and two-sample tests on proportions. If sample sizes are not large enough (Tamhane and Dunlop 2000, 300, 307) for sample proportions to be approximately normally distributed, r(warning) is set to 1. Otherwise, a note is displayed in the output, and r(warning) is set to 0.

❑

Clinical trials with repeated measures

In randomized controlled trials (RCTs), when comparing a standard treatment with an experimental therapy, it is not unusual for the study design to allow for repeated measurements of the outcome. Typically, one or more measurements are taken at baseline immediately before randomization, and additional measurements are taken at regular intervals during follow-up. Depending on the analysis method planned and the correlations between measurements at different time points, there can be a great increase in efficiency (variance reduction) from such designs over a simple study with only one measurement of the outcome.

Frison and Pocock (1992) discuss three methods used in RCTs to compare two treatments by using a continuous outcome measured at different times on each patient.

Posttreatment means (POST) uses the mean of each patient's follow-up measurements as the summary measurement. It compares the two groups by using a simple t test. This method ignores any baseline measurements.

Mean changes (CHANGE) uses each patient's difference between the mean of the follow-up measurements and the mean of baseline measurements as the summary measurement. It compares the two groups by using a simple t test.

Analysis of covariance (ANCOVA) uses the mean baseline measurement for each patient as a covariate in a linear model for treatment comparisons of follow-up means.

method() specifies which of these three analyses is planned to be used. sampsi will calculate the decrease in variance of the estimate of treatment effect from the number of measurements at baseline, the number of measurements during follow-up, and the correlations between measurements at different times, and use the calculation to estimate power or sample size, or both.

▷ Example 5

We are designing a clinical trial comparing a new medication for the treatment of angina to a placebo. We are planning on performing an exercise stress test on each patient four times during the study: once at time of treatment randomization and three more times at 4, 6, and 8 weeks after randomization. From each test, we will measure the time in seconds from the beginning of the test until the patient is unable to continue because of angina pain. From a previous pilot study, we estimated the means (sd#s) for the new drug and the placebo group to be 498 seconds (20.2) and 485 seconds (19.5), respectively, and an overall correlation at follow-up of 0.7. We will analyze these data by comparing each patient's difference between the mean of posttreatment measurements and the mean of baseline measurements, that is, the change method. To compute the number of subjects needed for allocation to each treatment group for $\alpha = 0.05$ (two-sided) and power of 90%, we issue the following command:

```
. sampsi 498 485, sd1(20.2) sd2(19.5) method(change) pre(1) post(3) r1(.7)

Estimated sample size for two samples with repeated measures

Assumptions:
                                             alpha =    0.0500  (two-sided)
                                             power =    0.9000
                                                m1 =       498
                                                m2 =       485
                                               sd1 =      20.2
                                               sd2 =      19.5
                                             n2/n1 =      1.00
                      number of follow-up measurements =       3
      correlation between follow-up measurements =   0.700
             number of baseline measurements =       1
    correlation between baseline & follow-up =   0.700

Method: CHANGE
     relative efficiency =      2.500
         adjustment to sd =      0.632
             adjusted sd1 =     12.776
             adjusted sd2 =     12.333

Estimated required sample sizes:
                   n1 =        20
                   n2 =        20
```

The output from `sampsi` for repeated measurements includes the specified parameters used to estimate the sample sizes or power, the relative efficiency of the design, and the adjustment to the standard deviation. These last two are the inverse and the square root of the calculated improvement in the variance compared with a similar study where only one measurement is planned.

We see that we need to allocate 20 subjects to each treatment group. Assume that we have funds to enroll only 30 patients into our study. If we randomly assigned 15 patients to each treatment group, what would be the expected power of our study, assuming that all other parameters remain the same?

```
. sampsi 498 485, sd1(20.2) sd2(19.5) meth(change) pre(1) post(3) r1(.7) n1(15)
> n2(15)

Estimated power for two samples with repeated measures

Assumptions:
                                             alpha =    0.0500  (two-sided)
                                                m1 =       498
                                                m2 =       485
                                               sd1 =      20.2
                                               sd2 =      19.5
                                      sample size n1 =        15
                                                n2 =        15
                                             n2/n1 =      1.00
                      number of follow-up measurements =       3
      correlation between follow-up measurements =   0.700
             number of baseline measurements =       1
    correlation between baseline & follow-up =   0.700
Method: CHANGE
     relative efficiency =      2.500
         adjustment to sd =      0.632
             adjusted sd1 =     12.776
             adjusted sd2 =     12.333

Estimated power:
                power =      0.809
```

If we enroll 30 patients into our study instead of the recommended 40, the power of the study decreases from 90% to approximately 81%.

◁

Saved results

sampsi saves the following in r():

Scalars

r(N_1)	sample size n_1
r(N_2)	sample size n_2
r(power)	power
r(adj)	adjustment to the SE
r(warning)	0 if assumptions are satisfied and 1 otherwise

Methods and formulas

sampsi is implemented as an ado-file.

In the following formulas, α is the significance level, $1 - \beta$ is the power, $z_{1-\alpha/2}$ is the $(1 - \alpha/2)$ quantile of the normal distribution, and $r = n_2/n_1$ is the ratio of sample sizes. The formulas below are for two-sided tests. The formulas for one-sided tests can be obtained by replacing $z_{1-\alpha/2}$ with $z_{1-\alpha}$.

1. The required sample sizes for a two-sample test of equality of means (assuming normality) are

$$n_1 = \frac{(\sigma_1^2 + \sigma_2^2/r)(z_{1-\alpha/2} + z_{1-\beta})^2}{(\mu_1 - \mu_2)^2}$$

and $n_2 = rn_1$ (Rosner 2006, 332).

2. For a one-sample test of a mean where the null hypothesis is $\mu = \mu_0$ and the alternative hypothesis is $\mu = \mu_A$, the required sample size (assuming normality) is

$$n = \left\{ \frac{(z_{1-\alpha/2} + z_{1-\beta})\sigma}{\mu_A - \mu_0} \right\}^2$$

(Pagano and Gauvreau 2000, 247–248).

3. The required sample sizes for a two-sample test of equality of proportions (using a normal approximation with a continuity correction) are

$$n_1 = \frac{n'}{4} \left[1 + \left\{ 1 + \frac{2(r+1)}{n'r\,|p_1 - p_2|} \right\}^{1/2} \right]^2$$

$$n_2 = rn_1$$

where

$$n' = \frac{\left[z_{1-\alpha/2}\{(r+1)\overline{pq}\}^{1/2} + z_{1-\beta}\left(rp_1q_1 + p_2q_2\right)^{1/2} \right]^2}{r(p_1 - p_2)^2}$$

and $\overline{p} = (p_1 + rp_2)/(r + 1)$ and $\overline{q} = 1 - \overline{p}$ (Fleiss, Levin, and Paik 2003, 76).

Without a continuity correction, the sample sizes are

$$n_1 = n'$$

$$n_2 = rn_1$$

where n' is defined above.

4. For a one-sample test of proportion where the null hypothesis is $p = p_0$ and the alternative hypothesis is $p = p_A$, the required sample size (using a normal approximation) is

$$n = \left[\frac{z_{1-\alpha/2}\{p_0(1 - p_0)\}^{1/2} + z_{1-\beta}\{p_A(1 - p_A)\}^{1/2}}{p_A - p_0} \right]^2$$

(Pagano and Gauvreau 2000, 332).

5. For repeated measurements, Frison and Pocock (1992) discuss three methods for use in randomized clinical trials to compare two treatments using a continuous outcome measured at different times on each patient. Each uses the average of baseline measurements, $\overline{x}_0$, and follow-up measurements, $\overline{x}_1$:

POST outcome is $\overline{x}_1$, where the analysis is by simple t test.

CHANGE outcome is $\overline{x}_1 - \overline{x}_0$, where the analysis is by simple t test.

ANCOVA outcome is $\overline{x}_1 - \beta\overline{x}_0$, where the β is estimated by analysis of covariance, correcting for the average at baseline.

ANCOVA will always be the most efficient of the three approaches. β is set so that $\beta\overline{x}_0$ accounts for the largest possible variation of $\overline{x}_1$.

For a study with one measurement each at baseline and follow-up, CHANGE will be more efficient than POST, provided that the correlation between measurements at baseline and measurements at follow-up is more than 0.5. POST ignores all baseline measurements, which tends to make it unpopular. CHANGE is the method most commonly used. With more than one baseline measurement, CHANGE and ANCOVA tend to produce similar sample sizes and power.

The improvements in variance of the estimate of treatment effect over a study with only one measurement depend on the number of measurements p at baseline; the number of measurements during follow-up; and the correlations between measurements at baseline $\overline{\rho}_{\mathrm{pre}}$, between measurements at follow-up $\overline{\rho}_{\mathrm{post}}$, and between measurements at baseline and measurements at follow-up $\overline{\rho}_{\mathrm{mix}}$. The improvements in variance for the POST method are given by

$$\frac{1 + (r - 1)\overline{\rho}_{\mathrm{post}}}{r}$$

for the CHANGE method by

$$\frac{1 + (r - 1)\overline{\rho}_{\mathrm{post}}}{r} + \frac{1 + (p - 1)\overline{\rho}_{\mathrm{pre}}}{p} - 2\overline{\rho}_{\mathrm{mix}}$$

and for the ANCOVA method by

$$\frac{1 + (r - 1)\overline{\rho}_{\mathrm{post}}}{r} - \frac{\overline{\rho}_{\mathrm{mix}}^2 p}{1 + (p - 1)\overline{\rho}_{\mathrm{pre}}}$$

Often the three correlations are assumed equal. In data from several trials, Frison and Pocock found that $\overline{\rho}_{\mathrm{pre}}$ and $\overline{\rho}_{\mathrm{post}}$ typically had values around 0.7, whereas $\overline{\rho}_{\mathrm{mix}}$ was nearer 0.5. This finding is consistent with the common finding that measurements closer in time are more strongly correlated.

Power calculations are based on estimates of one variance at all time points.

Acknowledgments

`sampsi` is based on the `sampsiz` command written by Joseph Hilbe of Arizona State University (Hilbe 1993). Paul Seed of Maternal and Fetal Research Unit, St. Thomas' Hospital (Seed 1997, 1998), expanded the command to allow for repeated measurements.

References

Fleiss, J. L., B. Levin, and M. C. Paik. 2003. *Statistical Methods for Rates and Proportions*. 3rd ed. New York: Wiley.

Frison, L., and S. J. Pocock. 1992. Repeated measures in clinical trials: Analysis using mean summary statistics and its implications for design. *Statistics in Medicine* 11: 1685–1704.

Hilbe, J. M. 1993. sg15: Sample size determination for means and proportions. *Stata Technical Bulletin* 11: 17–20. Reprinted in *Stata Technical Bulletin Reprints*, vol. 2, pp. 145–149. College Station, TX: Stata Press.

Pagano, M., and K. Gauvreau. 2000. *Principles of Biostatistics*. 2nd ed. Belmont, CA: Duxbury.

Rosner, B. 2006. *Fundamentals of Biostatistics*. 6th ed. Belmont, CA: Duxbury.

Royston, P., and A. Babiker. 2002. A menu-driven facility for complex sample size calculation in randomized controlled trials with a survival or a binary outcome. *Stata Journal* 2: 151–163.

Seed, P. T. 1997. sbe18: Sample size calculations for clinical trials with repeated measures data. *Stata Technical Bulletin* 40: 16–18. Reprinted in *Stata Technical Bulletin Reprints*, vol. 7, pp. 121–125. College Station, TX: Stata Press.

———. 1998. sbe18.1: Update of sampsi. *Stata Technical Bulletin* 45: 21. Reprinted in *Stata Technical Bulletin Reprints*, vol. 8, p. 84. College Station, TX: Stata Press.

Tamhane, A. C., and D. D. Dunlop. 2000. *Statistics and Data Analysis: From Elementary to Intermediate*. Upper Saddle River, NJ: Prentice Hall.

Also see

[ST] **stpower** — Sample-size, power, and effect-size determination for survival analysis

Title

saved results — Saved results

Syntax

List results from general commands, stored in r()

> <u>retu</u>rn <u>li</u>st [, all]

List results from estimation commands, stored in e()

> <u>ere</u>turn <u>li</u>st [, all]

List results from parsing commands, stored in s()

> <u>sret</u>urn <u>li</u>st

Description

Results of calculations are saved by many Stata commands so that they can be easily accessed and substituted into later commands.

return list lists results stored in r().

ereturn list lists results stored in e().

sreturn list lists results stored in s().

This entry discusses using saved results. Programmers wishing to save results should see [P] **return** and [P] **ereturn**.

Option

all is for use with return list and ereturn list. all specifies that hidden and historical saved results be listed along with the usual saved results. This option is seldom used. See *Using hidden and historical saved results* and *Programming hidden and historical saved results* under *Remarks* of [P] **return** for more information. These sections are written in terms of return list, but everything said there applies equally to ereturn list.

all is not allowed with sreturn list because s() does not allow hidden or historical results.

Remarks

Stata commands are classified as being

r-class	general commands that save results in r()
e-class	estimation commands that save results in e()
s-class	parsing commands that save results in s()
n-class	commands that do not save in r(), e(), or s()

There is also a c-class, c(), containing the values of system parameters and settings, along with certain constants, such as the value of pi; see [P] **creturn**. A program, however, cannot be c-class.

You can look at the *Saved results* section of the manual entry of a command to determine whether it is r-, e-, s-, or n-class, but it is easy enough to guess.

Commands producing statistical results are either r-class or e-class. They are e-class if they present estimation results and r-class otherwise. s-class is a class used by programmers and is primarily used in subprograms performing parsing. n-class commands explicitly state where the result is to go. For instance, generate and replace are n-class because their syntax is generate *varname* = ... and replace *varname* = ...

After executing a command, you can type return list, ereturn list, or sreturn list to see what has been saved.

▷ Example 1

```
. use http://www.stata-press.com/data/r12/auto4
(1978 Automobile Data)

. describe
Contains data from http://www.stata-press.com/data/r12/auto4.dta
  obs:           74                         1978 Automobile Data
 vars:            6                         6 Apr 2011 00:20
 size:        2,072

              storage   display    value
variable name   type    format     label     variable label

price           int     %8.0gc               Price
weight          int     %8.0gc               Weight (lbs.)
mpg             int     %8.0g                Mileage (mpg)
make            str18   %-18s                Make and Model
length          int     %8.0g                Length (in.)
rep78           int     %8.0g                Repair Record 1978

Sorted by:

. return list
scalars:
            r(changed) =  0
              r(width) =  28
                  r(k) =  6
                  r(N) =  74
```

To view all saved results, including those that are historical or hidden, specify the all option.

```
. return list, all
scalars:
            r(changed) =  0
              r(width) =  28
                  r(k) =  6
                  r(N) =  74
Historical; used before Stata 12, may exist only under version control
scalars:
           r(widthmax) =  1048576
             r(k_max) =  2048
             r(N_max) =  2147483647
```

r(widthmax), r(k_max), and r(N_max) are historical saved results. They are no longer relevant because Stata dynamically adjusts memory beginning with Stata 12.

◁

❑ Technical note

In the above example, we stated that r(widthmax) and r(N_max) are no longer relevant. In fact, they are not useful. Stata no longer has a fixed memory size, so the methods used to calculate r(widthmax) and r(N_max) are no longer appropriate.

❑

▷ Example 2

You can use saved results in expressions.

```
. summarize mpg
    Variable |       Obs        Mean    Std. Dev.        Min         Max
-------------+--------------------------------------------------------
         mpg |        74     21.2973    5.785503         12          41
. return list
scalars:
                  r(N) =  74
              r(sum_w) =  74
               r(mean) =  21.2972972972973
                r(Var) =  33.47204738985561
                 r(sd) =  5.785503209735141
                r(min) =  12
                r(max) =  41
                r(sum) =  1576
. generate double mpgstd = (mpg-r(mean))/r(sd)
. summarize mpgstd
    Variable |       Obs        Mean    Std. Dev.        Min         Max
-------------+--------------------------------------------------------
      mpgstd |        74    -1.64e-16           1   -1.606999     3.40553
```

Be careful to use results stored in r() soon because they will be replaced the next time you execute another r-class command. For instance, although r(mean) was 21.3 (approximately) after summarize mpg, it is −1.64e–16 now because you just ran summarize with mpgstd.

◁

▷ Example 3

e-class is really no different from r-class, except for where results are stored and that, when an estimation command stores results, it tends to store a lot of them:

```
. regress mpg weight length
  (output omitted )
. ereturn list
scalars:
                  e(N) =  74
               e(df_m) =  2
               e(df_r) =  71
                  e(F) =  69.34050004300227
                 e(r2) =  .6613903979336323
               e(rmse) =  3.413681741382589
                e(mss) =  1616.08062422659
                e(rss) =  827.3788352328695
               e(r2_a) =  .6518520992838754
                 e(ll) =  -194.3267619410807
               e(ll_0) =  -234.3943376482347
               e(rank) =  3
```

```
macros:
              e(cmdline) : "regress mpg weight length"
                e(title) : "Linear regression"
            e(marginsok) : "XB default"
                  e(vce) : "ols"
               e(depvar) : "mpg"
                  e(cmd) : "regress"
           e(properties) : "b V"
              e(predict) : "regres_p"
                e(model) : "ols"
            e(estat_cmd) : "regress_estat"
matrices:
                    e(b) :  1 x 3
                    e(V) :  3 x 3
functions:
                e(sample)
```

These e-class results will stick around until you run another estimation command. Typing return
list and ereturn list is the easy way to find out what a command stores.

◁

Both r- and e-class results come in four flavors: scalars, macros, matrices, and functions. (s-class
results come in only one flavor—macros—and as earlier noted, s-class is used solely by programmers,
so ignore it.)

Scalars are just that—numbers by any other name. You can subsequently refer to r(mean) or
e(rmse) in numeric expressions and obtain the result to full precision.

Macros are strings. For instance, e(depvar) contains "mpg". You can refer to it, too, in subsequent
expressions, but really that would be of most use to programmers, who will refer to it using constructs
like "`e(depvar)'". In any case, macros are macros, and you obtain their contents just as you
would a local macro, by enclosing their name in single quotes. The name here is the full name, so
`e(depvar)' is mpg.

Matrices are matrices, and all estimation commands store e(b) and e(V) containing the coefficient
vector and variance–covariance matrix of the estimates (VCE).

Functions are saved by e-class commands only, and the only function existing is e(sample).
e(sample) evaluates to 1 (meaning true) if the observation was used in the previous estimation and
to 0 (meaning false) otherwise.

❑ Technical note

Say that some command set r(scalar) and r(macro), the first being stored as a scalar and
the second as a macro. In theory, in subsequent use you are supposed to refer to r(scalar) and
`r(macro)'. In fact, however, you can refer to either one with or without quotes, so you could refer
to `r(scalar)' and r(macro). Programmers sometimes do this.

When you refer to r(scalar), you are referring to the full double-precision saved result. Think
of r(scalar) without quotes as a function returning the value of the saved result scalar. When you
refer to r(scalar) in quotes, Stata understands `r(scalar)' to mean "substitute the printed result
of evaluating r(scalar)". Pretend that r(scalar) equals the number 23. Then `r(scalar)' is
23, the character 2 followed by 3.

Referring to r(scalar) in quotes is sometimes useful. Say that you want to use the immediate
command ci with r(scalar). The immediate command ci requires its arguments to be numbers—
numeric literals in programmer's jargon—and it will not take an expression. Thus you could not type

'ci r(scalar) ...'. You could, however, type 'ci 'r(scalar)' ...' because 'r(scalar)' is just a numeric literal.

For r(macro), you are supposed to refer to it in quotes: 'r(macro)'. If, however, you omit the quotes in an expression context, Stata evaluates the macro and then pretends that it is the result of function-returning-string. There are side effects of this, the most important being that the result is trimmed to 80 characters.

Referring to r(macro) without quotes is never a good idea; the feature was included merely for completeness.

You can even refer to r(matrix) in quotes (assume that r(matrix) is a matrix). 'r(matrix)' does not result in the matrix being substituted; it returns the word matrix. Programmers sometimes find that useful.

❏

References

Jann, B. 2005. Making regression tables from stored estimates. *Stata Journal* 5: 288–308.

——. 2007. Making regression tables simplified. *Stata Journal* 7: 227–244.

Also see

[P] **ereturn** — Post the estimation results

[P] **return** — Return saved results

[U] **18.8 Accessing results calculated by other programs**

[U] **18.9 Accessing results calculated by estimation commands**

Title

> **scobit** — Skewed logistic regression

Syntax

scobit *depvar* [*indepvars*] [*if*] [*in*] [*weight*] [, *options*]

options	Description
Model	
<u>nocon</u>stant	suppress constant term
<u>off</u>set(*varname*)	include *varname* in model with coefficient constrained to 1
asis	retain perfect predictor variables
<u>constraints</u>(*constraints*)	apply specified linear constraints
<u>coll</u>inear	keep collinear variables
SE/Robust	
vce(*vcetype*)	*vcetype* may be oim, <u>r</u>obust, <u>c</u>luster *clustvar*, opg, <u>boot</u>strap, or <u>jack</u>knife
Reporting	
<u>level</u>(#)	set confidence level; default is level(95)
or	report odds ratios
<u>nocns</u>report	do not display constraints
display_options	control column formats, row spacing, line width, and display of omitted variables and base and empty cells
Maximization	
maximize_options	control the maximization process
<u>coefl</u>egend	display legend instead of statistics

indepvars may contain factor variables; see [U] **11.4.3 Factor variables**.
bootstrap, by, jackknife, nestreg, rolling, statsby, stepwise, and svy are allowed; see
[U] **11.1.10 Prefix commands**.
Weights are not allowed with the bootstrap prefix; see [R] **bootstrap**.
vce() and weights are not allowed with the svy prefix; see [SVY] **svy**.
fweights, iweights, and pweights are allowed; see [U] **11.1.6 weight**.
coeflegend does not appear in the dialog box.
See [U] **20 Estimation and postestimation commands** for more capabilities of estimation commands.

Menu

Statistics > Binary outcomes > Skewed logit regression

Description

scobit fits a maximum-likelihood skewed logit model.

See [R] **logistic** for a list of related estimation commands.

Options

⌐ Model ⌐

noconstant, offset(*varname*), constraints(*constraints*), collinear; see [R] **estimation options**.

asis forces retention of perfect predictor variables and their associated perfectly predicted observations and may produce instabilities in maximization; see [R] **probit**.

⌐ SE/Robust ⌐

vce(*vcetype*) specifies the type of standard error reported, which includes types that are derived from asymptotic theory, that are robust to some kinds of misspecification, that allow for intragroup correlation, and that use bootstrap or jackknife methods; see [R] ***vce_option***.

⌐ Reporting ⌐

level(*#*); see [R] **estimation options**.

or reports the estimated coefficients transformed to odds ratios, that is, e^b rather than b. Standard errors and confidence intervals are similarly transformed. This option affects how results are displayed, not how they are estimated. or may be specified at estimation or when replaying previously estimated results.

nocnsreport; see [R] **estimation options**.

display_options: noomitted, vsquish, noemptycells, baselevels, allbaselevels, cformat(% *fmt*), pformat(% *fmt*), sformat(% *fmt*), and nolstretch; see [R] **estimation options**.

⌐ Maximization ⌐

maximize_options: difficult, technique(*algorithm_spec*), iterate(*#*), [no]log, trace, gradient, showstep, hessian, showtolerance, tolerance(*#*), ltolerance(*#*), nrtolerance(*#*), nonrtolerance, and from(*init_specs*); see [R] **maximize**. These options are seldom used.

Setting the optimization type to technique(bhhh) resets the default *vcetype* to vce(opg).

The following option is available with scobit but is not shown in the dialog box:

coeflegend; see [R] **estimation options**.

Remarks

Remarks are presented under the following headings:

> *Skewed logistic model*
> *Robust standard errors*

Skewed logistic model

scobit fits maximum likelihood models with dichotomous dependent variables coded as 0/1 (or, more precisely, coded as 0 and not 0).

▷ Example 1

We have data on the make, weight, and mileage rating of 22 foreign and 52 domestic automobiles. We wish to fit a model explaining whether a car is foreign based on its mileage. Here is an overview of our data:

```
. use http://www.stata-press.com/data/r12/auto
(1978 Automobile Data)

. keep make mpg weight foreign

. describe
Contains data from http://www.stata-press.com/data/r12/auto.dta
  obs:            74                          1978 Automobile Data
  vars:            4                          13 Apr 2011 17:45
  size:         1,702                         (_dta has notes)

              storage   display    value
variable name   type    format     label      variable label

make            str18   %-18s                 Make and Model
mpg             int     %8.0g                 Mileage (mpg)
weight          int     %8.0gc                Weight (lbs.)
foreign         byte    %8.0g      origin     Car type

Sorted by:  foreign
     Note:  dataset has changed since last saved

. inspect foreign
foreign:  Car type                          Number of Observations

                                    Total    Integers    Nonintegers
  |  #                 Negative        -          -           -
  |  #                 Zero           52         52           -
  |  #                 Positive       22         22           -
  |  #                             _____    _____      _____
  |  #     #           Total          74         74           -
  |  #     #           Missing         -
  |_____           _____
  0             1                       74
     (2 unique values)

        foreign is labeled and all values are documented in the label.
```

The variable foreign takes on two unique values, 0 and 1. The value 0 denotes a domestic car, and 1 denotes a foreign car.

The model that we wish to fit is

$$\Pr(\texttt{foreign} = 1) = F(\beta_0 + \beta_1\texttt{mpg})$$

where $F(z) = 1 - 1/\{1 + \exp(z)\}^{\alpha}$.

To fit this model, we type

```
. scobit foreign mpg

Fitting logistic model:
Iteration 0:   log likelihood =  -45.03321
Iteration 1:   log likelihood = -39.380959
Iteration 2:   log likelihood = -39.288802
Iteration 3:   log likelihood =  -39.28864
Iteration 4:   log likelihood =  -39.28864

Fitting full model:
Iteration 0:   log likelihood =  -39.28864
Iteration 1:   log likelihood = -39.286393
Iteration 2:   log likelihood = -39.284415
Iteration 3:   log likelihood = -39.284234
Iteration 4:   log likelihood = -39.284197
Iteration 5:   log likelihood = -39.284196
```

Skewed logistic regression					Number of obs		=	74
					Zero outcomes		=	52
Log likelihood = -39.2842					Nonzero outcomes		=	22

foreign	Coef.	Std. Err.	z	P>|z|	[95% Conf. Interval]	
mpg	.1813879	.2407362	0.75	0.451	-.2904463	.6532222
_cons	-4.274883	1.399305	-3.06	0.002	-7.017471	-1.532295
/lnalpha	-.4450405	3.879885	-0.11	0.909	-8.049476	7.159395
alpha	.6407983	2.486224			.0003193	1286.133

```
Likelihood-ratio test of alpha=1:   chi2(1) =    0.01    Prob > chi2 = 0.9249
Note: likelihood-ratio tests are recommended for inference with scobit models.
```

We find that cars yielding better gas mileage are less likely to be foreign. The likelihood-ratio test at the bottom of the output indicates that the model is not significantly different from a logit model. Therefore, we should use the more parsimonious model.

◁

❑ Technical note

Stata interprets a value of 0 as a negative outcome (failure) and treats all other values (except missing) as positive outcomes (successes). Thus if the dependent variable takes on the values 0 and 1, then 0 is interpreted as failure and 1 as success. If the dependent variable takes on the values 0, 1, and 2, then 0 is still interpreted as failure, but both 1 and 2 are treated as successes.

Formally, when we type scobit y x, Stata fits the model

$$\Pr(y_j \neq 0 \mid \mathbf{x}_j) = 1 - 1 \Big/ \Big\{1 + \exp(\mathbf{x}_j\boldsymbol{\beta})\Big\}^\alpha$$

❑

Robust standard errors

If you specify the vce(robust) option, scobit reports robust standard errors as described in
[U] **20.20 Obtaining robust variance estimates**. For the model of foreign on mpg, the robust
calculation increases the standard error of the coefficient on mpg by around 25%:

```
. scobit foreign mpg, vce(robust) nolog
```

Skewed logistic regression   Number of obs   =   74
                                              Zero outcomes   =   52
Log pseudolikelihood = -39.2842   Nonzero outcomes =   22

foreign	Coef.	Robust Std. Err.	z	P>|z|	[95% Conf. Interval]	
mpg	.1813879	.3028487	0.60	0.549	-.4121847	.7749606
_cons	-4.274883	1.335521	-3.20	0.001	-6.892455	-1.657311
/lnalpha	-.4450405	4.71561	-0.09	0.925	-9.687466	8.797385
alpha	.6407983	3.021755			.0000621	6616.919

Without vce(robust), the standard error for the coefficient on mpg was reported to be 0.241, with
a resulting confidence interval of $[-0.29, 0.65]$.

Specifying the vce(cluster *clustvar*) option relaxes the independence assumption required by
the skewed logit estimator to being just independence between clusters. To demonstrate this, we will
switch to a different dataset.

▷ Example 2

We are studying the unionization of women in the United States and have a dataset with 26,200
observations on 4,434 women between 1970 and 1988. For our purposes, we will use the variables
age (the women were 14–26 in 1968 and the data thus span the age range of 16–46), grade (years
of schooling completed, ranging from 0 to 18), not_smsa (28% of the person-time was spent living
outside an SMSA—standard metropolitan statistical area), south (41% of the person-time was in the
South), and year. Each of these variables is included in the regression as a covariate along with the
interaction between south and year. This interaction, along with the south and year variables, is
specified in the scobit command using factor-variables notation, south##c.year. We also have
variable union. Overall, 22% of the person-time is marked as time under union membership and
44% of these women have belonged to a union.

We fit the following model, ignoring that women are observed an average of 5.9 times each in
these data:

```
. use http://www.stata-press.com/data/r12/union, clear
(NLS Women 14-24 in 1968)
. scobit union age grade not_smsa south##c.year, nrtol(1e-3)
```

(*output omitted*)

Skewed logistic regression

Number of obs =	26200
Zero outcomes =	20389
Nonzero outcomes =	5811

Log likelihood = -13540.61

union	Coef.	Std. Err.	z	P>|z|	[95% Conf. Interval]
age	.0185365	.0043615	4.25	0.000	.0099881 .0270849
grade	.0452803	.0057124	7.93	0.000	.0340842 .0564764
not_smsa	-.1886849	.0317802	-5.94	0.000	-.250973 -.1263968
1.south	-1.422381	.3949298	-3.60	0.000	-2.196429 -.6483327
year	-.0133017	.0049575	-2.68	0.007	-.0230182 -.0035853
south#c.year					
1	.0105663	.0049233	2.15	0.032	.0009168 .0202158
_cons	-10.19247	63.69015	-0.16	0.873	-135.0229 114.6379
/lnalpha	8.972796	63.68825	0.14	0.888	-115.8539 133.7995
alpha	7885.616	502221.1			4.85e-51 1.28e+58

Likelihood-ratio test of alpha=1:   chi2(1) =   3.76   Prob > chi2 = 0.0524
Note: likelihood-ratio tests are recommended for inference with scobit models.

The reported standard errors in this model are probably meaningless. Women are observed repeatedly, so the observations are not independent. Looking at the coefficients, we find a large southern effect against unionization and a different time trend for the south. The vce(cluster *clustvar*) option provides a way to fit this model and obtains correct standard errors:

```
. scobit union age grade not_smsa south##c.year, vce(cluster id) nrtol(1e-3)
```

(*output omitted*)

Skewed logistic regression

Number of obs =	26200
Zero outcomes =	20389
Nonzero outcomes =	5811

Log pseudolikelihood = -13540.61

(Std. Err. adjusted for 4434 clusters in idcode)

union	Coef.	Robust Std. Err.	z	P>|z|	[95% Conf. Interval]
age	.0185365	.0084867	2.18	0.029	.0019029 .0351701
grade	.0452803	.0125764	3.60	0.000	.0206311 .0699296
not_smsa	-.1886849	.0642035	-2.94	0.003	-.3145214 -.0628484
1.south	-1.422381	.5064916	-2.81	0.005	-2.415086 -.4296756
year	-.0133017	.0090621	-1.47	0.142	-.0310632 .0044597
south#c.year					
1	.0105663	.0063172	1.67	0.094	-.0018152 .0229478
_cons	-10.19247	.9458356	-10.78	0.000	-12.04627 -8.338666
/lnalpha	8.972796	.7483321	11.99	0.000	7.506092 10.4395
alpha	7885.616	5901.06			1819.09 34183.54

scobit, vce(cluster *clustvar*) is robust to assumptions about within-cluster correlation. That is, it inefficiently sums within cluster for the standard error calculation rather than attempting to exploit what might be assumed about the within-cluster correlation (as do the xtgee population-averaged models; see [XT] **xtgee**).

◁

❏ Technical note

The scobit model can be difficult to fit because of the functional form. Often it requires many iterations, or the optimizer prints out warning and informative messages during the optimization. For example, without the nrtol(1e-3) option, the model using the union dataset will not converge. See [R] **maximize** for details about the optimizer.

❏

❏ Technical note

The main reason for using scobit rather that logit is that the effects of the regressors on the probability of success are not constrained to be the largest when the probability is 0.5. Rather, the independent variables might show their largest impact when the probability of success is 0.3 or 0.6. This added flexibility results because the scobit function, unlike the logit function, can be skewed and is not constrained to be mirror symmetric about the 0.5 probability of success.

As Nagler (1994) pointed out, the point of maximum impact is constrained under the scobit model to fall within the interval $(0, 1 - e^{(-1)})$ or approximately $(0, 0.63)$. Achen (2002) notes that if we believe the maximum impact to be outside that range, we can instead estimate the "power logit" model by simply reversing the 0s and 1s of our outcome variable and estimating a scobit model on failure, rather than success. We would need to reverse the signs of the coefficients if we wanted to interpret them in terms of impact on success, or we could leave them as they are and interpret them in terms of impact on failure. The important thing to remember is that the scobit model, unlike the logit model, is not invariant to the choice of which result is assigned to success.

❏

Saved results

scobit saves the following in e():

Scalars
e(N)	number of observations
e(k)	number of parameters
e(k_eq)	number of equations in e(b)
e(k_aux)	number of auxiliary parameters
e(k_dv)	number of dependent variables
e(ll)	log likelihood
e(ll_c)	log likelihood, comparison model
e(N_f)	number of failures (zero outcomes)
e(N_s)	number of successes (nonzero outcomes)
e(alpha)	alpha
e(N_clust)	number of clusters
e(chi2)	χ^2
e(chi2_c)	χ^2 for comparison test
e(p)	significance
e(rank)	rank of e(V)
e(ic)	number of iterations
e(rc)	return code
e(converged)	1 if converged, 0 otherwise

Macros
e(cmd)	scobit
e(cmdline)	command as typed
e(depvar)	name of dependent variable
e(wtype)	weight type
e(wexp)	weight expression
e(title)	title in estimation output
e(clustvar)	name of cluster variable
e(offset)	linear offset variable
e(chi2type)	Wald or LR; type of model χ^2 test
e(chi2_ct)	Wald or LR; type of model χ^2 test corresponding to e(chi2_c)
e(vce)	*vcetype* specified in vce()
e(vcetype)	title used to label Std. Err.
e(opt)	type of optimization
e(which)	max or min; whether optimizer is to perform maximization or minimization
e(ml_method)	type of ml method
e(user)	name of likelihood-evaluator program
e(technique)	maximization technique
e(properties)	b V
e(predict)	program used to implement predict
e(footnote)	program used to implement the footnote display
e(asbalanced)	factor variables fvset as asbalanced
e(asobserved)	factor variables fvset as asobserved

Matrices
e(b)	coefficient vector
e(Cns)	constraints matrix
e(ilog)	iteration log (up to 20 iterations)
e(gradient)	gradient vector
e(V)	variance–covariance matrix of the estimators
e(V_modelbased)	model-based variance

Functions
e(sample)	marks estimation sample

Methods and formulas

scobit is implemented as an ado-file.

Skewed logit analysis is an alternative to logit that relaxes the assumption that individuals with initial probability of 0.5 are most sensitive to changes in independent variables.

The log-likelihood function for skewed logit is

$$\ln L = \sum_{j \in S} w_j \ln F(\mathbf{x}_j \mathbf{b}) + \sum_{j \notin S} w_j \ln\{1 - F(\mathbf{x}_j \mathbf{b})\}$$

where S is the set of all observations j such that $y_j \neq 0$, $F(z) = 1 - 1/\{1 + \exp(z)\}^{\alpha}$, and w_j denotes the optional weights. $\ln L$ is maximized as described in [R] **maximize**.

This command supports the Huber/White/sandwich estimator of the variance and its clustered version using vce(robust) and vce(cluster *clustvar*), respectively. See [P] **_robust**, particularly *Maximum likelihood estimators* and *Methods and formulas*.

scobit also supports estimation with survey data. For details on VCEs with survey data, see [SVY] **variance estimation**.

References

Achen, C. H. 2002. Toward a new political methodology: Microfoundations and ART. *Annual Review of Political Science* 5: 423–450.

Nagler, J. 1994. Scobit: An alternative estimator to logit and probit. *American Journal of Political Science* 38: 230–255.

Also see

[R] **scobit postestimation** — Postestimation tools for scobit

[R] **cloglog** — Complementary log-log regression

[R] **glm** — Generalized linear models

[R] **logistic** — Logistic regression, reporting odds ratios

[SVY] **svy estimation** — Estimation commands for survey data

[U] **20 Estimation and postestimation commands**

Title

> **scobit postestimation** — Postestimation tools for scobit

Description

The following postestimation commands are available after `scobit`:

Command	Description
contrast	contrasts and ANOVA-style joint tests of estimates
estat	AIC, BIC, VCE, and estimation sample summary
estat (svy)	postestimation statistics for survey data
estimates	cataloging estimation results
lincom	point estimates, standard errors, testing, and inference for linear combinations of coefficients
lrtest[1]	likelihood-ratio test
margins	marginal means, predictive margins, marginal effects, and average marginal effects
marginsplot	graph the results from margins (profile plots, interaction plots, etc.)
nlcom	point estimates, standard errors, testing, and inference for nonlinear combinations of coefficients
predict	predictions, residuals, influence statistics, and other diagnostic measures
predictnl	point estimates, standard errors, testing, and inference for generalized predictions
pwcompare	pairwise comparisons of estimates
suest	seemingly unrelated estimation
test	Wald tests of simple and composite linear hypotheses
testnl	Wald tests of nonlinear hypotheses

[1] `lrtest` is not appropriate with svy estimation results.

See the corresponding entries in the *Base Reference Manual* for details, but see [SVY] **estat** for details about `estat` (svy).

Syntax for predict

> predict [*type*] *newvar* [*if*] [*in*] [, *statistic* <u>nooff</u>set]

> predict [*type*] { *stub** | *newvar*_{reg} *newvar*_{lnalpha} } [*if*] [*in*] , <u>sc</u>ores

statistic	Description
Main	
<u>p</u>r	probability of a positive outcome; the default
xb	$x_j b$, linear prediction
stdp	standard error of the linear prediction

These statistics are available both in and out of sample; type `predict ... if e(sample) ...` if wanted only for the estimation sample.

Menu

Statistics > Postestimation > Predictions, residuals, etc.

Options for predict

⌐ Main ⌐

pr, the default, calculates the probability of a positive outcome.

xb calculates the linear prediction.

stdp calculates the standard error of the linear prediction.

nooffset is relevant only if you specified offset(*varname*) for scobit. It modifies the calculations made by predict so that they ignore the offset variable; the linear prediction is treated as $\mathbf{x}_j\mathbf{b}$ rather than as $\mathbf{x}_j\mathbf{b} + \text{offset}_j$.

scores calculates equation-level score variables.

The first new variable will contain $\partial\ln L/\partial(\mathbf{x}_j\boldsymbol{\beta})$.

The second new variable will contain $\partial\ln L/\partial\ln\alpha$.

Remarks

Once you have fit a model, you can obtain the predicted probabilities by using the predict command for both the estimation sample and other samples; see [U] **20 Estimation and postestimation commands** and [R] **predict**. Here we will make only a few additional comments.

predict without arguments calculates the predicted probability of a positive outcome. With the xb option, it calculates the linear combination $\mathbf{x}_j\mathbf{b}$, where $\mathbf{x}_j$ are the independent variables in the jth observation and $\mathbf{b}$ is the estimated parameter vector.

With the stdp option, predict calculates the standard error of the prediction, which is *not* adjusted for replicated covariate patterns in the data.

▷ Example 1

In example 1 of [R] **scobit**, we fit the model scobit foreign mpg. To obtain predicted probabilities, we type

```
. use http://www.stata-press.com/data/r12/auto
(1978 Automobile Data)
. keep make mpg weight foreign
. scobit foreign mpg
 (output omitted )
. predict p
(option pr assumed; Pr(foreign))
. summarize foreign p
```

Variable	Obs	Mean	Std. Dev.	Min	Max
foreign	74	.2972973	.4601885	0	1
p	74	.2974049	.182352	.0714664	.871624

◁

Methods and formulas

All postestimation commands listed above are implemented as ado-files.

Also see

[R] **scobit** — Skewed logistic regression

[U] **20 Estimation and postestimation commands**

Title

> **sdtest** — Variance-comparison tests

Syntax

One-sample variance-comparison test

> sdtest *varname* == # [*if*] [*in*] [, $\underline{\text{l}}$evel(#)]

Two-sample variance-comparison test

> sdtest *varname*$_1$ == *varname*$_2$ [*if*] [*in*] [, $\underline{\text{l}}$evel(#)]

Two-group variance-comparison test

> sdtest *varname* [*if*] [*in*] , by(*groupvar*) [$\underline{\text{l}}$evel(#)]

Immediate form of one-sample variance-comparison test

> sdtesti #$_{\text{obs}}$ { #$_{\text{mean}}$ | . } #$_{\text{sd}}$ #$_{\text{val}}$ [, $\underline{\text{l}}$evel(#)]

Immediate form of two-sample variance-comparison test

> sdtesti #$_{\text{obs},1}$ { #$_{\text{mean},1}$ | . } #$_{\text{sd},1}$ #$_{\text{obs},2}$ { #$_{\text{mean},2}$ | . } #$_{\text{sd},2}$ [, $\underline{\text{l}}$evel(#)]

Robust tests for equality of variances

> robvar *varname* [*if*] [*in*] , by(*groupvar*)

> by is allowed with sdtest and robvar; see [D] **by**.

Menu

one-sample

Statistics > Summaries, tables, and tests > Classical tests of hypotheses > One-sample variance-comparison test

two-sample

Statistics > Summaries, tables, and tests > Classical tests of hypotheses > Two-sample variance-comparison test

two-group

Statistics > Summaries, tables, and tests > Classical tests of hypotheses > Two-group variance-comparison test

immediate command: one-sample

Statistics > Summaries, tables, and tests > Classical tests of hypotheses > One-sample variance-comparison calculator

immediate command: two-sample

Statistics > Summaries, tables, and tests > Classical tests of hypotheses > Two-sample variance-comparison calculator

robvar

Statistics > Summaries, tables, and tests > Classical tests of hypotheses > Robust equal-variance test

Description

sdtest performs tests on the equality of standard deviations (variances). In the first form, sdtest tests that the standard deviation of *varname* is #. In the second form, sdtest tests that *varname$_1$* and *varname$_2$* have the same standard deviation. In the third form, sdtest performs the same test, using the standard deviations of the two groups defined by *groupvar*.

sdtesti is the immediate form of sdtest; see [U] **19 Immediate commands**.

Both the traditional F test for the homogeneity of variances and Bartlett's generalization of this test to K samples are sensitive to the assumption that the data are drawn from an underlying Gaussian distribution. See, for example, the cautionary results discussed by Markowski and Markowski (1990). Levene (1960) proposed a test statistic for equality of variance that was found to be robust under nonnormality. Then Brown and Forsythe (1974) proposed alternative formulations of Levene's test statistic that use more robust estimators of central tendency in place of the mean. These reformulations were demonstrated to be more robust than Levene's test when dealing with skewed populations.

robvar reports Levene's robust test statistic (W_0) for the equality of variances between the groups defined by *groupvar* and the two statistics proposed by Brown and Forsythe that replace the mean in Levene's formula with alternative location estimators. The first alternative (W_{50}) replaces the mean with the median. The second alternative replaces the mean with the 10% trimmed mean (W_{10}).

Options

level(#) specifies the confidence level, as a percentage, for confidence intervals of the means. The default is level(95) or as set by set level; see [U] **20.7 Specifying the width of confidence intervals**.

by(*groupvar*) specifies the *groupvar* that defines the groups to be compared. For sdtest, there should be two groups, but for robvar there may be more than two groups. Do not confuse the by() option with the by prefix; both may be specified.

Remarks

Remarks are presented under the following headings:

> *Basic form*
> *Immediate form*
> *Robust test*

Basic form

sdtest performs two different statistical tests: one testing equality of variances and the other testing that the standard deviation is equal to a known constant. Which test it performs is determined by whether you type a variable name or a number to the right of the equal sign.

▷ Example 1: One-sample test of variance

We have a sample of 74 automobiles. For each automobile, we know the mileage rating. We wish to test whether the overall standard deviation is 5 mpg:

```
. use http://www.stata-press.com/data/r12/auto
(1978 Automobile Data)
. sdtest mpg == 5
One-sample test of variance
```

Variable	Obs	Mean	Std. Err.	Std. Dev.	[95% Conf. Interval]	
mpg	74	21.2973	.6725511	5.785503	19.9569	22.63769

```
      sd = sd(mpg)                                    c = chi2 =  97.7384
  Ho: sd = 5                              degrees of freedom =       73

    Ha: sd < 5                Ha: sd != 5                 Ha: sd > 5
  Pr(C < c) = 0.9717      2*Pr(C > c) = 0.0565      Pr(C > c) = 0.0283
```

◁

▷ Example 2: Variance ratio test

We are testing the effectiveness of a new fuel additive. We run an experiment on 12 cars, running each without and with the additive. The data can be found in [R] **ttest**. The results for each car are stored in the variables mpg1 and mpg2:

```
. use http://www.stata-press.com/data/r12/fuel
. sdtest mpg1==mpg2
Variance ratio test
```

Variable	Obs	Mean	Std. Err.	Std. Dev.	[95% Conf. Interval]	
mpg1	12	21	.7881701	2.730301	19.26525	22.73475
mpg2	12	22.75	.9384465	3.250874	20.68449	24.81551
combined	24	21.875	.6264476	3.068954	20.57909	23.17091

```
     ratio = sd(mpg1) / sd(mpg2)                            f =   0.7054
  Ho: ratio = 1                           degrees of freedom =   11, 11

    Ha: ratio < 1              Ha: ratio != 1              Ha: ratio > 1
  Pr(F < f) = 0.2862      2*Pr(F < f) = 0.5725      Pr(F > f) = 0.7138
```

We cannot reject the hypothesis that the standard deviations are the same.

In [R] **ttest**, we draw an important distinction between paired and unpaired data, which, in this example, means whether there are 12 cars in a before-and-after experiment or 24 different cars. For sdtest, on the other hand, there is no distinction. If the data had been unpaired and stored as described in [R] **ttest**, we could have typed sdtest mpg, by(treated), and the results would have been the same.

◁

Immediate form

▷ Example 3: sdtesti

Immediate commands are used not with data, but with reported summary statistics. For instance, to test whether a variable on which we have 75 observations and a reported standard deviation of 6.5 comes from a population with underlying standard deviation 6, we would type

```
. sdtesti 75 . 6.5 6
One-sample test of variance
```

	Obs	Mean	Std. Err.	Std. Dev.	[95% Conf. Interval]
x	75	.	.7505553	6.5	. .

```
       sd = sd(x)                                    c = chi2 =   86.8472
Ho: sd = 6                              degrees of freedom =        74
      Ha: sd < 6                  Ha: sd != 6                  Ha: sd > 6
   Pr(C < c) = 0.8542       2*Pr(C > c) = 0.2916       Pr(C > c) = 0.1458
```

The mean plays no role in the calculation, so it may be omitted.

To test whether the variable comes from a population with the same standard deviation as another for which we have a calculated standard deviation of 7.5 over 65 observations, we would type

```
. sdtesti 75 . 6.5  65 . 7.5
Variance ratio test
```

	Obs	Mean	Std. Err.	Std. Dev.	[95% Conf. Interval]
x	75	.	.7505553	6.5	. .
y	65	.	.9302605	7.5	. .
combined	140	.	.	.	. .

```
    ratio = sd(x) / sd(y)                              f =    0.7511
Ho: ratio = 1                           degrees of freedom =    74, 64
      Ha: ratio < 1               Ha: ratio != 1               Ha: ratio > 1
   Pr(F < f) = 0.1172       2*Pr(F < f) = 0.2344       Pr(F > f) = 0.8828
```

◁

Robust test

▷ Example 4: robvar

We wish to test whether the standard deviation of the length of stay for patients hospitalized for a given medical procedure differs by gender. Our data consist of observations on the length of hospital stay for 1778 patients: 884 males and 894 females. Length of stay, lengthstay, is highly skewed (skewness coefficient = 4.912591) and thus violates Bartlett's normality assumption. Therefore, we use robvar to compare the variances.

```
. use http://www.stata-press.com/data/r12/stay
. robvar lengthstay, by(sex)
```

	Summary of Length of stay in days		
sex	Mean	Std. Dev.	Freq.
male	9.0874434	9.7884747	884
female	8.800671	9.1081478	894
Total	8.9432508	9.4509466	1778

```
W0  = .55505315   df(1, 1776)   Pr > F = .45635888
W50 = .42714734   df(1, 1776)   Pr > F = .51347664
W10 = .44577674   df(1, 1776)   Pr > F = .50443411
```

For these data, we cannot reject the null hypothesis that the variances are equal. However, Bartlett's test yields a significance probability of 0.0319 because of the pronounced skewness of the data. ◁

❏ Technical note

robvar implements both the conventional Levene's test centered at the mean and a median-centered test. In a simulation study, Conover, Johnson, and Johnson (1981) compare the properties of the two tests and recommend using the median test for asymmetric data, although for small sample sizes the test is somewhat conservative. See Carroll and Schneider (1985) for an explanation of why both mean- and median-centered tests have approximately the same level for symmetric distributions, but for asymmetric distributions the median test is closer to the correct level.

❏

Saved results

sdtest and sdtesti save the following in r():

Scalars

r(N)	number of observations
r(p_l)	lower one-sided p-value
r(p_u)	upper one-sided p-value
r(p)	two-sided p-value
r(F)	F statistic
r(sd)	standard deviation
r(sd_1)	standard deviation for first variable
r(sd_2)	standard deviation for second variable
r(df)	degrees of freedom
r(df_1)	numerator degrees of freedom
r(df_2)	denominator degrees of freedom
r(chi2)	χ^2

robvar saves the following in r():

Scalars

r(N)	number of observations
r(w50)	Brown and Forsythe's F statistic (median)
r(p_w50)	Brown and Forsythe's p-value
r(w0)	Levene's F statistic
r(p_w0)	Levene's p-value
r(w10)	Brown and Forsythe's F statistic (trimmed mean)
r(p_w10)	Brown and Forsythe's p-value (trimmed mean)
r(df_1)	numerator degrees of freedom
r(df_2)	denominator degrees of freedom

Methods and formulas

`sdtest`, `sdtesti`, and `robvar` are implemented as ado-files.

See Armitage et al. (2002, 149–153) or Bland (2000, 171–172) for an introduction and explanation of the calculation of these tests.

The test for $\sigma = \sigma_0$ is given by

$$\chi^2 = \frac{(n-1)s^2}{\sigma_0^2}$$

which is distributed as χ^2 with $n - 1$ degrees of freedom.

The test for $\sigma_x^2 = \sigma_y^2$ is given by

$$F = \frac{s_x^2}{s_y^2}$$

which is distributed as F with $n_x - 1$ and $n_y - 1$ degrees of freedom.

`robvar` is also implemented as an ado-file.

Let X_{ij} be the jth observation of X for the ith group. Let $Z_{ij} = |X_{ij} - \overline{X}_i|$, where $\overline{X}_i$ is the mean of X in the ith group. Levene's test statistic is

$$W_0 = \frac{\sum_i n_i (\overline{Z}_i - \overline{Z})^2 / (g - 1)}{\sum_i \sum_j (Z_{ij} - \overline{Z}_i)^2 / \sum_i (n_i - 1)}$$

where n_i is the number of observations in group i and g is the number of groups. W_{50} is obtained by replacing $\overline{X}_i$ with the ith group median of X_{ij}, whereas W_{10} is obtained by replacing $\overline{X}_i$ with the 10% trimmed mean for group i.

References

Armitage, P., G. Berry, and J. N. S. Matthews. 2002. *Statistical Methods in Medical Research*. 4th ed. Oxford: Blackwell.

Bland, M. 2000. *An Introduction to Medical Statistics*. 3rd ed. Oxford: Oxford University Press.

Brown, M. B., and A. B. Forsythe. 1974. Robust tests for the equality of variances. *Journal of the American Statistical Association* 69: 364–367.

Carroll, R. J., and H. Schneider. 1985. A note on Levene's tests for equality of variances. *Statistics and Probability Letters* 3: 191–194.

Cleves, M. A. 1995. sg35: Robust tests for the equality of variances. *Stata Technical Bulletin* 25: 13–15. Reprinted in *Stata Technical Bulletin Reprints*, vol. 5, pp. 91–93. College Station, TX: Stata Press.

——. 2000. sg35.2: Robust tests for the equality of variances update to Stata 6. *Stata Technical Bulletin* 53: 17–18. Reprinted in *Stata Technical Bulletin Reprints*, vol. 9, pp. 158–159. College Station, TX: Stata Press.

Conover, W. J., M. E. Johnson, and M. M. Johnson. 1981. A comparative study of tests for homogeneity of variances, with applications to the outer continental shelf bidding data. *Technometrics* 23: 351–361.

Gastwirth, J. L., Y. R. Gel, and W. Miao. 2009. The impact of Levene's test of equality of variances on statistical theory and practice. *Statistical Science* 24: 343–360.

Levene, H. 1960. Robust tests for equality of variances. In *Contributions to Probability and Statistics: Essays in Honor of Harold Hotelling*, ed. I. Olkin, S. G. Ghurye, W. Hoeffding, W. G. Madow, and H. B. Mann, 278–292. Menlo Park, CA: Stanford University Press.

Markowski, C. A., and E. P. Markowski. 1990. Conditions for the effectiveness of a preliminary test of variance. *American Statistician* 44: 322–326.

Seed, P. T. 2000. sbe33: Comparing several methods of measuring the same quantity. *Stata Technical Bulletin* 55: 2–9. Reprinted in *Stata Technical Bulletin Reprints*, vol. 10, pp. 73–82. College Station, TX: Stata Press.

Tobías, A. 1998. gr28: A graphical procedure to test equality of variances. *Stata Technical Bulletin* 42: 4–6. Reprinted in *Stata Technical Bulletin Reprints*, vol. 7, pp. 68–70. College Station, TX: Stata Press.

Also see

[R] **ttest** — Mean-comparison tests

Title

search — Search Stata documentation

Syntax

search *word* [*word* ...] [, *search_options*]

set searchdefault { local | net | all } [, permanently]

findit *word* [*word* ...]

search_options	Description
local	search using Stata's keyword database; the default
net	search across materials available via Stata's net command
all	search across both the local keyword database and the net material
author	search by author's name
entry	search by entry ID
exact	search across both the local keyword database and the net materials; prevents matching on abbreviations
faq	search the FAQs posted to the Stata website
historical	search entries that are of historical interest only
or	list an entry if *any* of the words typed after search are associated with the entry
manual	search the entries in the *Stata Documentation*
sj	search the entries in the *Stata Journal* and the STB

Menu

Help > Search...

Description

search searches a keyword database and the Internet.

Capitalization of the words following search is irrelevant, as is the inclusion or exclusion of special characters such as commas and hyphens.

set searchdefault affects the default behavior of the search command. local is the default.

findit is equivalent to search *word* [*word* ...], all. findit results are displayed in the Viewer. findit is the best way to search for information on a topic across all sources, including the online help, the FAQs at the Stata website, the *Stata Journal*, and all Stata-related Internet sources including user-written additions. From findit, you can click to go to a source or to install additions.

See [R] **hsearch** for a command that searches help files.

Options for search

local, the default (unless changed by set searchdefault), specifies that the search be performed using only Stata's keyword database.

net specifies that the search be performed across the materials available via Stata's net command. Using search *word* [*word* ...], net is equivalent to typing net search *word* [*word* ...] (without options); see [R] **net search**.

all specifies that the search be performed across both the local keyword database and the net materials.

author specifies that the search be performed on the basis of author's name rather than keywords. A search with the author option is performed on the local keyword database only.

entry specifies that the search be performed on the basis of entry IDs rather than keywords. A search with the entry option is performed on the local keyword database only.

exact prevents matching on abbreviations. A search with the exact option is performed across both the local keyword database and the net materials.

faq limits the search to the FAQs posted on the Stata website: http://www.stata.com. A search with the faq option is performed on the local keyword database only.

historical adds to the search entries that are of historical interest only. By default, such entries are not listed. Past entries are classified as historical if they discuss a feature that later became an official part of Stata. Updates to historical entries will always be found, even if historical is not specified. A search with the historical option is performed on the local keyword database only.

or specifies that an entry be listed if any of the words typed after search are associated with the entry. The default is to list the entry only if all the words specified are associated with the entry. A search with the or option is performed on the local keyword database only.

manual limits the search to entries in the *Stata Documentation*; that is, the search is limited to the *User's Guide* and all the reference manuals. A search with the manual option is performed on the local keyword database only.

sj limits the search to entries in the *Stata Journal* and its predecessor, the *Stata Technical Bulletin*; see [R] **sj**. A search with the sj option is performed on the local keyword database only.

Option for set searchdefault

permanently specifies that, in addition to making the change right now, the searchdefault setting be remembered and become the default setting when you invoke Stata.

Remarks

Remarks are presented under the following headings:

> *Introduction*
> *Internet searches*
> *Author searches*
> *Entry ID searches*
> *Return codes*

Introduction

See [U] **4 Stata's help and search facilities** for a tutorial introduction to search. search is one of Stata's most useful commands. To understand the advanced features of search, you need to know how it works.

search has a database—files—containing the titles, etc., of every entry in the *User's Guide*, the *Base Reference Manual*, the *Data-Management Reference Manual*, the *Graphics Reference Manual*, the *Longitudinal-Data/Panel-Data Reference Manual*, the *Multiple-Imputation Reference Manual*, the *Multivariate Statistics Reference Manual*, the *Programming Reference Manual*, the *Structural Equation Modeling Reference Manual*, the *Survey Data Reference Manual*, the *Survival Analysis and Epidemiological Tables Reference Manual*, the *Time-Series Reference Manual*, the *Mata Reference Manual*, undocumented help files, NetCourses, Stata Press books, FAQs posted on the Stata website, selected articles on StataCorp's official blog, selected user-written FAQs and examples, and the articles in the *Stata Journal* and the *Stata Technical Bulletin*. In these files is a list of words, called keywords, associated with each entry.

When you type search *xyz*, search reads the database and compares the list of keywords with *xyz*. If it finds *xyz* in the list or a keyword that allows an abbreviation of *xyz*, it displays the entry.

When you type search *xyz abc*, search does the same thing but displays an entry only if it contains both keywords. The order does not matter, so you can search linear regression or search regression linear.

Obviously, how many entries search finds depends on how the search database was constructed. We have included a plethora of keywords under the theory that, for a given request, it is better to list too much rather than risk listing nothing at all. Still, you are in the position of guessing the keywords. Do you look up normality test, normality tests, or tests of normality? Well, normality test would be best, but all would work. In general, use the singular, and strike the unnecessary words. For guidelines for specifying keywords, see [U] **4.6 More on search**.

set searchdefault allows you to specify where search searches. set searchdefault local, the default, restricts search to using only Stata's keyword database. set searchdefault net restricts search to searching only the Internet. set searchdefault all indicates that both the keyword database and the Internet are to be searched.

Internet searches

search with the net option searches the Internet for user-written additions to Stata, including, but not limited to, user-written additions published in the *Stata Journal* (SJ) and the *Stata Technical Bulletin* (STB). search *keywords*, net performs the same search as the command net search (with no options); see [R] **net search**.

```
. search random effect, net

Keyword search
      Keywords:  random effect
        Search:  (1) Web resources from Stata and from other users

Web resources from Stata and other users
(contacting http://www.stata.com)
135 packages found (Stata Journal and STB listed first)
-----------------------------------------------------

st0201 from http://www.stata-journal.com/software/sj10-3
    SJ10-3 st0201.  metaan: Random-effects meta-analysis / metaan:
    Random-effects meta-analysis / by Evangelos Kontopantelis, / National
    Primary Care Research and Development Centre (NPCRDC), / University of
    Manchester, Manchester, UK / David Reeves, / National Primary Care
```

```
st0175 from http://www.stata-journal.com/software/sj9-4
    SJ9-4 st0175.  A menu-driven facility for sample-size... / A menu-driven
    facility for sample-size calculation in / novel multiarm, multistage
    randomized controlled / trials with a time-to-event outcome / by
    Friederike M.-S. Barthel, GlaxoSmithKline / Patrick Royston, UK Medical

sbe24_3 from http://www.stata-journal.com/software/sj9-2
    SJ9-2 sbe24_3.  Update: metan: fixed- and random-effects... / Update:
    metan: fixed- and random-effects meta-analysis / by Ross J. Harris, Roger
    M. Harbord, and Jonathan A. C. Sterne, / Department of Social Medicine,
    University of Bristol / Jonathan J. Deeks, Department of Primary Care

st0156 from http://www.stata-journal.com/software/sj9-1
    SJ9-1 st0156.  Multivariate random-effects meta-analysis / Multivariate
    random-effects meta-analysis / by Ian R. White, MRC Biostatistics Unit, /
    Institute of Public Health, UK / Support:  ian.white@mrc-bsu.cam.ac.uk /
    After installation, type help mvmeta and mvmeta_make
```

  (*output omitted*)

(end of search)

Author searches

search ordinarily compares the words following search with the keywords for the entry. If you specify the author option, however, it compares the words with the author's name. In the search database, we have filled in author names for all SJ and STB inserts.

For instance, in [R] **kdensity** in this manual you will discover that Isaías H. Salgado-Ugarte wrote the first version of Stata's kdensity command and published it in the STB. Assume that you read his original insert and found the discussion useful. You might now wonder what else he has written in the SJ or STB. To find out, you type

```
. search Salgado-Ugarte, author
```
  (*output omitted*)

Names like Salgado-Ugarte are confusing to many people. search does not require you to specify the entire name; what you type is compared with each "word" of the name and, if any part matches, the entry is listed. The hyphen is a special character, and you can omit it. Thus you can obtain the same list by looking up Salgado, Ugarte, or Salgado Ugarte without the hyphen.

Actually, to find all entries written by Salgado-Ugarte, you need to type

```
. search Salgado-Ugarte, author historical
```
  (*output omitted*)

Prior inserts in the SJ or STB that provide a feature that later was superseded by a built-in feature of Stata are marked as historical in the search database and, by default, are not listed. The historical option ensures that all entries are listed.

Entry ID searches

If you specify the entry option, search compares what you have typed with the entry ID. The entry ID is not the title—it is the reference listed to the left of the title that tells you where to look. For instance, in

```
[R]     regress . . . . . . . . . . . . . . . . . . . . . Linear regression
        (help regress)
```

[R] **regress** is the entry ID. This is a reference, of course, to this manual. In

```
FAQ         . . . . . . . . . . . Analysis of multiple failure-time survival data
            . . . . . . . . . . . . . . . . . . . . . . . . . . . . . . M. Cleves
            11/99   How do I analyze multiple failure-time data using Stata?
                    http://www.stata.com/support/faqs/stat/stmfail.html
```

"FAQ" is the entry ID. In

```
SJ-7-1   st0118  . . A survey on survey stat.: What is and can be done in Stata
         . . . . . . . . . . . . . . . . . . . . . . . F. Kreuter and R. Valliant
         Q1/07   SJ7(1):1--21                                      (no commands)
         discusses survey issues in analyzing complex survey
         data and describes some of Stata's capabilities for
         such analyses
```

"SJ-7-1" is the entry ID.

search with the entry option searches these entry IDs.

Thus you could generate a table of contents for the *User's Guide* by typing

```
. search [U], entry
(output omitted)
```

You could generate a table of contents for *Stata Journal*, Volume 1, Issue 1, by typing

```
. search sj-1-1, entry
(output omitted)
```

To generate a table of contents for the 26th issue of the STB, you would type

```
. search STB-26, entry historical
(output omitted)
```

The historical option here is possibly important. STB-26 was published in July 1995, and perhaps some of its inserts have already been marked historical.

You could obtain a list of all inserts associated with sg53 by typing

```
. search sg53, entry historical
(output omitted)
```

Again we include the historical option in case any of the relevant inserts have been marked historical.

Return codes

In addition to indexing the entries in the *User's Guide* and all the *Reference* manuals, search also can be used to search return codes.

To see information on return code 131, type

```
. search rc 131
[P]     error . . . . . . . . . . . . . . . . . . . . . . . Return code 131
        not possible with test;
        You requested a test of a hypothesis that is nonlinear in the
        variables.  test tests only linear hypotheses.  Use testnl.
```

If you want a list of all Stata return codes, type

```
. search error, entry
(output omitted)
```

Methods and formulas

findit is implemented as an ado-file.

Acknowledgment

findit grew from a suggestion by Nicholas J. Cox, Durham University.

Also see

[R] **hsearch** — Search help files

[R] **help** — Display online help

[R] **net search** — Search the Internet for installable packages

[U] **4 Stata's help and search facilities**

Title

> **serrbar** — Graph standard error bar chart

Syntax

serrbar *mvar svar xvar* [*if*] [*in*] [, *options*]

options	Description
Main	
scale(*#*)	scale length of graph bars; default is scale(1)
Error bars	
rcap_options	affect rendition of capped spikes
Plotted points	
mvopts(*scatter_options*)	affect rendition of plotted points
Add plots	
addplot(*plot*)	add other plots to generated graph
Y axis, X axis, Titles, Legend, Overall	
twoway_options	any options other than by() documented in [G-3] ***twoway_options***

Menu

Statistics > Other > Quality control > Standard error bar chart

Description

serrbar graphs $mvar \pm$ scale() $\times$ *svar* against *xvar*. Usually, but not necessarily, *mvar* and *svar* will contain means and standard errors or standard deviations of some variable so that a standard error bar chart is produced.

Options

⌐ Main ⌐

scale(*#*) controls the length of the bars. The upper and lower limits of the bars will be $mvar +$ scale() $\times$ *svar* and $mvar -$ scale() $\times$ *svar*. The default is scale(1).

⌐ Error bars ⌐

rcap_options affect the rendition of the plotted error bars (the capped spikes). See [G-2] **graph twoway rcap**.

⌐ Plotted points ⌐

mvopts(*scatter_options*) affects the rendition of the plotted points (*mvar* versus *xvar*). See [G-2] **graph twoway scatter**.

⌐ Add plots ⌐

addplot(*plot*) provides a way to add other plots to the generated graph; see [G-3] ***addplot_option***.

⌐ Y axis, X axis, Titles, Legend, Overall ⌐

twoway_options are any of the options documented in [G-3] ***twoway_options***, excluding by(). These include options for titling the graph (see [G-3] ***title_options***) and for saving the graph to disk (see [G-3] ***saving_option***).

Remarks

▷ Example 1

In quality-control applications, the three most commonly used variables with this command are the process mean, process standard deviation, and time. For instance, we have data on the average weights and standard deviations from an assembly line in San Francisco for the period January 8 to January 16. Our data are

```
. use http://www.stata-press.com/data/r12/assembly
. list, sep(0) divider
```

	date	mean	std
1.	108	192.22	3.94
2.	109	192.64	2.83
3.	110	192.37	4.58
4.	113	194.76	3.25
5.	114	192.69	2.89
6.	115	195.02	1.73
7.	116	193.40	2.62

We type serrbar mean std date, scale(2) but, after seeing the result, decide to make it fancier:

```
. serrbar mean std date, scale(2) title("Observed Weight Variation")
> sub("San Francisco plant, 1/8 to 1/16") yline(195) yaxis(1 2)
> ylab(195, axis(2)) ytitle("", axis(2))
```

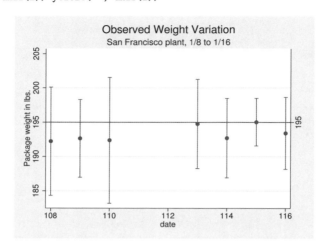

◁

Methods and formulas

serrbar is implemented as an ado-file.

Acknowledgment

serrbar was written by Nicholas J. Cox of Durham University.

Also see

[R] **qc** — Quality control charts

Title

> **set** — Overview of system parameters

Syntax

set [*setcommand* ...]

set typed without arguments is equivalent to query typed without arguments.

Description

This entry provides a reference to Stata's set commands. For many entries, more thorough information is provided elsewhere; see the Reference field in each entry below for the location of this information.

To reset system parameters to factory defaults, see [R] **set_defaults**.

Remarks

set adosize
    Syntax: set adosize # [, permanently]
    Default: 1,000
    Description: sets the maximum amount of memory that automatically loaded do-files
                may consume. $10 \leq \# \leq 10000$.
    Reference: [P] **sysdir**

set autotabgraphs (Windows only)
    Syntax: set autotabgraphs {on | off} [, permanently]
    Default: off
    Description: determines whether graphs are created as tabs within one window or as separate
                windows.

set cformat
    Syntax: set cformat [*fmt*] [, permanently]
    Description: specifies the output format of coefficients, standard errors, and confidence limits
                in coefficient tables. *fmt* is a numerical format; see [D] **format**.
    Reference: [R] **set cformat**

set checksum
    Syntax: set checksum {on | off} [, permanently]
    Default: off
    Description: determines whether files should be prevented from being downloaded from the
                Internet if checksums do not match.
    Reference: [D] **checksum**

`set conren` (Unix console only)

Syntax 1:	`set conren`
Syntax 2:	`set conren clear`
Syntax 3:	`set conren` $\big[\texttt{sf} \mid \texttt{bf} \mid \texttt{it}\big]$
	$\big\{\underline{\texttt{res}}\texttt{ult} \mid \big[\texttt{txt} \mid \texttt{text}\big] \mid \underline{\texttt{input}} \mid \underline{\texttt{err}}\texttt{or} \mid \underline{\texttt{link}} \mid \underline{\texttt{hi}}\texttt{lite}\big\}$
	$\big[char\big[char\dots\big]\big]$
Syntax 4:	`set conren` $\{\texttt{ulon} \mid \underline{\texttt{ul}}\texttt{off}\}$ $\big[char\,\big[char\dots\big]\big]$
Syntax 5:	`set conren reset` $\big[char\,\big[char\dots\big]\big]$
Description:	can possibly make the output on your screen appear prettier.
	`set conren` displays a list of the currently defined display codes.
	`set conren clear` clears all codes.
	`set conren` followed by a font type (`bf`, `sf`, or `it`) and display context (`result`, `error`, `link`, or `hilite`) and then followed by a series of space-separated characters sets the code for the specified font type and display context. If the font type is omitted, the code is set to the same specified code for all three font types.
	`set conren ulon` and `set conren uloff` set the codes for turning on and off underlining.
	`set conren reset` sets the code that will turn off all display and underlining codes.
Reference:	[GSU] **conren**

`set copycolor` (Mac and Windows only)

Syntax:	`set copycolor` $\{\underline{\texttt{auto}}\texttt{matic} \mid \texttt{asis} \mid \texttt{gs1} \mid \texttt{gs2} \mid \texttt{gs3}\}$ $\big[\,, \underline{\texttt{perm}}\texttt{anently}\big]$
Default:	`automatic`
Description:	determines how colors are handled when graphs are copied to the Clipboard.
Reference:	[G-2] **set printcolor**

`set dockable` (Windows only)

Syntax:	`set dockable` $\{\texttt{on} \mid \texttt{off}\}$ $\big[\,, \underline{\texttt{perm}}\texttt{anently}\big]$
Default:	`on`
Description:	determines whether to enable the use of dockable window characteristics, including the ability to dock or tab a window into another window.

`set dockingguides` (Windows only)

Syntax:	`set `$\underline{\texttt{dockingg}}$`uides` $\{\texttt{on} \mid \texttt{off}\}$ $\big[\,, \underline{\texttt{perm}}\texttt{anently}\big]$
Default:	`on`
Description:	determines whether to enable the use of dockable guides when repositioning a dockable window.

`set doublebuffer` (Windows only)

Syntax:	`set doublebuffer` $\{\texttt{on} \mid \texttt{off}\}$ $\big[\,, \underline{\texttt{perm}}\texttt{anently}\big]$
Default:	`on`
Description:	enables or disables double buffering of the Results, Viewer, and Data Editor windows. Double buffering prevents the windows from flickering when redrawn or resized. Users who encounter performance problems such as the Results window outputting very slowly should disable double buffering.

set dp
    Syntax: set dp {comma | period} [, permanently]
    Default: period
    Description: determines whether a period or a comma is to be used as the decimal point.
    Reference: [D] **format**

set emptycells
    Syntax: set emptycells {keep | drop} [, permanently]
    Default: keep
    Description: sets what to do with empty cells in interactions.
    Reference: [R] **set emptycells**

set eolchar (Mac only)
    Syntax: set eolchar {mac | unix} [, permanently]
    Default: unix
    Description: sets the default end-of-line delimiter for text files created in Stata.

set fastscroll (Unix and Windows only)
    Syntax: set fastscroll {on | off} [, permanently]
    Default: on
    Description: sets the scrolling method for new output in the Results window. Setting
                  fastscroll to on is faster but can be jumpy. Setting fastscroll to off
                  is slower but smoother.

set floatresults (Windows only)
    Syntax: set floatresults {on | off}
    Default: off
    Description: determines whether to enable floating window behavior for the Results window. The
                  term "float" in this context means the Results window will always float over the
                  main Stata window; the Results window can never be placed behind the main Stata
                  window. There is no permanently option because permanently is implied.

set floatwindows (Windows only)
    Syntax: set floatwindows {on | off}
    Default: off
    Description: determines whether to enable floating window behavior for dialog boxes and dockable
                  window. The term "float" in this context means that a window will always float
                  over the main Stata window; these windows cannot be placed behind the main Stata
                  window. There is no permanently option because permanently is implied.

set graphics
    Syntax: set graphics {on | off}
    Default: on; default is off for console Stata
    Description: determines whether graphs are displayed on your monitor.
    Reference: [G-2] **set graphics**

`set httpproxy`
    Syntax: `set httpproxy {on | off} [ , init]`
    Default: `off`
    Description: turns on/off the use of a proxy server. There is no `permanently` option because
               `permanently` is implied.
    Reference: [R] **netio**

`set httpproxyauth`
    Syntax: `set httpproxyauth {on | off}`
    Default: `off`
    Description: determines whether authorization is required for the proxy server.
               There is no `permanently` option because `permanently` is implied.
    Reference: [R] **netio**

`set httpproxyhost`
    Syntax: `set httpproxyhost ["]name["]`
    Description: sets the name of a host to be used as a proxy server. There is no `permanently`
               option because `permanently` is implied.
    Reference: [R] **netio**

`set httpproxyport`
    Syntax: `set httpproxyport #`
    Default: 8080 if Stata cannot autodetect the proper setting for your computer.
    Description: sets the port number for a proxy server. There is no `permanently` option
               because `permanently` is implied.
    Reference: [R] **netio**

`set httpproxypw`
    Syntax: `set httpproxypw ["]password["]`
    Description: sets the appropriate password. There is no `permanently` option because
               `permanently` is implied.
    Reference: [R] **netio**

`set httpproxyuser`
    Syntax: `set httpproxyuser ["]name["]`
    Description: sets the appropriate user ID. There is no `permanently` option because
               `permanently` is implied.
    Reference: [R] **netio**

`set include_bitmap` (Mac only)
    Syntax: `set include_bitmap {on | off} [ , permanently]`
    Default: `on`
    Description: sets the output behavior when copying an image to the Clipboard.

set level
    Syntax: set <u>l</u>evel # $\left[\,,\ \underline{\text{perm}}\text{anently}\right]$
    Default: 95
    Description: sets the default significance level for confidence intervals for all commands
                    that report confidence intervals. $10.00 \leq \# \leq 99.99$, and # can have at
                    most two digits after the decimal point.
    Reference: [R] **level**

set linegap
    Syntax: set <u>linegap</u> #
    Default: 1
    Description: sets the space between lines, in pixels, in the Results window. There is no
                    permanently option because permanently is implied.

set linesize
    Syntax: set <u>lin</u>esize #
    Default: 1 less than the full width of the screen
    Description: sets the line width, in characters, for both the screen and the log file.
    Reference: [R] **log**

set locksplitters (Windows only)
    Syntax: set <u>locksplitters</u> $\left\{\text{on}\,|\,\text{off}\right\}$ $\left[\,,\ \underline{\text{perm}}\text{anently}\right]$
    Default: off
    Description: determines whether splitters should be locked so that docked windows
                    cannot be resized.

set logtype
    Syntax: set <u>log</u>type $\left\{\underline{\text{t}}\text{ext}\,|\,\underline{\text{sm}}\text{cl}\right\}$ $\left[\,,\ \underline{\text{perm}}\text{anently}\right]$
    Default: smcl
    Description: sets the default log filetype.
    Reference: [R] **log**

set lstretch
    Syntax: set lstretch $\left\{\text{on}\,|\,\text{off}\right\}$ $\left[\,,\ \underline{\text{perm}}\text{anently}\right]$
    Description: specifies whether to automatically widen the coefficient table up to the width of
                    the Results window to accommodate longer variable names.

set matacache, set matafavor, set matalibs, set matalnum, set matamofirst,
    set mataoptimize, and set matastrict; see [M-3] **mata set**.

set matsize
    Syntax: set <u>mat</u>size # $\left[\,,\ \underline{\text{perm}}\text{anently}\right]$
    Default: 400 for Stata/MP, Stata/SE, and Stata/IC; 40 for Small Stata
    Description: sets the maximum number of variables that can be included in any estimation
                    command. This setting cannot be changed in Small Stata.
                    $10 \leq \# \leq 11000$ for Stata/MP and Stata/SE; $10 \leq \# \leq 800$ for Stata/IC.
    Reference: [R] **matsize**

set max_memory
    Syntax: set max_memory #[b | k | m | g] [, permanently]
    Default: . (all the memory the operating system will supply)
    Description: specifies the maximum amount of memory Stata can use to store your data.
                $2 \times \text{segmentsize} \le \# \le .$
    Reference: [D] **memory**

set maxdb
    Syntax: set maxdb # [, permanently]
    Default: 50
    Description: sets the maximum number of dialog boxes whose contents are remembered
                from one invocation to the next during a session. $5 \le \# \le 1000$
    Reference: [R] **db**

set maxiter
    Syntax: set maxiter # [, permanently]
    Default: 16000
    Description: sets the default maximum number of iterations for estimation commands.
                $0 \le \# \le 16000$
    Reference: [R] **maximize**

set maxvar
    Syntax: set maxvar # [, permanently]
    Default: 5000 for Stata/MP and Stata/SE, 2048 for Stata/IC, and 99 for Small Stata
    Description: sets the maximum number of variables. This can be changed only in Stata/MP and
                Stata/SE. $2048 \le \# \le 32767$
    Reference: [D] **memory**

set min_memory
    Syntax: set min_memory #[b | k | m | g] [, permanently]
    Default: 0
    Description: specifies an amount of memory Stata will not fall below. This setting affects
                efficiency, not the size of datasets you can analyze. $0 \le \# \le \text{max\_memory}$
    Reference: [D] **memory**

set more
    Syntax: set more {on | off} [, permanently]
    Default: on
    Description: pauses when —more— is displayed, continuing only when the user presses a key.
    Reference: [R] **more**

set niceness
    Syntax: set niceness # [, permanently]
    Default: 5
    Description: affects how soon Stata gives back unused segments to the operating system.
                $0 \le \# \le 10$
    Reference: [D] **memory**

set notifyuser (Mac only)
    Syntax: set notifyuser {on | off} [, permanently]
    Default: on
    Description: sets the default Notification Manager behavior in Stata.

set obs
    Syntax: set obs #
    Default: current number of observations
    Description: changes the number of observations in the current dataset. # must be at least
                    as large as the current number of observations. If there are variables in memory,
                    the values of all new observations are set to *missing*.
    Reference: [D] **obs**

set odbcmgr (Unix only)
    Syntax: set odbcmgr {iodbc | unixodbc} [, permanently]
    Default: iodbc
    Description: determines whether iODBC or unixODBC is your ODBC driver manager.
    Reference: [D] **odbc**

set output
    Syntax: set output {proc | inform | error}
    Default: proc
    Description: specifies the output to be displayed. proc means display all output; inform
                    suppresses procedure output but displays informative messages and error messages;
                    error suppresses all output except error messages. set output is seldom used.
    Reference: [P] **quietly**

set pagesize
    Syntax: set pagesize #
    Default: 2 less than the physical number of lines on the screen
    Description: sets the number of lines between —more— messages.
    Reference: [R] **more**

set pformat
    Syntax: set pformat [*fmt*] [, permanently]
    Description: specifies the output format of *p*-values in coefficient tables.
                    fmt is a numerical format; see [D] **format**.
    Reference: [R] **set cformat**

set pinnable (Windows only)
    Syntax: set pinnable {on | off} [, permanently]
    Default: on
    Description: determines whether to enable the use of pinnable window characteristics for certain
                    windows in Stata.

set playsnd (Mac only)
Syntax: set playsnd {on | off} [, permanently]
Default: on
Description: sets the sound behavior for the Notification Manager behavior in Stata.

set printcolor
Syntax: set printcolor {automatic | asis | gs1 | gs2 | gs3} [, permanently]
Default: automatic
Description: determines how colors are handled when graphs are printed.
Reference: [G-2] **set printcolor**

set processors
Syntax: set processors #
Description: sets the number of processors or cores that Stata/MP will use. The default
              is the number of processors available on the computer, or the number of
              processors allowed by Stata/MP's license, whichever is less.

set reventries
Syntax: set reventries # [, permanently]
Default: 5000
Description: sets the number of scrollback lines available in the Review window.
              $5 \leq \# \leq 32000$.

set revkeyboard (Mac only)
Syntax: set revkeyboard {on | off} [, permanently]
Default: on
Description: sets the keyboard navigation behavior for the Review window. on indicates
              that you can use the keyboard to navigate and enter items from the Review
              window into the Command window. off indicates that all keyboard input be
              directed at the Command window; items can be entered from the Review
              window only by using the mouse.

set rmsg
Syntax: set rmsg {on | off} [, permanently]
Default: off
Description: indicates whether a return message telling the execution time is to be displayed at
              the completion of each command.
Reference: [P] **rmsg**

set scheme
Syntax: set scheme schemename [, permanently]
Default: s2color
Description: determines the overall look for graphs.
Reference: [G-2] **set scheme**

```
set scrollbufsize
```
  Syntax: set scrollbufsize #
  Default: 200000
  Description: sets the scrollback buffer size, in bytes, for the Results window;
         may be set between 10,000 and 2,000,000.

```
set searchdefault
```
  Syntax: set searchdefault {local | net | all} [, permanently]
  Default: local
  Description: sets the default behavior of the search command. set searchdefault local
         restricts search to use only Stata's keyword database. set searchdefault net
         restricts search to searching only the Internet. set searchdefault all
         indicates that both the keyword database and the Internet are to be searched.
  Reference: [R] **search**

```
set seed
```
  Syntax: set seed {# | code}
  Default: 123456789
  Description: specifies initial value of the random-number seed used by the runiform() function.
  Reference: [R] **set seed**

```
set segmentsize
```
  Syntax: set segmentsize #[b | k | m | g] [, permanently]
  Default: 32m for 64-bit machines; 16m for 32-bit machines
  Description: Stata allocates memory for data in units of segmentsize. This setting changes the
         amount of memory in a single segment.
         $1m \leq \# \leq 32g$ for 64-bit machines; $1m \leq \# \leq 1g$ for 32-bit machines
  Reference: [D] **memory**

```
set sformat
```
  Syntax: set sformat [fmt] [, permanently]
  Description: specifies the output format of test statistics in coefficient tables.
         fmt is a numerical format; see [D] **format**.
  Reference: [R] **set cformat**

```
set showbaselevels
```
  Syntax: set showbaselevels [on | off | all] [, permanently]
  Description: specifies whether to display base levels of factor variables and their interactions
         in coefficient tables.
  Reference: [R] **set showbaselevels**

```
set showemptycells
```
  Syntax: set showemptycells [on | off] [, permanently]
  Description: specifies whether to display empty cells in coefficient tables.
  Reference: [R] **set showbaselevels**

set showomitted
   Syntax: set showomitted [on | off] [, permanently]
   Description: specifies whether to display omitted coefficients in coefficient tables.
   Reference: [R] **set showbaselevels**

set smoothfonts (Mac only)
   Syntax: set smoothfonts {on | off}
   Default: on
   Description: determines whether to use font smoothing (antialiased text) in the Results, Viewer,
                 and Data Editor windows.

set timeout1
   Syntax: set timeout1 *#seconds* [, permanently]
   Default: 30
   Description: sets the number of seconds Stata will wait for a remote host to respond to an initial
                 contact before giving up. In general, users should not modify this value unless
                 instructed to do so by Stata Technical Services.
   Reference: [R] **netio**

set timeout2
   Syntax: set timeout2 *#seconds* [, permanently]
   Default: 180
   Description: sets the number of seconds Stata will keep trying to get information from a remote
                 host after initial contact before giving up. In general, users should not modify this
                 value unless instructed to do so by Stata Technical Services.
   Reference: [R] **netio**

set trace
   Syntax: set trace {on | off}
   Default: off
   Description: determines whether to trace the execution of programs for debugging.
   Reference: [P] **trace**

set tracedepth
   Syntax: set tracedepth #
   Default: 32000 (equivalent to ∞)
   Description: if trace is set on, traces execution of programs and nested programs up to
                 tracedepth. For example, if tracedepth is 2, the current program and any
                 subroutine called would be traced, but subroutines of subroutines would not
                 be traced.
   Reference: [P] **trace**

set traceexpand
   Syntax: set traceexpand {on | off} [, permanently]
   Default: on
   Description: if trace is set on, shows lines both before and after macro expansion. If
                 traceexpand is set off, only the line before macro expansion is shown.
   Reference: [P] **trace**

`set tracehilite`

Syntax: set <u>trace</u>hilite "*pattern*" $\left[\, , \, \text{word} \right]$
Default: ""
Description: highlights *pattern* in the trace output.
Reference: [P] **trace**

`set traceindent`

Syntax: set <u>trace</u>indent $\left\{ \text{on} \, | \, \text{off} \right\}$ $\left[\, , \, \underline{\text{perma}}\text{nently} \right]$
Default: on
Description: if `trace` is set on, indents displayed lines according to their nesting level. The lines of the main program are not indented. Two spaces of indentation are used for each level of nested subroutine.
Reference: [P] **trace**

`set tracenumber`

Syntax: set <u>trace</u>number $\left\{ \text{on} \, | \, \text{off} \right\}$ $\left[\, , \, \underline{\text{perma}}\text{nently} \right]$
Default: off
Description: if `trace` is set on, shows the nesting level numerically in front of the line. Lines of the main program are preceded by 01, lines of subroutines called by the main program are preceded by 02, etc.
Reference: [P] **trace**

`set tracesep`

Syntax: set <u>trace</u>sep $\left\{ \text{on} \, | \, \text{off} \right\}$ $\left[\, , \, \underline{\text{perma}}\text{nently} \right]$
Default: on
Description: if `trace` is set on, displays a horizontal separator line that displays the name of the subroutine whenever a subroutine is called or exits.
Reference: [P] **trace**

`set type`

Syntax: set <u>type</u> $\left\{ \text{float} \, | \, \text{double} \right\}$ $\left[\, , \, \underline{\text{perma}}\text{nently} \right]$
Default: float
Description: specifies the default storage type assigned to new variables.
Reference: [D] **generate**

`set update_interval` (Mac and Windows only)

Syntax: set update_interval #
Default: 7
Description: sets the number of days to elapse before performing the next automatic update query.
Reference: [R] **update**

`set update_prompt` (Mac and Windows only)

Syntax: set update_prompt $\left\{ \text{on} \, | \, \text{off} \right\}$
Default: on
Description: determines wheter a dialog is to be displayed before performing an automatic update query. There is no `permanently` option because `permanently` is implied.
Reference: [R] **update**

set update_query (Mac and Windows only)
   Syntax: set update_query {on | off}
   Default: on
   Description: determines whether update query is to be automatically performed when Stata
                is launched. There is no permanently option because permanently is implied.
   Reference: [R] **update**

set varabbrev
   Syntax: set varabbrev {on | off} [, permanently]
   Default: on
   Description: indicates whether Stata should allow variable abbreviations.
   Reference: [P] **varabbrev**

set varkeyboard (Mac only)
   Syntax: set varkeyboard {on | off} [, permanently]
   Default: on
   Description: sets the keyboard navigation behavior for the Variables window. on indicates
                that you can use the keyboard to navigate and enter items from the Variables
                window into the Command window. off indicates that all keyboard input be
                directed at the Command window; items can be entered from the Variables
                window only by using the mouse.

Also see

[R] **query** — Display system parameters

[R] **set_defaults** — Reset system parameters to original Stata defaults

[M-3] **mata set** — Set and display Mata system parameters

Title

set cformat — Format settings for coefficient tables

Syntax

set cformat [*fmt*] [, <u>perm</u>anently]

set pformat [*fmt*] [, <u>perm</u>anently]

set sformat [*fmt*] [, <u>perm</u>anently]

where *fmt* is a numerical format.

Description

set cformat specifies the output format of coefficients, standard errors, and confidence limits in coefficient tables.

set pformat specifies the output format of p-values in coefficient tables.

set sformat specifies the output format of test statistics in coefficient tables.

Option

permanently specifies that, in addition to making the change right now, the setting be remembered and become the default setting when you invoke Stata.

Remarks

The formatting of the numbers in the coefficient table can be controlled by using the set cformat, set pformat, and set sformat commands or by using the cformat(%*fmt*), pformat(%*fmt*), and sformat(%*fmt*) options at the time of estimation or on replay of the estimation command. See [R] **estimation options**.

▷ Example 1

We use `auto.dta` to illustrate.

```
. use http://www.stata-press.com/data/r12/auto
(1978 Automobile Data)

. regress mpg weight displacement
```

Source	SS	df	MS
Model	1595.40969	2	797.704846
Residual	848.049768	71	11.9443629
Total	2443.45946	73	33.4720474

Number of obs	=	74
F(2, 71)	=	66.79
Prob > F	=	0.0000
R-squared	=	0.6529
Adj R-squared	=	0.6432
Root MSE	=	3.4561

mpg	Coef.	Std. Err.	t	P>|t|	[95% Conf. Interval]	
weight	-.0065671	.0011662	-5.63	0.000	-.0088925	-.0042417
displacement	.0052808	.0098696	0.54	0.594	-.0143986	.0249602
_cons	40.08452	2.02011	19.84	0.000	36.05654	44.11251

```
. set cformat %9.2f

. regress mpg weight displacement
```

Source	SS	df	MS
Model	1595.40969	2	797.704846
Residual	848.049768	71	11.9443629
Total	2443.45946	73	33.4720474

Number of obs	=	74
F(2, 71)	=	66.79
Prob > F	=	0.0000
R-squared	=	0.6529
Adj R-squared	=	0.6432
Root MSE	=	3.4561

mpg	Coef.	Std. Err.	t	P>|t|	[95% Conf. Interval]	
weight	-0.01	0.00	-5.63	0.000	-0.01	-0.00
displacement	0.01	0.01	0.54	0.594	-0.01	0.02
_cons	40.08	2.02	19.84	0.000	36.06	44.11

```
. regress mpg weight displacement, cformat(%9.3f)
```

Source	SS	df	MS
Model	1595.40969	2	797.704846
Residual	848.049768	71	11.9443629
Total	2443.45946	73	33.4720474

Number of obs	=	74
F(2, 71)	=	66.79
Prob > F	=	0.0000
R-squared	=	0.6529
Adj R-squared	=	0.6432
Root MSE	=	3.4561

mpg	Coef.	Std. Err.	t	P>|t|	[95% Conf. Interval]	
weight	-0.007	0.001	-5.63	0.000	-0.009	-0.004
displacement	0.005	0.010	0.54	0.594	-0.014	0.025
_cons	40.085	2.020	19.84	0.000	36.057	44.113

To reset the cformat setting to its command-specific default, type

```
. set cformat
. regress mpg weight displacement
```

Source	SS	df	MS	
Model	1595.40969	2	797.704846	
Residual	848.049768	71	11.9443629	
Total	2443.45946	73	33.4720474	

Number of obs	=	74
F(2, 71)	=	66.79
Prob > F	=	0.0000
R-squared	=	0.6529
Adj R-squared	=	0.6432
Root MSE	=	3.4561

mpg	Coef.	Std. Err.	t	P>|t|	[95% Conf. Interval]	
weight	-.0065671	.0011662	-5.63	0.000	-.0088925	-.0042417
displacement	.0052808	.0098696	0.54	0.594	-.0143986	.0249602
_cons	40.08452	2.02011	19.84	0.000	36.05654	44.11251

◁

Also see

[R] **estimation options** — Estimation options

[R] **set** — Overview of system parameters

[R] **query** — Display system parameters

[U] **20.8 Formatting the coefficient table**

Title

> **set_defaults** — Reset system parameters to original Stata defaults

Syntax

set_defaults { *category* | _all } [, <u>perm</u>anently]

where *category* is one of <u>mem</u>ory | <u>out</u>put | <u>inter</u>face | <u>graph</u>ics | <u>eff</u>iciency |

<u>net</u>work | <u>up</u>date | trace | mata | <u>oth</u>er

Description

set_defaults resets settings made by set to the original default settings that were shipped with Stata.

set_defaults _all resets all the categories, whereas set defaults *category* resets only the settings for the specified category.

Option

permanently specifies that, in addition to making the change right now, the settings be remembered and become the default settings when you invoke Stata.

Remarks

▷ Example 1

To assist us in debugging a new command, we modified some of the trace settings. To return them to their original values, we type

```
. set_defaults trace
-> set trace off
-> set tracedepth 32000
-> set traceexpand on
-> set tracesep on
-> set traceindent on
-> set tracenumber off
-> set tracehilite ""
(preferences reset)
```

◁

Methods and formulas

set_defaults is implemented as an ado-file.

Also see

[R] **query** — Display system parameters

[R] **set** — Overview of system parameters

[M-3] **mata set** — Set and display Mata system parameters

Title

set emptycells — Set what to do with empty cells in interactions

Syntax

set emptycells { keep | drop } [, underline{perma}nently]

Description

set emptycells allows you to control how Stata handles interaction terms with empty cells. Stata can keep empty cells or drop them. The default is to keep empty cells.

Option

permanently specifies that, in addition to making the change right now, the setting be remembered and become the default setting when you invoke Stata.

Remarks

By default, Stata keeps empty cells so they can be reported in the coefficient table. For example, type

```
. use http://www.stata-press.com/data/r12/auto
. regress mpg rep78#foreign, baselevels
```

and you will see a regression of mpg on 10 indicator variables because rep78 takes on 5 values and foreign takes on 2 values in the auto dataset. Two of those cells will be reported as empty because the data contain no observations of foreign cars with a rep78 value of 1 or 2.

Many real datasets contain a large number of empty cells, and this could cause the "matsize too small" error, r(908). In that case, type

```
. set emptycells drop
```

to get Stata to drop empty cells from the list of coefficients. If you commonly fit models with empty cells, you can permanently set Stata to drop empty cells by typing the following:

```
. set emptycells drop, permanently
```

Also see

[R] **set** — Overview of system parameters

Title

> **set seed** — Specify initial value of random-number seed

Syntax

set <u>seed</u> #

set <u>seed</u> *statecode*

where

is any number between 0 and 2^{31-1} (2,147,483,647), and

statecode is a random-number state previously obtained from `creturn` value `c(seed)`.

Description

`set seed` # specifies the initial value of the random-number seed used by the random-number functions, such as `runiform()` and `rnormal()`.

`set seed` *statecode* resets the state of the random-number functions to the value specified, which is a state previously obtained from `creturn` value `c(seed)`.

Remarks

Remarks are presented under the following headings:

> *Examples*
> *Setting the seed*
> *How to choose a seed*
> *Do not set the seed too often*
> *Preserving and restoring the random-number generator state*

Examples

1. Specify initial value of random-number seed

   `. set seed 339487731`

2. Create variable u containing uniformly distributed pseudorandom numbers on the interval $[0, 1)$

   `. generate u = runiform()`

3. Create variable z containing normally distributed random numbers with mean 0 and standard deviation 1

   `. generate z = rnormal()`

4. Obtain state of pseudorandom-number generator and store it in a local macro named `state`

   `. local state = c(seed)`

5. Restore pseudorandom-number generator state to that previously stored in local macro named `state`

   `. set seed 'state'`

Setting the seed

Stata's random-number generation functions, such as `runiform()` and `rnormal()`, do not really produce random numbers. These functions are deterministic algorithms that produce numbers that can pass for random. `runiform()` produces numbers that can pass for independent draws from a rectangular distribution over $[0, 1)$; `rnormal()` produces numbers that can pass for independent draws from $N(0, 1)$. Stata's random-number functions are formally called pseudorandom-number functions.

The sequences these functions produce are determined by the seed, which is just a number and which is set to 123456789 every time Stata is launched. This means that `runiform()` produces the same sequence each time you start Stata. The first time you use `runiform()` after Stata is launched, `runiform()` returns 0.136984078446403146. The second time you use it, `runiform()` returns 0.643220667960122228. The third time you use it, ...

To obtain different sequences, you must specify different seeds using the `set seed` command. You might specify the seed 472195:

```
. set seed 472195
```

If you were now to use `runiform()`, the first call would return 0.247166610788553953, the second call would return 0.593119932804256678, and so on. Whenever you `set seed 472195`, `runiform()` will return those numbers the first two times you use it.

Thus you set the seed to obtain different pseudorandom sequences from the pseudorandom-number functions.

If you record the seed you set, pseudorandom results such as results from a simulation or imputed values from `mi impute` can be reproduced later. Whatever you do after setting the seed, if you set the seed to the same value and repeat what you did, you will obtain the same results.

How to choose a seed

Your best choice for the seed is an element chosen randomly from the set $\{0, 1, \ldots, 2{,}147{,}483{,}647\}$. We recommend that, but that is difficult to achieve because finding easy-to-access, truly random sources is difficult.

One person we know uses digits from the serial numbers from dollar bills he finds in his wallet. Of course, the numbers he obtains are not really random, but they are good enough, and they are probably a good deal more random than the seeds most people choose. Some people use dates and times, although we recommend against that because, over the day, it just gets later and later, and that is a pattern. Others try to make up a random number, figuring if they include enough digits, the result just has to be random. This is a variation on the five-second rule for dropped food, and we admit to using both of these rules.

It does not really matter how you set the seed, as long as there is no obvious pattern in the seeds that you set and as long as you do not set the seed too often during a session.

Nonetheless, here are two methods that we have seen used but you should not use:

1. The first time you set the seed, you set the number 1. The next time, you set 2, and then 3, and so on. Variations on this included setting 1001, 1002, 1003, . . . , or setting 1001, 2001, 3001, and so on.

   Do not follow any of these procedures. The seeds you set must not exhibit a pattern.

2. To set the seed, you obtain a pseudorandom number from `runiform()` and then use the digits from that to form the seed.

   This is a bad idea because the pseudorandom-number generator can converge to a cycle. If you obtained the pseudorandom-number generator unrelated to those in Stata, this would work well, but then you would have to find a rule to set the first generator's seed. In any case, the pseudorandom-number generators in Stata are all closely related, and so you must not follow this procedure.

Choosing seeds that do not exhibit a pattern is of great importance. That the seeds satisfy the other properties of randomness is minor by comparison.

Do not set the seed too often

We cannot emphasize this enough: Do not set the seed too often.

To see why this is such a bad idea, consider the limiting case: You set the seed, draw one pseudorandom number, reset the seed, draw again, and so continue. The pseudorandom numbers you obtain will be nothing more than the seeds you run through a mathematical function. The results you obtain will not pass for random unless the seeds you choose pass for random. If you already had such numbers, why are you even bothering to use the pseudorandom-number generator?

The definition of too often is more than once per problem.

If you are running a simulation of 10,000 replications, set the seed at the start of the simulation and do not reset it until the 10,000th replication is finished. The pseudorandom-number generators provided by Stata have long periods. The longer you go between setting the seed, the more random-like are the numbers produced.

It is sometimes useful later to be able to reproduce in isolation any one of the replications, and so you might be tempted to set the seed to a known value for each of the replications. We negatively mentioned setting the seed to 1, 2, ..., and it is in exactly such situations that we have seen this done. The advantage, however, is that you could reproduce the fifth replication merely by setting the seed to 5 and then repeating whatever it is that is to be replicated. If this is your goal, you do not need to reset the seed. You can record the state of the random-number generator, save the state with your replication results, and then use the recorded states later to reproduce whichever of the replications that you wish. This will be discussed in *Preserving and restoring the random-number generator state*.

There is another reason you might be tempted to set the seed more than once per problem. It sometimes happens that you run a simulation, let's say for 5,000 replications, and then you decide you should have run it for 10,000 replications. Instead of running all 10,000 replications afresh, you decide to save time by running another 5,000 replications and then combining those results with your previous 5,000 results. That is okay. We at StataCorp do this kind of thing. If you do this, it is important that you set the seed especially well, particularly if you repeat this process to add yet another 5,000 replications. It is also important that in each run there be a large enough number of replications, which is say thousands of them.

Even so, do not do this: You want 500,000 replications. To obtain them, you run in batches of 1,000, setting the seed 500 times. Unless you have a truly random source for the seeds, it is unlikely you can produce a patternless sequence of 500 seeds. The fact that you ran 1,000 replications in between choosing the seeds does not mitigate the requirement that there be no pattern to the seeds you set.

In all cases, the best solution is to set the seed only once and then use the method we suggest in the next section.

Preserving and restoring the random-number generator state

In the previous section, we discussed the case in which you might be tempted to set the seed more frequently than otherwise necessary, either to save time or to be able to rerun any one of the replications. In such cases, there is an alternative to setting a new seed: recording the state of the pseudorandom-number generator and then restoring the state later should the need arise.

The state of the random-number generator is a string that looks like this:

```
Xb5804563c43f462544a474abacbdd93d00021fb3
```

You can obtain the state from c(seed):

```
. display c(seed)
Xb5804563c43f462544a474abacbdd93d00021fb3
```

The name c(seed) is unfortunate because it suggests that Xb5804563c43f462544a474abacbdd93d00021fb3 is nothing more than a seed such as 1073741823 in a different guise. It is not. A better name for c(seed) would have been c(rng_state). The state string specifies an entry point into the sequence produced by the pseudorandom-number generator. Let us explain.

The best way to use a pseudorandom-number generator would be to choose a seed once, draw random numbers until you use up the generator, and then get a new generator and choose a new key. Pseudorandom-number generators have a period, after which they repeat the original sequence. That is what we mean by using up a generator. The period of the pseudorandom-number generator that Stata is currently using is over 2^{123}. Stata uses the KISS generator. It is difficult to imagine that you could ever use up KISS.

The string reported by c(seed) reports an encoded form of the information necessary for Stata to reestablish exactly where it is located in the pseudorandom-number generator's sequence.

We are not seriously suggesting you choose only one seed over your entire lifetime, but let's look at how you might do that. Sometime after birth, when you needed your first random number, you would set your seed,

```
. set seed 1073741823
```

On that day, you would draw, say, 10,000 pseudorandom numbers, perhaps to impute some missing values. Being done for the day, you type

```
. display c(seed)
X15b512f3b2143ab434f1c92f4e7058e400023bc3
```

The next day, after launching Stata, you type

```
. set seed X15b512f3b2143ab434f1c92f4e7058e400023bc3
```

When you type set seed followed by a state string rather than a number, instead of setting the seed, Stata reestablishes the previous state. Thus the next time you draw a pseudorandom number, Stata will produce the 10,001st result after setting seed 1073741823. Let's assume that you draw 100,000 numbers this day. Done for the day, you display c(seed).

```
. display c(seed)
X5d13d693a72ad0602b093cc4f61e07a500020381
```

On the third day, after setting the seed to the string above, you will be in a position to draw the 110,001st pseudorandom number.

In this way, you would eat your way though the 2^{123} random numbers, but you would be unlikely ever to make it to the end. Assuming you did this every day for 100 years, to arrive at the end of the sequence you would need to consume 2.9e+32 pseudorandom numbers per day.

We do not expect you to set the seed just once in your life, but using the state string makes it easy to set the seed just once for a problem.

When we do simulations at StataCorp, we record c(seed) for each replication. Just like everybody else, we record results from replications as observations in datasets; we just happen to have an extra variable in the dataset, namely, a string variable named state. That string is filled in observation by observation from the then-current values of c(seed), which is a function and so can be used in any context that a function can be used in Stata.

Anytime we want to reproduce a particular replication, we thus have the information we need to reset the pseudorandom-number generator, and having it in the dataset is convenient because we had to go there anyway to determine which replication we wanted to reproduce.

In addition to recording each of the state strings for each replication, we record the closing value of c(seed) as a note, which is easy enough to do:

```
. note: closing state 'c(seed)'
```

If we want to add more replications later, we have a state string that we can use to continue from where we left off.

Also see

[R] **set** — Overview of system parameters

[D] **functions** — Functions

Title

> **set showbaselevels** — Display settings for coefficient tables

Syntax

> set showbaselevels [on | off | all] [, permanently]
>
> set showemptycells [on | off] [, permanently]
>
> set showomitted [on | off] [, permanently]

Description

set showbaselevels specifies whether to display base levels of factor variables and their interactions in coefficient tables. set showbaselevels on specifies that base levels be reported for factor variables and for interactions whose bases cannot be inferred from their component factor variables. set showbaselevels all specifies that all base levels of factor variables and interactions be reported.

set showemptycells specifies whether to display empty cells in coefficient tables.

set showomitted specifies whether to display omitted coefficients in coefficient tables.

Option

permanently specifies that, in addition to making the change right now, the setting be remembered and become the default setting when you invoke Stata.

Remarks

▷ Example 1

We illustrate these three set commands using cholesterol2.dta.

```
. use http://www.stata-press.com/data/r12/cholesterol2
(Artificial cholesterol data, empty cells)

. generate x = race

. regress chol race##agegrp x
note: 2.race#2.agegrp identifies no observations in the sample
note: x omitted because of collinearity
```

Source	SS	df	MS		Number of obs =	70
					F(13, 56) =	13.51
Model	15751.6113	13	1211.66241		Prob > F =	0.0000
Residual	5022.71559	56	89.6913498		R-squared =	0.7582
					Adj R-squared =	0.7021
Total	20774.3269	69	301.077201		Root MSE =	9.4706

chol	Coef.	Std. Err.	t	P>|t|	[95% Conf. Interval]	
race						
2	12.84185	5.989703	2.14	0.036	.8430383	24.84067
3	-.167627	5.989703	-0.03	0.978	-12.16644	11.83119
agegrp						
2	17.24681	5.989703	2.88	0.006	5.247991	29.24562
3	31.43847	5.989703	5.25	0.000	19.43966	43.43729
4	34.86613	5.989703	5.82	0.000	22.86732	46.86495
5	44.43374	5.989703	7.42	0.000	32.43492	56.43256
race#agegrp						
2 2	0	(empty)				
2 3	-22.83983	8.470719	-2.70	0.009	-39.80872	-5.870939
2 4	-14.67558	8.470719	-1.73	0.089	-31.64447	2.293306
2 5	-10.51115	8.470719	-1.24	0.220	-27.48004	6.457735
3 2	-6.054425	8.470719	-0.71	0.478	-23.02331	10.91446
3 3	-11.48083	8.470719	-1.36	0.181	-28.44971	5.488063
3 4	-.6796112	8.470719	-0.08	0.936	-17.6485	16.28928
3 5	-1.578052	8.470719	-0.19	0.853	-18.54694	15.39084
x	0	(omitted)				
_cons	175.2309	4.235359	41.37	0.000	166.7464	183.7153

```
. set showemptycells off

. set showomitted off

. set showbaselevels all
```

```
. regress chol race##agegrp x
note: 2.race#2.agegrp identifies no observations in the sample
note: x omitted because of collinearity
```

Source	SS	df	MS
Model	15751.6113	13	1211.66241
Residual	5022.71559	56	89.6913498
Total	20774.3269	69	301.077201

```
Number of obs =      70
F( 13,   56) =   13.51
Prob > F      =  0.0000
R-squared     =  0.7582
Adj R-squared =  0.7021
Root MSE      =  9.4706
```

chol	Coef.	Std. Err.	t	P>|t|	[95% Conf. Interval]	
race						
1	0	(base)				
2	12.84185	5.989703	2.14	0.036	.8430383	24.84067
3	-.167627	5.989703	-0.03	0.978	-12.16644	11.83119
agegrp						
1	0	(base)				
2	17.24681	5.989703	2.88	0.006	5.247991	29.24562
3	31.43847	5.989703	5.25	0.000	19.43966	43.43729
4	34.86613	5.989703	5.82	0.000	22.86732	46.86495
5	44.43374	5.989703	7.42	0.000	32.43492	56.43256
race#agegrp						
1 1	0	(base)				
1 2	0	(base)				
1 3	0	(base)				
1 4	0	(base)				
1 5	0	(base)				
2 1	0	(base)				
2 3	-22.83983	8.470719	-2.70	0.009	-39.80872	-5.870939
2 4	-14.67558	8.470719	-1.73	0.089	-31.64447	2.293306
2 5	-10.51115	8.470719	-1.24	0.220	-27.48004	6.457735
3 1	0	(base)				
3 2	-6.054425	8.470719	-0.71	0.478	-23.02331	10.91446
3 3	-11.48083	8.470719	-1.36	0.181	-28.44971	5.488063
3 4	-.6796112	8.470719	-0.08	0.936	-17.6485	16.28928
3 5	-1.578052	8.470719	-0.19	0.853	-18.54694	15.39084
_cons	175.2309	4.235359	41.37	0.000	166.7464	183.7153

To restore the display of empty cells, omitted predictors, and baselevels to their command-specific default behavior, type

```
. set showemptycells

. set showomitted

. set showbaselevels

. regress chol race##agegrp x
note: 2.race#2.agegrp identifies no observations in the sample
note: x omitted because of collinearity
```

Source	SS	df	MS			
Model	15751.6113	13	1211.66241			
Residual	5022.71559	56	89.6913498			
Total	20774.3269	69	301.077201			

Number of obs = 70
F(13, 56) = 13.51
Prob > F = 0.0000
R-squared = 0.7582
Adj R-squared = 0.7021
Root MSE = 9.4706

chol	Coef.	Std. Err.	t	P>|t|	[95% Conf.	Interval]
race						
2	12.84185	5.989703	2.14	0.036	.8430383	24.84067
3	-.167627	5.989703	-0.03	0.978	-12.16644	11.83119
agegrp						
2	17.24681	5.989703	2.88	0.006	5.247991	29.24562
3	31.43847	5.989703	5.25	0.000	19.43966	43.43729
4	34.86613	5.989703	5.82	0.000	22.86732	46.86495
5	44.43374	5.989703	7.42	0.000	32.43492	56.43256
race#agegrp						
2 2	0	(empty)				
2 3	-22.83983	8.470719	-2.70	0.009	-39.80872	-5.870939
2 4	-14.67558	8.470719	-1.73	0.089	-31.64447	2.293306
2 5	-10.51115	8.470719	-1.24	0.220	-27.48004	6.457735
3 2	-6.054425	8.470719	-0.71	0.478	-23.02331	10.91446
3 3	-11.48083	8.470719	-1.36	0.181	-28.44971	5.488063
3 4	-.6796112	8.470719	-0.08	0.936	-17.6485	16.28928
3 5	-1.578052	8.470719	-0.19	0.853	-18.54694	15.39084
x	0	(omitted)				
_cons	175.2309	4.235359	41.37	0.000	166.7464	183.7153

◁

Also see

[R] **set** — Overview of system parameters

[R] **query** — Display system parameters

Title

signrank — Equality tests on matched data

Syntax

Wilcoxon matched-pairs signed-ranks test

signrank *varname* = *exp* [*if*] [*in*]

Sign test of matched pairs

signtest *varname* = *exp* [*if*] [*in*]

by is allowed with signrank and signtest; see [D] **by**.

Menu

signrank

Statistics > Nonparametric analysis > Tests of hypotheses > Wilcoxon matched-pairs signed-rank test

signtest

Statistics > Nonparametric analysis > Tests of hypotheses > Test equality of matched pairs

Description

signrank tests the equality of matched pairs of observations by using the Wilcoxon matched-pairs signed-ranks test (Wilcoxon 1945). The null hypothesis is that both distributions are the same.

signtest also tests the equality of matched pairs of observations (Arbuthnott [1710], but better explained by Snedecor and Cochran [1989]) by calculating the differences between *varname* and the expression. The null hypothesis is that the median of the differences is zero; no further assumptions are made about the distributions. This, in turn, is equivalent to the hypothesis that the true proportion of positive (negative) signs is one-half.

For equality tests on unmatched data, see [R] **ranksum**.

Remarks

▷ Example 1: signrank

We are testing the effectiveness of a new fuel additive. We run an experiment with 12 cars. We first run each car without the fuel treatment and measure the mileage. We then add the fuel treatment and repeat the experiment. The results of the experiment are

Without treatment	With treatment	Without treatment	With treatment
20	24	18	17
23	25	24	28
21	21	20	24
25	22	24	27
18	23	23	21
17	18	19	23

We create two variables called mpg1 and mpg2, representing mileage without and with the treatment, respectively. We can test the null hypothesis that the treatment had no effect by typing

```
. use http://www.stata-press.com/data/r12/fuel
. signrank mpg1=mpg2
Wilcoxon signed-rank test
```

sign	obs	sum ranks	expected
positive	3	13.5	38.5
negative	8	63.5	38.5
zero	1	1	1
all	12	78	78

```
unadjusted variance       162.50
adjustment for ties        -1.62
adjustment for zeros       -0.25

adjusted variance         160.62

Ho: mpg1 = mpg2
          z =   -1.973
   Prob > |z| =   0.0485
```

The output indicates that we can reject the null hypothesis at any level above 4.85%.   ◁

▷ Example 2: signtest

signtest tests that the median of the differences is zero, making no further assumptions, whereas signrank assumed that the distributions are equal as well. Using the data above, we type

```
. signtest mpg1=mpg2
Sign test
```

sign	observed	expected
positive	3	5.5
negative	8	5.5
zero	1	1
all	12	12

```
One-sided tests:
  Ho: median of mpg1 - mpg2 = 0 vs.
  Ha: median of mpg1 - mpg2 > 0
      Pr(#positive >= 3) =
          Binomial(n = 11, x >= 3, p = 0.5) =  0.9673
  Ho: median of mpg1 - mpg2 = 0 vs.
  Ha: median of mpg1 - mpg2 < 0
      Pr(#negative >= 8) =
          Binomial(n = 11, x >= 8, p = 0.5) =  0.1133
```

```
Two-sided test:
  Ho: median of mpg1 - mpg2 = 0 vs.
  Ha: median of mpg1 - mpg2 != 0
       Pr(#positive >= 8 or #negative >= 8) =
           min(1, 2*Binomial(n = 11, x >= 8, p = 0.5)) =  0.2266
```

The summary table indicates that there were three comparisons for which mpg1 exceeded mpg2, eight comparisons for which mpg2 exceeded mpg1, and one comparison for which they were the same.

The output below the summary table is based on the binomial distribution. The significance of the one-sided test, where the alternative hypothesis is that the median of mpg2 − mpg1 is greater than zero, is 0.1133. The significance of the two-sided test, where the alternative hypothesis is simply that the median of the differences is different from zero, is $0.2266 = 2 \times 0.1133$.   ◁

Saved results

signrank saves the following in r():

Scalars

r(N_neg)	number of negative comparisons	r(sum_neg)	sum of the negative ranks
r(N_pos)	number of positive comparisons	r(z)	z statistic
r(N_tie)	number of tied comparisons	r(Var_a)	adjusted variance
r(sum_pos)	sum of the positive ranks		

signtest saves the following in r():

Scalars

r(N_neg)	number of negative comparisons	r(p_2)	two-sided probability
r(N_pos)	number of positive comparisons	r(p_neg)	one-sided probability of negative comparison
r(N_tie)	number of tied comparisons	r(p_pos)	one-sided probability of positive comparison

Methods and formulas

signrank and signtest are implemented as ado-files.

For a practical introduction to these techniques with an emphasis on examples rather than theory, see Bland (2000) or Sprent and Smeeton (2007). For a summary of these tests, see Snedecor and Cochran (1989).

Methods and formulas are presented under the following headings:

> *signrank*
> *signtest*

signrank

Both the sign test and Wilcoxon signed-rank tests test the null hypothesis that the distribution of a random variable $D = varname - exp$ has median zero. The sign test makes no additional assumptions, but the Wilcoxon signed-rank test makes the additional assumption that the distribution of D is symmetric. If $D = X_1 - X_2$, where X_1 and X_2 have the same distribution, then it follows that the distribution of D is symmetric about zero. Thus the Wilcoxon signed-rank test is often described as a test of the hypothesis that two distributions are the same, that is, $X_1 \sim X_2$.

Let d_j denote the difference for any matched pair of observations,

$$d_j = x_{1j} - x_{2j} = varname - exp$$

for $j = 1, 2, \ldots, n$.

Rank the absolute values of the differences, $|d_j|$, and assign any tied values the average rank. Consider the signs of d_j, and let

$$r_j = \text{sign}(d_j) \, \text{rank}(|d_j|)$$

be the signed ranks. The test statistic is

$$T_{\text{obs}} = \sum_{j=1}^{n} r_j = (\text{sum of ranks for } + \text{ signs}) - (\text{sum of ranks for } - \text{ signs})$$

The null hypothesis is that the distribution of d_j is symmetric about 0. Hence the likelihood is unchanged if we flip signs on the d_j, and thus the randomization datasets are the 2^n possible sign changes for the d_j. Thus the randomization distribution of our test statistic T can be computed by considering all the 2^n possible values of

$$T = \sum_{j=1}^{n} S_j r_j$$

where the r_j are the observed signed ranks (considered fixed) and S_j is either $+1$ or -1.

With this distribution, the mean and variance of T are given by

$$E(T) = 0 \qquad \text{and} \qquad \text{Var}_{\text{adj}}(T) = \sum_{j=1}^{n} r_j^2$$

The test statistic for the Wilcoxon signed-rank test is often expressed (equivalently) as the sum of the positive signed-ranks, T_+, where

$$E(T_+) = \frac{n(n+1)}{4} \qquad \text{and} \qquad \text{Var}_{\text{adj}}(T_+) = \frac{1}{4} \sum_{j=1}^{n} r_j^2$$

Zeros and ties do not affect the theory above, and the exact variance is still given by the above formula for $\text{Var}_{\text{adj}}(T_+)$. When $d_j = 0$ is observed, d_j will always be zero in each of the randomization datasets (using $\text{sign}(0) = 0$). When there are ties, you can assign averaged ranks for each group of ties and then treat them the same as the other ranks.

The "unadjusted variance" reported by signrank is the variance that the randomization distribution would have had if there had been no ties or zeros:

$$\text{Var}_{\text{unadj}}(T_+) = \frac{1}{4} \sum_{j=1}^{n} j^2 = \frac{n(n+1)(2n+1)}{24}$$

The adjustment for zeros is the change in the variance when the ranks for the zeros are signed to make $r_j = 0$,

$$\Delta \text{Var}_{\text{zero adj}}(T_+) = -\frac{1}{4} \sum_{j=1}^{n_0} j^2 = -\frac{n_0(n_0+1)(2n_0+1)}{24}$$

where n_0 is the number of zeros. The adjustment for ties is the change in the variance when the ranks (for nonzero observations) are replaced by averaged ranks:

$$\Delta\text{Var}_{\text{ties adj}}(T_+) = \text{Var}_{\text{adj}}(T_+) - \text{Var}_{\text{unadj}}(T_+) - \Delta\text{Var}_{\text{zero adj}}(T_+)$$

A normal approximation is used to calculate

$$z = \frac{T_+ - E(T_+)}{\sqrt{\text{Var}_{\text{adj}}(T_+)}}$$

signtest

The test statistic for the sign test is the number n_+ of differences

$$d_j = x_{1j} - x_{2j} = varname - exp$$

greater than zero. Assuming that the probability of a difference being equal to zero is exactly zero, then, under the null hypothesis, $n_+ \sim \text{binomial}(n, p = 1/2)$, where n is the total number of observations.

But what if some differences are zero? This question has a ready answer if you view the test from the perspective of Fisher's Principle of Randomization (Fisher 1935). Fisher's idea (stated in a modern way) was to look at a family of transformations of the observed data such that the a priori likelihood (under the null hypothesis) of the transformed data is the same as the likelihood of the observed data. The distribution of the test statistic is then produced by calculating its value for each of the transformed "randomization" datasets, assuming that each dataset is equally likely.

For the sign test, the "data" are simply the set of signs of the differences. Under the null hypothesis of the sign test, the probability that d_j is less than zero is equal to the probability that d_j is greater than zero. Thus you can transform the observed signs by flipping any number of them, and the set of signs will have the same likelihood. The 2^n possible sign changes form the family of randomization datasets. If you have no zeros, this procedure again leads to $n_+ \sim \text{binomial}(n, p = 1/2)$.

If you do have zeros, changing their signs leaves them as zeros. So, if you observe n_0 zeros, each of the 2^n sign-change datasets will also have n_0 zeros. Hence, the values of n_+ calculated over the sign-change datasets range from 0 to $n - n_0$, and the "randomization" distribution of n_+ is $\text{binomial}(n - n_0, p = 1/2)$.

The work of Arbuthnott (1710) and later eighteenth-century contributions is discussed by Hald (2003, chap. 17).

Frank Wilcoxon (1892–1965) was born in Ireland to American parents. After working in various occupations (including merchant seaman, oil-well pump attendant, and tree surgeon), he settled in chemistry, gaining degrees from Rutgers and Cornell and employment from various companies. Working mainly on the development of fungicides and insecticides, Wilcoxon became interested in statistics in 1925 and made several key contributions to nonparametric methods. After retiring from industry, he taught statistics at Florida State until his death.

References

Arbuthnott, J. 1710. An argument for divine providence, taken from the constant regularity observed in the births of both sexes. *Philosophical Transaction of the Royal Society of London* 27: 186–190.

Bland, M. 2000. *An Introduction to Medical Statistics.* 3rd ed. Oxford: Oxford University Press.

Bradley, R. A. 2001. Frank Wilcoxon. In *Statisticians of the Centuries*, ed. C. C. Heyde and E. Seneta, 420–424. New York: Springer.

Fisher, R. A. 1935. *The Design of Experiments.* Edinburgh: Oliver & Boyd.

Hald, A. 2003. *A History of Probability and Statistics and Their Applications before 1750.* New York: Wiley.

Kaiser, J. 2007. An exact and a Monte Carlo proposal to the Fisher–Pitman permutation tests for paired replicates and for independent samples. *Stata Journal* 7: 402–412.

Newson, R. 2006. Confidence intervals for rank statistics: Somers' D and extensions. *Stata Journal* 6: 309–334.

Snedecor, G. W., and W. G. Cochran. 1989. *Statistical Methods.* 8th ed. Ames, IA: Iowa State University Press.

Sprent, P., and N. C. Smeeton. 2007. *Applied Nonparametric Statistical Methods.* 4th ed. Boca Raton, FL: Chapman & Hall/CRC.

Sribney, W. M. 1995. crc40: Correcting for ties and zeros in sign and rank tests. *Stata Technical Bulletin* 26: 2–4. Reprinted in *Stata Technical Bulletin Reprints*, vol. 5, pp. 5–8. College Station, TX: Stata Press.

Wilcoxon, F. 1945. Individual comparisons by ranking methods. *Biometrics* 1: 80–83.

Also see

[R] **ranksum** — Equality tests on unmatched data

[R] **ttest** — Mean-comparison tests

Title

> **simulate** — Monte Carlo simulations

Syntax

> simulate [*exp_list*] , <u>rep</u>s(*#*) [*options*] : *command*

options	Description
<u>nodots</u>	suppress replication dots
<u>noi</u>sily	display any output from *command*
<u>trace</u>	trace *command*
<u>sa</u>ving(*filename*, ...)	save results to *filename*
<u>nol</u>egend	suppress table legend
<u>v</u>erbose	display the full table legend
seed(*#*)	set random-number seed to *#*

All weight types supported by *command* are allowed; see [U] **11.1.6 weight**.

exp_list contains	(*name*: *elist*)
	elist
	eexp
elist contains	*newvar* = (*exp*)
	(*exp*)
eexp is	*specname*
	[*eqno*]*specname*
specname is	_b
	_b[]
	_se
	_se[]
eqno is	# #
	name

exp is a standard Stata expression; see [U] **13 Functions and expressions**.

Distinguish between [ ], which are to be typed, and [ ], which indicate optional arguments.

Description

simulate eases the programming task of performing Monte Carlo–type simulations. Typing

> . simulate *exp_list*, reps(*#*): *command*

runs *command* for *#* replications and collects the results in *exp_list*.

command defines the command that performs one simulation. Most Stata commands and user-written programs can be used with simulate, as long as they follow standard Stata syntax; see [U] **11 Language syntax**. The by prefix may not be part of *command*.

exp_list specifies the expression to be calculated from the execution of *command*. If no expressions are given, *exp_list* assumes a default, depending upon whether *command* changes results in e() or r(). If *command* changes results in e(), the default is _b. If *command* changes results in r() (but not e()), the default is all the scalars posted to r(). It is an error not to specify an expression in *exp_list* otherwise.

Options

reps(#) is required—it specifies the number of replications to be performed.

nodots suppresses display of the replication dots. By default, one dot character is displayed for each successful replication. A red 'x' is displayed if *command* returns an error or if one of the values in *exp_list* is missing.

noisily requests that any output from *command* be displayed. This option implies the nodots option.

trace causes a trace of the execution of *command* to be displayed. This option implies the noisily option.

saving(*filename*[, *suboptions*]) creates a Stata data file (.dta file) consisting of (for each statistic in *exp_list*) a variable containing the simulated values.

   double specifies that the results for each replication be stored as doubles, meaning 8-byte reals. By default, they are stored as floats, meaning 4-byte reals.

   every(#) specifies that results be written to disk every #th replication. every() should be specified only in conjunction with saving() when *command* takes a long time for each replication. This will allow recovery of partial results should some other software crash your computer. See [P] **postfile**.

   replace specifies that *filename* be overwritten if it exists.

nolegend suppresses display of the table legend. The table legend identifies the rows of the table with the expressions they represent.

verbose requests that the full table legend be displayed. By default, coefficients and standard errors are not displayed.

seed(#) sets the random-number seed. Specifying this option is equivalent to typing the following command before calling simulate:

   . set seed #

Remarks

For an introduction to Monte Carlo methods, see Cameron and Trivedi (2010, chap. 4). White (2010) provides a command for analyzing results of simulation studies.

▷ Example 1

We have a dataset containing means and variances of 100-observation samples from a lognormal distribution (as a first step in evaluating, say, the coverage of a 95%, *t*-based confidence interval). Then we perform the experiment 1,000 times.

The following command definition will generate 100 independent observations from a lognormal distribution and compute the summary statistics for this sample.

```
program lnsim, rclass
        version 12
        drop _all
        set obs 100
        gen z = exp(rnormal())
        summarize z
        return scalar mean = r(mean)
        return scalar Var  = r(Var)
end
```

We can save 1,000 simulated means and variances from lnsim by typing

```
. set seed 1234

. simulate mean=r(mean) var=r(Var), reps(1000) nodots: lnsim
        command:  lnsim
           mean:  r(mean)
            var:  r(Var)

. describe *

              storage  display    value
variable name  type    format     label     variable label

mean           float   %9.0g                 r(mean)
var            float   %9.0g                 r(Var)

. summarize
    Variable |      Obs       Mean    Std. Dev.       Min        Max

        mean |     1000    1.638466    .214371    1.095099    2.887392
         var |     1000     4.63856   6.428406       .8626    175.3746
```

◁

❏ Technical note

Before executing our lnsim simulator, we can verify that it works by executing it interactively.

```
. set seed 1234

. lnsim
obs was 0, now 100
    Variable |      Obs       Mean    Std. Dev.       Min        Max

           z |      100    1.597757    1.734328    .0625807    12.71548

. return list

scalars:
            r(Var) =  3.007893773683719
           r(mean) =  1.59775722913444
```

❏

▷ Example 2

Consider a more complicated problem. Let's experiment with fitting $y_j = a + bx_j + u_j$ when the true model has $a = 1$, $b = 2$, $u_j = z_j + cx_j$, and when z_j is $N(0, 1)$. We will save the parameter estimates and standard errors and experiment with varying c. x_j will be fixed across experiments but will originally be generated as $N(0, 1)$. We begin by interactively making the true data:

```
. drop _all
. set obs 100
obs was 0, now 100
. set seed 54321
. gen x = rnormal()
. gen true_y = 1+2*x
. save truth
file truth.dta saved
```

Our program is

```
program hetero1
        version 12
        args c
        use truth, clear
        gen y = true_y + (rnormal() + 'c'*x)
        regress y x
end
```

Note the use of 'c' in our statement for generating y. c is a local macro generated from args c and thus refers to the first argument supplied to hetero1. If we want $c = 3$ for our experiment, we type

```
. simulate _b _se, reps(10000): hetero1 3
```
   (*output omitted*)

Our program hetero1 could, however, be more efficient because it rereads the file truth once every replication. It would be better if we could read the data just once. In fact, if we read in the data right before running simulate, we really should not have to reread for each subsequent replication. A faster version reads

```
program hetero2
        version 12
        args c
        capture drop y
        gen y = true_y + (rnormal() + 'c'*x)
        regress y x
end
```

Requiring that the current dataset has the variables true_y and x may become inconvenient. Another improvement would be to require that the user supply variable names, such as in

```
program hetero3
        version 12
        args truey x c
        capture drop y
        gen y = 'truey' + (rnormal() + 'c'*'x')
        regress y x
end
```

Thus we can type

```
. simulate _b _se, reps(10000): hetero3 true_y x 3
```
   (*output omitted*)

◁

▷ Example 3

Now let's consider the problem of simulating the ratio of two medians. Suppose that each sample of size n_i comes from a normal population with a mean μ_i and standard deviation σ_i, where $i = 1, 2$. We write the program below and save it as a text file called `myratio.ado` (see [U] **17 Ado-files**). Our program is an `rclass` command that requires six arguments as input, identified by the local macros `n1`, `mu1`, `sigma1`, `n2`, `mu2`, and `sigma2`, which correspond to n_1, μ_1, σ_1, n_2, μ_2, and σ_2, respectively. With these arguments, `myratio` will generate the data for the two samples, use `summarize` to compute the two medians and save the ratio of the medians in `r(ratio)`.

```
program myratio, rclass
        version 12
        args n1 mu1 sigma1 n2 mu2 sigma2
        //  generate the data
        drop _all
        local N = 'n1'+'n2'
        set obs 'N'
        tempvar y
        generate 'y' = rnormal()
        replace 'y' = cond(_n<='n1','mu1'+'y'*'sigma1','mu2'+'y'*'sigma2')
        //  calculate the medians
        tempname m1
        summarize 'y' if _n<='n1', detail
        scalar 'm1' = r(p50)
        summarize 'y' if _n>'n1', detail
        //  save the results
        return scalar ratio = 'm1' / r(p50)
end
```

The result of running our simulation is

```
. set seed 19192
. simulate ratio=r(ratio), reps(1000) nodots: myratio 5 3 1 10 3 2
        command:  myratio 5 3 1 10 3 2
          ratio:  r(ratio)
. summarize
```

Variable	Obs	Mean	Std. Dev.	Min	Max
ratio	1000	1.08571	.4427828	.3834799	6.742217

◁

❏ Technical note

Stata lets us do simulations of simulations and simulations of bootstraps. Stata's `bootstrap` command (see [R] **bootstrap**) works much like `simulate`, except that it feeds the user-written program a bootstrap sample. Say that we want to evaluate the bootstrap estimator of the standard error of the median when applied to lognormally distributed data. We want to perform a simulation, resulting in a dataset of medians and bootstrap estimated standard errors.

As background, `summarize` (see [R] **summarize**) calculates summary statistics, leaving the mean in `r(mean)` and the standard deviation in `r(sd)`. `summarize` with the `detail` option also calculates summary statistics, but more of them, and leaves the median in `r(p50)`.

Thus our plan is to perform simulations by randomly drawing a dataset: we calculate the median of our random sample, we use `bootstrap` to obtain a dataset of medians calculated from bootstrap samples of our random sample, the standard deviation of those medians is our estimate of the standard error, and the summary statistics are saved in the results of `summarize`.

Our simulator is

```
program define bsse, rclass
        version 12
        drop _all
        set obs 100
        gen x = rnormal()
        tempfile bsfile
        bootstrap midp=r(p50), rep(100) saving('bsfile'): summarize x, detail
        use 'bsfile', clear
        summarize midp
        return scalar mean = r(mean)
        return scalar sd   = r(sd)
end
```

We can obtain final results, running our simulation 1,000 times, by typing

```
. set seed 48901

. simulate med=r(mean) bs_se=r(sd), reps(1000): bsse

      command:  bsse
          med:  r(mean)
        bs_se:  r(sd)

Simulations (1000)
    ──────┼─── 1 ───┼─── 2 ───┼─── 3 ───┼─── 4 ───┼─── 5
    .................................................         50
    .................................................        100
    .................................................        150
    .................................................        200
    .................................................        250
    .................................................        300
    .................................................        350
    .................................................        400
    .................................................        450
    .................................................        500
    .................................................        550
    .................................................        600
    .................................................        650
    .................................................        700
    .................................................        750
    .................................................        800
    .................................................        850
    .................................................        900
    .................................................        950
    .................................................       1000

. summarize
```

Variable	Obs	Mean	Std. Dev.	Min	Max
med	1000	-.0008696	.1210451	-.3132536	.4058724
bs_se	1000	.126236	.029646	.0326791	.2596813

This is a case where the simulation dots (drawn by default, unless the nodots option is specified) will give us an idea of how long this simulation will take to finish as it runs.   ❑

Methods and formulas

simulate is implemented as an ado-file.

References

Cameron, A. C., and P. K. Trivedi. 2010. *Microeconometrics Using Stata*. Rev. ed. College Station, TX: Stata Press.

Gould, W. W. 1994. ssi6.1: Simplified Monte Carlo simulations. *Stata Technical Bulletin* 20: 22–24. Reprinted in *Stata Technical Bulletin Reprints*, vol. 4, pp. 207–210. College Station, TX: Stata Press.

Hamilton, L. C. 2009. *Statistics with Stata (Updated for Version 10)*. Belmont, CA: Brooks/Cole.

Hilbe, J. M. 2010. Creating synthetic discrete-response regression models. *Stata Journal* 10: 104–124.

Weesie, J. 1998. ip25: Parameterized Monte Carlo simulations: Enhancement to the simulation command. *Stata Technical Bulletin* 43: 13–15. Reprinted in *Stata Technical Bulletin Reprints*, vol. 8, pp. 75–77. College Station, TX: Stata Press.

White, I. R. 2010. simsum: Analyses of simulation studies including Monte Carlo error. *Stata Journal* 10: 369–385.

Also see

[R] **bootstrap** — Bootstrap sampling and estimation

[R] **jackknife** — Jackknife estimation

[R] **permute** — Monte Carlo permutation tests

Title

sj — Stata Journal and STB installation instructions

Description

The *Stata Journal* (SJ) is a quarterly journal containing articles about statistics, data analysis, teaching methods, and effective use of Stata's language. The SJ publishes reviewed papers together with shorter notes and comments, regular columns, tips, book reviews, and other material of interest to researchers applying statistics in a variety of disciplines. You can read all about the *Stata Journal* at http://www.stata-journal.com.

The *Stata Journal* is a printed and electronic journal with corresponding software. If you want the journal, you must subscribe, but the software is available for no charge from our website at http://www.stata-journal.com. PDF copies of SJ articles that are older than three years are available for download for no charge at http://www.stata-journal.com/archives.html. More recent articles may be individually purchased.

The predecessor to the *Stata Journal* was the *Stata Technical Bulletin* (STB). The STB was also a printed and electronic journal with corresponding software. PDF copies of all STB journals are available for download for no charge at http://www.stata-press.com/journals/stbj.html. The STB software is available for no charge from our website at http://www.stata.com.

Below are instructions for installing the *Stata Journal* and the *Stata Technical Bulletin* software from our website.

Remarks

Remarks are presented under the following headings:

> *Installing the Stata Journal software*
> > *Obtaining from the Internet by pointing and clicking*
> > *Obtaining from the Internet via command mode*
> *Installing the STB software*
> > *Obtaining from the Internet by pointing and clicking*
> > *Obtaining from the Internet via command mode*

Installing the Stata Journal software

Each issue of the *Stata Journal* is labeled Volume #, Number #. Volume 1 refers to the first year of publication, Volume 2 to the second, and so on. Issues are numbered 1, 2, 3, and 4 within each year. The first issue of the *Journal* was published in the fourth quarter of 2001, and that issue is numbered Volume 1, Number 1. For installation purposes, we refer to this issue as sj1-1.

The articles, columns, notes, and comments that make up the *Stata Journal* are assigned a letter-and-number code, called an insert tag, such as st0001, an0034, or ds0011. The letters represent a category: st is the statistics category, an is the announcements category, etc. The numbers are assigned sequentially, so st0001 is the first article in the statistics category.

Sometimes inserts are subsequently updated, either to fix bugs or to add new features. A number such as st0001_1 indicates that this article, column, note, or comment is an update to the original st0001 article. Updates are complete; that is, installing st0001_1 provides all the features of the original article and more.

The *Stata Journal* software may be obtained by pointing and clicking or by using command mode.

The sections below detail how to install an insert. In all cases, pretend that you wish to install insert st0001_1 from `sj2-2`.

Obtaining from the Internet by pointing and clicking

1. Select **Help > SJ and User-written Programs**.

2. Click on *Stata Journal*.

3. Click on *sj2-2*.

4. Click on *st0001_1*.

5. Click on *(click here to install)*.

Obtaining from the Internet via command mode

Type the following:

```
. net from http://www.stata-journal.com/software
. net cd sj2-2
. net describe st0001_1
. net install st0001_1
```

The above could be shortened to

```
. net from http://www.stata-journal.com/software/sj2-2
. net describe st0001_1
. net install st0001_1
```

Alternatively, you could type

```
. net sj 2-2
. net describe st0001_1
. net install st0001_1
```

but going about it the long way is more entertaining, at least the first time.

Installing the STB software

Each issue of the STB is numbered. STB-1 refers to the first issue (published May 1991), STB-2 refers to the second (published July 1991), and so on.

An issue of the STB consists of inserts—articles—and these are assigned letter-and-number combinations, such as sg84, dm80, sbe26.1, etc. The letters represent a category; for example, sg is the general statistics category and dm the data-management category. The numbers are assigned sequentially, so sbe39 is the 39th insert in the biostatistics and epidemiology series.

Insert sbe39, it turns out, provides a method of accounting for publication bias in meta-analysis; it adds a new command called `metatrim` to Stata. If you installed sbe39, you would have that command and its online help. Insert sbe39 was published in STB-57 (September 2000). Obtaining `metatrim` simply requires going to STB-57 and getting sbe39.

Sometimes inserts were subsequently updated, either to fix bugs or to add new features. sbe39 was updated: the first update is sbe39.1 and the second is sbe39.2. You could install insert sbe39.2, and it would not matter whether you had previously installed sbe39.1. Updates are complete: installing sbe39.2 provides all the features of the original insert and more.

For computer naming purposes, insert sbe39.2 is referred to as `sbe39_2`. When referred to in normal text, however, the insert is still called sbe39.2 because that looks nicer.

Inserts are easily available from the Internet. Inserts may be obtained by pointing and clicking or by using command mode.

The sections below detail how to install an insert. In all cases, pretend that you wish to install insert sbe39.2 from STB-61.

Obtaining from the Internet by pointing and clicking

1. Select **Help > SJ and User-written Programs**.

2. Click on *STB*.

3. Click on *stb61*.

4. Click on *sbe39_2*.

5. Click on *(click here to install)*.

Obtaining from the Internet via command mode

Type the following:

```
. net from http://www.stata.com
. net cd stb
. net cd stb61
. net describe sbe39_2
. net install sbe39_2
```

The above could be shortened to

```
. net from http://www.stata.com/stb/stb61
. net describe sbe39_2
. net install sbe39_2
```

but going about it the long way is more entertaining, at least the first time.

Also see

[R] **search** — Search Stata documentation

[R] **net** — Install and manage user-written additions from the Internet

[R] **net search** — Search the Internet for installable packages

[R] **update** — Update Stata

[U] **3.5 The Stata Journal**

[U] **28 Using the Internet to keep up to date**

[GSM] **19 Updating and extending Stata—Internet functionality**

[GSU] **19 Updating and extending Stata—Internet functionality**

[GSW] **19 Updating and extending Stata—Internet functionality**

Title

> **sktest** — Skewness and kurtosis test for normality

Syntax

> sktest *varlist* [*if*] [*in*] [*weight*] [, <u>no</u>adjust]
>
> aweights and fweights are allowed; see [U] **11.1.6 weight**.

Menu

Statistics > Summaries, tables, and tests > Distributional plots and tests > Skewness and kurtosis normality test

Description

For each variable in *varlist*, sktest presents a test for normality based on skewness and another based on kurtosis and then combines the two tests into an overall test statistic. sktest requires a minimum of 8 observations to make its calculations. See [MV] **mvtest normality** for multivariate tests of normality.

Option

⌐ Main ⌐

<u>no</u>adjust suppresses the empirical adjustment made by Royston (1991c) to the overall χ^2 and its significance level and presents the unaltered test as described by D'Agostino, Belanger, and D'Agostino (1990).

Remarks

Also see [R] **swilk** for the Shapiro–Wilk and Shapiro–Francia tests for normality. Those tests are, in general, preferred for nonaggregated data (Gould and Rogers 1991; Gould 1992; Royston 1991c). Moreover, a normal quantile plot should be used with any test for normality; see [R] **diagnostic plots** for more information.

▷ Example 1

Using our automobile dataset, we will test whether the variables mpg and trunk are normally distributed:

```
. use http://www.stata-press.com/data/r12/auto
(1978 Automobile Data)
. sktest mpg trunk
```

			Skewness/Kurtosis tests for Normality		joint	
Variable	Obs	Pr(Skewness)	Pr(Kurtosis)	adj chi2(2)	Prob>chi2
mpg	74	0.0015	0.0804	10.95	0.0042
trunk	74	0.9115	0.0445	4.19	0.1228

We can reject the hypothesis that mpg is normally distributed, but we cannot reject the hypothesis that trunk is normally distributed, at least at the 12% level. The kurtosis for trunk is 2.19, as can be verified by issuing the command

```
. summarize trunk, detail
```
   (*output omitted*)

and the *p*-value of 0.0445 shown in the table above indicates that it is significantly different from the kurtosis of a normal distribution at the 5% significance level. However, on the basis of skewness alone, we cannot reject the hypothesis that trunk is normally distributed.

◁

❑ Technical note

sktest implements the test as described by D'Agostino, Belanger, and D'Agostino (1990) but with the adjustment made by Royston (1991c). In the above example, if we had specified the noadjust option, the χ^2 values would have been 13.13 for mpg and 4.05 for trunk. With the adjustment, the χ^2 value might show as '.'. This result should be interpreted as an absurdly large number; the data are most certainly not normal.

❑

Saved results

sktest saves the following in r():

Scalars

r(chi2)	χ^2
r(P_skew)	Pr(skewness)
r(P_kurt)	Pr(kurtosis)
r(P_chi2)	Prob > chi2

Matrices

r(N)	matrix of observations
r(Utest)	matrix of test results, one row per variable

Methods and formulas

sktest is implemented as an ado-file.

sktest implements the test described by D'Agostino, Belanger, and D'Agostino (1990) with the empirical correction developed by Royston (1991c).

Let g_1 denote the coefficient of skewness and b_2 denote the coefficient of kurtosis as calculated by summarize, and let n denote the sample size. If weights are specified, then g_1, b_2, and n denote the weighted coefficients of skewness and kurtosis and weighted sample size, respectively. See [R] **summarize** for the formulas for skewness and kurtosis.

To perform the test of skewness, we compute

$$Y = g_1 \left\{ \frac{(n+1)(n+3)}{6(n-2)} \right\}^{1/2}$$

$$\beta_2(g_1) = \frac{3(n^2 + 27n - 70)(n+1)(n+3)}{(n-2)(n+5)(n+7)(n+9)}$$

$$W^2 = -1 + [2\{\beta_2(g_1) - 1\}]^{1/2}$$

and

$$\alpha = \{2/(W^2 - 1)\}^{1/2}$$

Then the distribution of the test statistic

$$Z_1 = \frac{1}{\sqrt{\ln W}} \ln \left[Y/\alpha + \{(Y/\alpha)^2 + 1\}^{1/2} \right]$$

is approximately standard normal under the null hypothesis that the data are distributed normally.

To perform the test of kurtosis, we compute

$$E(b_2) = \frac{3(n-1)}{n+1}$$

$$\mathrm{var}(b_2) = \frac{24n(n-2)(n-3)}{(n+1)^2(n+3)(n+5)}$$

$$X = \{b_2 - E(b_2)\} / \sqrt{\mathrm{var}(b_2)}$$

$$\sqrt{\beta_1(b_2)} = \frac{6(n^2 - 5n + 2)}{(n+7)(n+9)} \left\{ \frac{6(n+3)(n+5)}{n(n-2)(n-3)} \right\}^{1/2}$$

and

$$A = 6 + \frac{8}{\sqrt{\beta_1(b_2)}} \left[\frac{2}{\sqrt{\beta_1(b_2)}} + \left\{ 1 + \frac{4}{\beta_1(b_2)} \right\}^{1/2} \right]$$

Then the distribution of the test statistic

$$Z_2 = \frac{1}{\sqrt{2/(9A)}} \left[\left(1 - \frac{2}{9A} \right) - \left\{ \frac{1 - 2/A}{1 + X\sqrt{2/(A-4)}} \right\}^{1/3} \right]$$

is approximately standard normal under the null hypothesis that the data are distributed normally.

D'Agostino, Balanger, and D'Agostino Jr.'s omnibus test of normality uses the statistic

$$K^2 = Z_1^2 + Z_2^2$$

which has approximately a χ^2 distribution with 2 degrees of freedom under the null of normality.

Royston (1991c) proposed the following adjustment to the test of normality, which sktest uses by default. Let $\Phi(x)$ denote the cumulative standard normal distribution function for x, and let $\Phi^{-1}(p)$ denote the inverse cumulative standard normal function [that is, $x = \Phi^{-1}\{\Phi(x)\}$]. Define the following terms:

$$Z_c = -\Phi^{-1}\left\{\exp\left(-\frac{1}{2}K^2\right)\right\}$$
$$Z_t = 0.55n^{0.2} - 0.21$$
$$a_1 = (-5 + 3.46\ln n)\exp(-1.37\ln n)$$
$$b_1 = 1 + (0.854 - 0.148\ln n)\exp(-0.55\ln n)$$
$$a_2 = a_1 - \{2.13/(1 - 2.37\ln n)\}Z_t$$
and
$$b_2 = 2.13/(1 - 2.37\ln n) + b1$$

If $Z_c < -1$ set $Z = Z_c$; else if $Z_c < Z_t$ set $Z = a_1 + b_1 Z_c$; else set $Z = a_2 + b_2 Z_c$. Define $P = 1 - \Phi(Z)$. Then $K^2 = -2\ln P$ is approximately distributed χ^2 with 2 degrees of freedom.

The relative merits of the skewness and kurtosis test versus the Shapiro–Wilk and Shapiro–Francia tests have been a subject of debate. The interested reader is directed to the articles in the *Stata Technical Bulletin*. Our recommendation is to use the Shapiro–Francia test whenever possible, that is, whenever dealing with nonaggregated or ungrouped data (Gould and Rogers 1991; Gould 1992); see [R] **swilk**. If normality is rejected, use sktest to determine the source of the problem.

As both D'Agostino, Belanger, and D'Agostino (1990) and Royston (1991d) mention, researchers should also examine the normal quantile plot to determine normality rather than blindly relying on a few test statistics. See the qnorm command documented in [R] **diagnostic plots** for more information on normal quantile plots.

sktest is similar in spirit to the Jarque–Bera (1987) test of normality. The Jarque–Bera test statistic is also calculated from the sample skewness and kurtosis, though it is based on asymptotic standard errors with no corrections for sample size. In effect, sktest offers two adjustments for sample size, that of Royston (1991c) and that of D'Agostino, Belanger, and D'Agostino (1990).

Acknowledgments

sktest has benefited greatly by the comments and work of Patrick Royston of the MRC Clinical Trials Unit, London; at this point, the program should be viewed as due as much to Royston as to us, except, of course, for any errors. We are also indebted to Nicholas J. Cox, Durham University, for helpful comments.

References

D'Agostino, R. B., A. J. Belanger, and R. B. D'Agostino, Jr. 1990. A suggestion for using powerful and informative tests of normality. *American Statistician* 44: 316–321.

——. 1991. sg3.3: Comment on tests of normality. *Stata Technical Bulletin* 3: 20. Reprinted in *Stata Technical Bulletin Reprints*, vol. 1, pp. 105–106. College Station, TX: Stata Press.

Gould, W. W. 1991. sg3: Skewness and kurtosis tests of normality. *Stata Technical Bulletin* 1: 20–21. Reprinted in *Stata Technical Bulletin Reprints*, vol. 1, pp. 99–101. College Station, TX: Stata Press.

——. 1992. sg11.1: Quantile regression with bootstrapped standard errors. *Stata Technical Bulletin* 9: 19–21. Reprinted in *Stata Technical Bulletin Reprints*, vol. 2, pp. 137–139. College Station, TX: Stata Press.

Gould, W. W., and W. H. Rogers. 1991. sg3.4: Summary of tests of normality. *Stata Technical Bulletin* 3: 20–23. Reprinted in *Stata Technical Bulletin Reprints*, vol. 1, pp. 106–110. College Station, TX: Stata Press.

Jarque, C. M., and A. K. Bera. 1987. A test for normality of observations and regression residuals. *International Statistical Review* 2: 163–172.

Marchenko, Y. V., and M. G. Genton. 2010. A suite of commands for fitting the skew-normal and skew-t models. *Stata Journal* 10: 507–539.

Royston, P. 1991a. sg3.1: Tests for departure from normality. *Stata Technical Bulletin* 2: 16–17. Reprinted in *Stata Technical Bulletin Reprints*, vol. 1, pp. 101–104. College Station, TX: Stata Press.

——. 1991b. sg3.2: Shapiro–Wilk and Shapiro–Francia tests. *Stata Technical Bulletin* 3: 19. Reprinted in *Stata Technical Bulletin Reprints*, vol. 1, p. 105. College Station, TX: Stata Press.

——. 1991c. sg3.5: Comment on sg3.4 and an improved D'Agostino test. *Stata Technical Bulletin* 3: 23–24. Reprinted in *Stata Technical Bulletin Reprints*, vol. 1, pp. 110–112. College Station, TX: Stata Press.

——. 1991d. sg3.6: A response to sg3.3: Comment on tests of normality. *Stata Technical Bulletin* 4: 8–9. Reprinted in *Stata Technical Bulletin Reprints*, vol. 1, pp. 112–114. College Station, TX: Stata Press.

Also see

[R] **diagnostic plots** — Distributional diagnostic plots

[R] **ladder** — Ladder of powers

[R] **lv** — Letter-value displays

[R] **swilk** — Shapiro–Wilk and Shapiro–Francia tests for normality

[MV] **mvtest normality** — Multivariate normality tests

Title

> **slogit** — Stereotype logistic regression

Syntax

slogit *depvar* [*indepvars*] [*if*] [*in*] [*weight*] [, *options*]

options	Description
Model	
<u>dimen</u>sion(*#*)	dimension of the model; default is dimension(1)
<u>baseo</u>utcome(*# | lbl*)	set the base outcome to *#* or *lbl*; default is the last outcome
<u>const</u>raints(*numlist*)	apply specified linear constraints
<u>coll</u>inear	keep collinear variables
<u>nocorn</u>er	do not generate the corner constraints
SE/Robust	
vce(*vcetype*)	*vcetype* may be oim, <u>r</u>obust, <u>c</u>luster *clustvar*, opg, <u>boot</u>strap, or <u>jack</u>knife
Reporting	
<u>level</u>(*#*)	set confidence level; default is level(95)
<u>nocnsr</u>eport	do not display constraints
display_options	control column formats, row spacing, line width, and display of omitted variables and base and empty cells
Maximization	
maximize_options	control the maximization process; seldom used
<u>init</u>ialize(*initype*)	method of initializing scale parameters; *initype* can be constant, random, or svd; see *Options* for details
<u>nonorm</u>alize	do not normalize the numeric variables
<u>coefl</u>egend	display legend instead of statistics

indepvars may contain factor variables; see [U] **11.4.3 Factor variables**.
bootstrap, by, jackknife, rolling, statsby, and svy are allowed; see [U] **11.1.10 Prefix commands**.
Weights are not allowed with the bootstrap prefix; see [R] **bootstrap**.
vce() and weights are not allowed with the svy prefix; see [SVY] **svy**.
fweights, iweights, and pweights are allowed; see [U] **11.1.6 weight**.
coeflegend does not appear in the dialog box.
See [U] **20 Estimation and postestimation commands** for more capabilities of estimation commands.

Menu

Statistics > Categorical outcomes > Stereotype logistic regression

Description

slogit fits maximum-likelihood stereotype logistic regression models as developed by Anderson (1984). Like multinomial logistic and ordered logistic models, stereotype logistic models are for use with categorical dependent variables. In a multinomial logistic model, the categories cannot be ranked, whereas in an ordered logistic model the categories follow a natural ranking scheme. You can view stereotype logistic models as a compromise between those two models. You can use them when you are unsure of the relevance of the ordering, as is often the case when subjects are asked to assess or judge something. You can also use them in place of multinomial logistic models when you suspect that some of the alternatives are similar. Unlike ordered logistic models, stereotype logistic models do not impose the proportional-odds assumption.

Options

    ┌ Model └

dimension(#) specifies the dimension of the model, which is the number of equations required to describe the relationship between the dependent variable and the independent variables. The maximum dimension is $\min(m - 1, p)$, where m is the number of categories of the dependent variable and p is the number of independent variables in the model. The stereotype model with maximum dimension is a reparameterization of the multinomial logistic model.

baseoutcome(# | lbl) specifies the outcome level whose scale parameters and intercept are constrained to be zero. The base outcome may be specified as a number of a label. By default, slogit assumes that the outcome levels are ordered and uses the largest level of the dependent variable as the base outcome.

constraints(numlist), collinear; see [R] **estimation options**.

By default, the linear equality constraints suggested by Anderson (1984), termed the corner constraints, are generated for you. You can add constraints to these as needed, or you can turn off the corner constraints by specifying nocorner. These constraints are in addition to the constraints placed on the ϕ parameters corresponding to baseoutcome(#).

nocorner specifies that slogit not generate the corner constraints. If you specify nocorner, you must specify at least dimension() × dimension() constraints for the model to be identified.

    ┌ SE/Robust └

vce(vcetype) specifies the type of standard error reported, which includes types that are derived from asymptotic theory, that are robust to some kinds of misspecification, that allow for intragroup correlation, and that use bootstrap or jackknife methods; see [R] **vce_option**.

If specifying vce(bootstrap) or vce(jackknife), you must also specify baseoutcome().

    ┌ Reporting └

level(#); see [R] **estimation options**.

nocnsreport; see [R] **estimation options**.

display_options: noomitted, vsquish, noemptycells, baselevels, allbaselevels, cformat(%fmt), pformat(%fmt), sformat(%fmt), and nolstretch; see [R] **estimation options**.

⌐ Maximization ⌐

maximize_options: <u>dif</u>ficult, <u>techn</u>ique(*algorithm_spec*), <u>iter</u>ate(*#*), [<u>no</u>]<u>log</u>, <u>trac</u>e, gradient, showstep, <u>hess</u>ian, <u>showtol</u>erance, <u>tol</u>erance(*#*), <u>ltol</u>erance(*#*), <u>nrtol</u>erance(*#*), <u>nonrtol</u>erance, and from(*init_specs*); see [R] **maximize**. These options are seldom used.

Setting the optimization type to technique(bhhh) resets the default *vcetype* to vce(opg).

initialize(<u>constant</u> | <u>rand</u>om | svd) specifies how initial estimates are computed. The default, initialize(constant), is to set the scale parameters to the constant $\min(1/2, 1/d)$, where d is the dimension specified in dimension().

> initialize(random) requests that uniformly distributed random numbers between 0 and 1 be used as initial values for the scale parameters. If you specify this option, you should also use set seed to ensure that you can replicate your results; see [R] **set seed**.

> initialize(svd) requests that a singular value decomposition (SVD) be performed on the matrix of regression estimates from mlogit to reduce its rank to the dimension specified in dimension(). slogit uses the reduced-rank components of the SVD as initial estimates for the scale and regression coefficients. For details, see *Methods and formulas*.

nonormalize specifies that the numeric variables not be normalized. Normalization of the numeric variables improves numerical stability but consumes more memory in generating temporary double-precision variables. Variables that are of type byte are not normalized, and if initial estimates are specified using the from() option, normalization of variables is not performed. See *Methods and formulas* for more information.

The following option is available with slogit but is not shown in the dialog box:

coeflegend; see [R] **estimation options**.

Remarks

Remarks are presented under the following headings:

> *Introduction*
> *One-dimensional model*
> *Higher-dimension models*

Introduction

Stereotype logistic models are often used when subjects are requested to assess or judge something. For example, consider a survey in which consumers may be asked to rate the quality of a product on a scale from 1 to 5, with 1 indicating poor quality and 5 indicating excellent quality. If the categories are monotonically related to an underlying latent variable, the ordered logistic model is appropriate. However, suppose that consumers assess quality not just along one dimension, but rather weigh two or three latent factors. Stereotype logistic regression allows you to specify multiple equations to capture the effects of those latent variables, which you then parameterize in terms of observable characteristics. Unlike with multinomial logit, the number of equations you specify could be less than $m - 1$, where m is the number of categories of the dependent variable.

Stereotype logistic models are also used when categories may be indistinguishable. Suppose that a consumer must choose among A, B, C, or D. Multinomial logistic modeling assumes that the four choices are distinct in the sense that a consumer choosing one of the goods can distinguish its characteristics from the others. If goods A and B are in fact similar, consumers may be randomly picking between the two. One alternative is to combine the two categories and fit a three-category multinomial logistic model. A more flexible alternative is to use a stereotype logistic model.

In the multinomial logistic model, you estimate $m - 1$ parameter vectors $\widetilde{\boldsymbol{\beta}}_k$, $k = 1, \ldots, m - 1$, where m is the number of categories of the dependent variable. The stereotype logistic model is a restriction on the multinomial model in the sense that there are d parameter vectors, where d is between one and $\min(m - 1, p)$, and p is the number of regressors. The relationship between the stereotype model's coefficients $\boldsymbol{\beta}_j$, $j = 1, \ldots, d$, and the multinomial model's coefficients is $\widetilde{\boldsymbol{\beta}}_k = -\sum_{j=1}^{d} \phi_{jk}\boldsymbol{\beta}_j$. The ϕs are scale parameters to be estimated along with the $\boldsymbol{\beta}_j$s.

Given a row vector of covariates $\mathbf{x}$, let $\eta_k = \theta_k - \sum_{j=1}^{d} \phi_{jk}\mathbf{x}\boldsymbol{\beta}_j$. The probability of observing outcome k is

$$\Pr(Y_i = k) = \begin{cases} \dfrac{\exp(\eta_k)}{1 + \sum_{l=1}^{m-1} \exp(\eta_l)} & k < m \\[3ex] \dfrac{1}{1 + \sum_{l=1}^{m-1} \exp(\eta_l)} & k = m \end{cases}$$

This model includes a set of θ parameters so that each equation has an unrestricted constant term. If $d = m - 1$, the stereotype model is just a reparameterization of the multinomial logistic model. To identify the ϕs and the βs, you must place at least d^2 restrictions on the parameters. By default, slogit uses the "corner constraints" $\phi_{jj} = 1$ and $\phi_{jk} = 0$ for $j \neq k$, $k \leq d$, and $j \leq d$.

For a discussion of the stereotype logistic model, see Lunt (2005).

One-dimensional model

▷ Example 1

We have 2 years of repair rating data on the make, price, mileage rating, and gear ratio of 104 foreign and 44 domestic automobiles (with 13 missing values on repair rating). We wish to fit a stereotype logistic model to discriminate between the levels of repair rating using mileage, price, gear ratio, and origin of the manufacturer. Here is an overview of our data:

```
. use http://www.stata-press.com/data/r12/auto2yr
(Automobile Models)

. tabulate repair
```

repair	Freq.	Percent	Cum.
Poor	5	3.70	3.70
Fair	19	14.07	17.78
Average	57	42.22	60.00
Good	38	28.15	88.15
Excellent	16	11.85	100.00
Total	135	100.00	

The variable repair can take five values, 1, ..., 5, which represent the subjective rating of the car model's repair record as *Poor*, *Fair*, *Average*, *Good*, and *Excellent*.

We wish to fit the one-dimensional stereotype logistic model

$$\eta_k = \theta_k - \phi_k \left(\beta_1 \texttt{foreign} + \beta_2 \texttt{mpg} + \beta_3 \texttt{price} + \beta_4 \texttt{gratio} \right)$$

for $k < 5$ and $\eta_5 = 0$. To fit this model, we type

```
. slogit repair foreign mpg price gratio
Iteration 0:   log likelihood = -173.78178  (not concave)
Iteration 1:   log likelihood = -164.77316
Iteration 2:   log likelihood =  -161.7069
Iteration 3:   log likelihood = -159.76138
Iteration 4:   log likelihood = -159.34327
Iteration 5:   log likelihood = -159.25914
Iteration 6:   log likelihood = -159.25691
Iteration 7:   log likelihood = -159.25691
```

```
Stereotype logistic regression               Number of obs   =        135
                                              Wald chi2(4)    =       9.33
Log likelihood = -159.25691                   Prob > chi2     =     0.0535
 ( 1)   [phi1_1]_cons = 1
```

repair	Coef.	Std. Err.	z	P>|z|	[95% Conf. Interval]	
foreign	5.947382	2.094126	2.84	0.005	1.84297	10.05179
mpg	.1911968	.08554	2.24	0.025	.0235414	.3588521
price	-.0000576	.0001357	-0.42	0.671	-.0003236	.0002083
gratio	-4.307571	1.884713	-2.29	0.022	-8.00154	-.6136017
/phi1_1	1	(constrained)				
/phi1_2	1.262268	.3530565	3.58	0.000	.5702904	1.954247
/phi1_3	1.17593	.3169397	3.71	0.000	.5547394	1.79712
/phi1_4	.8657195	.2411228	3.59	0.000	.3931275	1.338311
/phi1_5	0	(base outcome)				
/theta1	-6.864749	4.21252	-1.63	0.103	-15.12114	1.391639
/theta2	-7.613977	4.861803	-1.57	0.117	-17.14294	1.914981
/theta3	-5.80655	4.987508	-1.16	0.244	-15.58189	3.968786
/theta4	-3.85724	3.824132	-1.01	0.313	-11.3524	3.637922
/theta5	0	(base outcome)				

```
(repair=Excellent is the base outcome)
```

The coefficient associated with the first scale parameter, ϕ_{11}, is 1, and its standard error and other statistics are missing. This is the corner constraint applied to the one-dimensional model; in the header, this constraint is listed as [phi1_1]_cons = 1. Also, the ϕ and θ parameters that are associated with the base outcome are identified. Keep in mind, though, that there are no coefficient estimates for [phi1_5]_cons or [theta5]_cons in the ereturn matrix e(b). The Wald statistic is for a test of the joint significance of the regression coefficients on foreign, mpg, price, and gratio.

The one-dimensional stereotype model restricts the multinomial logistic regression coefficients $\widetilde{\beta}_k$, $k = 1, \ldots, m - 1$ to be parallel; that is, $\widetilde{\beta}_k = -\phi_k\beta$. As Lunt (2001) discusses, in the one-dimensional stereotype model, one linear combination $x_i\beta$ best discriminates the outcomes of the dependent variable, and the scale parameters ϕ_k measure the distance between the outcome levels and the linear predictor. If $\phi_1 \geq \phi_2 \geq \cdots \phi_{m-1} \geq \phi_m \equiv 0$, the model suggests that the subjective assessment of the dependent variable is indeed ordered. Here the maximum likelihood estimates of the ϕs are not monotonic, as would be assumed in an ordered logit model.

We test that $\phi_1 = \phi_2$ by typing

```
. test [phi1_2]_cons = [phi1_1]_cons
 ( 1)   - [phi1_1]_cons + [phi1_2]_cons = 0
         chi2(  1) =      0.55
       Prob > chi2 =    0.4576
```

Because the two parameters are not statistically different, we decide to add a constraint to force $\phi_1 = \phi_2$:

```
. constraint 1 [phi1_2]_cons = [phi1_1]_cons
. slogit repair foreign mpg price gratio, constraint(1) nolog
```

Stereotype logistic regression				Number of obs	=	135
				Wald chi2(4)	=	21.28
Log likelihood = -159.65769				Prob > chi2	=	0.0003

```
( 1)   [phi1_1]_cons = 1
( 2)  - [phi1_1]_cons + [phi1_2]_cons = 0
```

repair	Coef.	Std. Err.	z	P>|z|	[95% Conf. Interval]	
foreign	7.166515	1.690177	4.24	0.000	3.853829	10.4792
mpg	.2340043	.0807042	2.90	0.004	.0758271	.3921816
price	-.000041	.0001618	-0.25	0.800	-.0003581	.000276
gratio	-5.218107	1.798717	-2.90	0.004	-8.743528	-1.692686
/phi1_1	1	(constrained)				
/phi1_2	1	(constrained)				
/phi1_3	.9751096	.1286563	7.58	0.000	.7229478	1.227271
/phi1_4	.7209343	.1220353	5.91	0.000	.4817494	.9601191
/phi1_5	0	(base outcome)				
/theta1	-8.293452	4.645182	-1.79	0.074	-17.39784	.8109368
/theta2	-6.958451	4.629292	-1.50	0.133	-16.0317	2.114795
/theta3	-5.620232	4.953981	-1.13	0.257	-15.32986	4.089392
/theta4	-3.745624	3.809189	-0.98	0.325	-11.2115	3.720249
/theta5	0	(base outcome)				

(repair=Excellent is the base outcome)

The ϕ estimates are now monotonically decreasing and the standard errors of the ϕs are small relative to the size of the estimates, so we conclude that, with the exception of outcomes *Poor* and *Fair*, the groups are distinguishable for the one-dimensional model and that the quality assessment can be ordered.

◁

Higher-dimension models

The stereotype logistic model is not limited to ordered categorical dependent variables; you can use it on nominal data to reduce the dimension of the regressions. Recall that a multinomial model fit to a categorical dependent variable with m levels will have $m - 1$ sets of regression coefficients. However, a model with fewer dimensions may fit the data equally well, suggesting that some of the categories are indistinguishable.

▷ Example 2

As discussed in [R] **mlogit**, we have data on the type of health insurance available to 616 psychologically depressed subjects in the United States (Tarlov et al. 1989; Wells et al. 1989). Patients may have either an indemnity (fee-for-service) plan or a prepaid plan, such as an HMO, or may be uninsured. Demographic variables include age, gender, race, and site.

First, we fit the saturated, two-dimensional model that is equivalent to a multinomial logistic model. We choose the base outcome to be 1 (indemnity insurance) because that is the default for mlogit.

```
. use http://www.stata-press.com/data/r12/sysdsn1
(Health insurance data)

. slogit insure age male nonwhite i.site, dim(2) base(1)

Iteration 0:   log likelihood = -534.36165
Iteration 1:   log likelihood = -534.36165

Stereotype logistic regression                Number of obs    =        615
                                               Wald chi2(10)    =      38.17
Log likelihood = -534.36165                    Prob > chi2      =     0.0000

 ( 1)  [phi1_2]_cons = 1
 ( 2)  [phi1_3]_cons = 0
 ( 3)  [phi2_2]_cons = 0
 ( 4)  [phi2_3]_cons = 1
```

insure	Coef.	Std. Err.	z	P>|z|	[95% Conf. Interval]	
dim1						
age	.011745	.0061946	1.90	0.058	-.0003962	.0238862
male	-.5616934	.2027465	-2.77	0.006	-.9590693	-.1643175
nonwhite	-.9747768	.2363213	-4.12	0.000	-1.437958	-.5115955
site						
2	-.1130359	.2101903	-0.54	0.591	-.5250013	.2989296
3	.5879879	.2279351	2.58	0.010	.1412433	1.034733
dim2						
age	.0077961	.0114418	0.68	0.496	-.0146294	.0302217
male	-.4518496	.3674867	-1.23	0.219	-1.17211	.268411
nonwhite	-.2170589	.4256361	-0.51	0.610	-1.05129	.6171725
site						
2	1.211563	.4705127	2.57	0.010	.2893747	2.133751
3	.2078123	.3662926	0.57	0.570	-.510108	.9257327
/phi1_1	0	(base outcome)				
/phi1_2	1	(constrained)				
/phi1_3	0	(omitted)				
/phi2_1	0	(base outcome)				
/phi2_2	0	(omitted)				
/phi2_3	1	(constrained)				
/theta1	0	(base outcome)				
/theta2	.2697127	.3284422	0.82	0.412	-.3740222	.9134476
/theta3	-1.286943	.5923219	-2.17	0.030	-2.447872	-.1260134

```
(insure=Indemnity is the base outcome)
```

For comparison, we also fit the model by using `mlogit`:

```
. mlogit insure age male nonwhite i.site, nolog
```

Multinomial logistic regression

			Number of obs	=	615
			LR chi2(10)	=	42.99
			Prob > chi2	=	0.0000
Log likelihood = -534.36165 | | | Pseudo R2 | = | 0.0387 |

insure	Coef.	Std. Err.	z	P>|z|	[95% Conf. Interval]
Indemnity	(base outcome)					
Prepaid						
age	-.011745	.0061946	-1.90	0.058	-.0238862	.0003962
male	.5616934	.2027465	2.77	0.006	.1643175	.9590693
nonwhite	.9747768	.2363213	4.12	0.000	.5115955	1.437958
site						
2	.1130359	.2101903	0.54	0.591	-.2989296	.5250013
3	-.5879879	.2279351	-2.58	0.010	-1.034733	-.1412433
_cons	.2697127	.3284422	0.82	0.412	-.3740222	.9134476
Uninsure						
age	-.0077961	.0114418	-0.68	0.496	-.0302217	.0146294
male	.4518496	.3674867	1.23	0.219	-.268411	1.17211
nonwhite	.2170589	.4256361	0.51	0.610	-.6171725	1.05129
site						
2	-1.211563	.4705127	-2.57	0.010	-2.133751	-.2893747
3	-.2078123	.3662926	-0.57	0.570	-.9257327	.510108
_cons	-1.286943	.5923219	-2.17	0.030	-2.447872	-.1260134

Apart from having opposite signs, the coefficients from the stereotype logistic model are identical to those from the multinomial logit model. Recall the definition of η_k given in the *Remarks*, particularly the minus sign in front of the summation. One other difference in the output is that the constant estimates labeled /theta in the `slogit` output are the constants labeled _cons in the `mlogit` output.

Next we examine the one-dimensional model.

```
. slogit insure age male nonwhite i.site, dim(1) base(1) nolog
Stereotype logistic regression                  Number of obs   =        615
                                                Wald chi2(5)    =      28.20
Log likelihood = -539.75205                     Prob > chi2     =     0.0000
 ( 1)  [phi1_2]_cons = 1
```

insure	Coef.	Std. Err.	z	P>|z|	[95% Conf. Interval]	
age	.0108366	.0061918	1.75	0.080	-.0012992	.0229723
male	-.5032537	.2078171	-2.42	0.015	-.9105678	-.0959396
nonwhite	-.9480351	.2340604	-4.05	0.000	-1.406785	-.489285
site						
2	-.2444316	.2246366	-1.09	0.277	-.6847113	.1958481
3	.556665	.2243799	2.48	0.013	.1168886	.9964415
/phi1_1	0	(base outcome)				
/phi1_2	1	(constrained)				
/phi1_3	.0383539	.4079705	0.09	0.925	-.7612535	.8379613
/theta1	0	(base outcome)				
/theta2	.187542	.3303847	0.57	0.570	-.4600001	.835084
/theta3	-1.860134	.2158898	-8.62	0.000	-2.28327	-1.436997

```
(insure=Indemnity is the base outcome)
```

We have reduced a two-dimensional multinomial model to one dimension, reducing the number of estimated parameters by four and decreasing the model likelihood by ≈ 5.4.

slogit does not report a model likelihood-ratio test. The test of $d = 1$ (a one-dimensional model) versus $d = 0$ (the null model) does not have an asymptotic χ^2 distribution because the unconstrained ϕ parameters (/phi1_3 in the previous example) cannot be identified if $\beta = 0$. More generally, this problem precludes testing any hierarchical model of dimension d versus $d - 1$. Of course, the likelihood-ratio test of a full-dimension model versus $d = 0$ is valid because the full model is just multinomial logistic, and all the ϕ parameters are fixed at 0 or 1.

◁

❑ Technical note

The stereotype model is a special case of the reduced-rank vector generalized linear model discussed by Yee and Hastie (2003). If we define $\eta_{ik} = \theta_k - \sum_{j=1}^{d} \phi_{jk} \mathbf{x}_i \beta_j$, for $k = 1, \ldots, m - 1$, we can write the expression in matrix notation as

$$\boldsymbol{\eta}_i = \boldsymbol{\theta} + \boldsymbol{\Phi} (\mathbf{x}_i \mathbf{B})'$$

where $\boldsymbol{\Phi}$ is a $(m - 1) \times d$ matrix containing the ϕ_{jk} parameters and $\mathbf{B}$ is a $p \times d$ matrix with columns containing the β_j parameters, $j = 1, \ldots, d$. The factorization $\boldsymbol{\Phi}\mathbf{B}'$ is not unique because $\boldsymbol{\Phi}\mathbf{B}' = \boldsymbol{\Phi}\mathbf{M}\mathbf{M}^{-1}\mathbf{B}'$ for any nonsingular $d \times d$ matrix $\mathbf{M}$. To avoid this identifiability problem, we choose $\mathbf{M} = \boldsymbol{\Phi}_1^{-1}$, where

$$\boldsymbol{\Phi} = \begin{pmatrix} \boldsymbol{\Phi}_1 \\ \boldsymbol{\Phi}_2 \end{pmatrix}$$

and $\mathbf{\Phi}_1$ is $d \times d$ of rank d so that

$$\mathbf{\Phi M} = \begin{pmatrix} \mathbf{I}_d \\ \mathbf{\Phi}_2 \mathbf{\Phi}_1^{-1} \end{pmatrix}$$

and $\mathbf{I}_d$ is a $d \times d$ identity matrix. Thus the corner constraints used by slogit are $\phi_{jj} \equiv 1$ and $\phi_{jk} \equiv 0$ for $j \neq k$ and $k, j \leq d$.

❑

Saved results

slogit saves the following in e():

Scalars

e(N)	number of observations
e(k)	number of parameters
e(k_indvars)	number of independent variables
e(k_out)	number of outcomes
e(k_eq)	number of equations in e(b)
e(k_eq_model)	number of equations in overall model test
e(df_m)	Wald test degrees of freedom
e(df_0)	null model degrees of freedom
e(k_dim)	model dimension
e(i_base)	base outcome index
e(ll)	log likelihood
e(ll_0)	null model log likelihood
e(N_clust)	number of clusters
e(chi2)	χ^2
e(p)	significance
e(ic)	number of iterations
e(rank)	rank of e(V)
e(rc)	return code
e(converged)	1 if converged, 0 otherwise

Macros
    e(cmd)   slogit
    e(cmdline)   command as typed
    e(depvar)   name of dependent variable
    e(indvars)   independent variables
    e(wtype)   weight type
    e(wexp)   weight expression
    e(title)   title in estimation output
    e(clustvar)   name of cluster variable
    e(out#)   outcome labels, $\# = 1,...,$ e(k_out)
    e(chi2type)   Wald; type of model χ^2 test
    e(labels)   outcome labels or numeric levels
    e(vce)   *vcetype* specified in vce()
    e(vcetype)   title used to label Std. Err.
    e(opt)   type of optimization
    e(which)   max or min; whether optimizer is to perform maximization or minimization
    e(ml_method)   type of ml method
    e(user)   name of likelihood-evaluator program
    e(technique)   maximization technique
    e(properties)   b V
    e(predict)   program used to implement predict
    e(marginsnotok)   predictions disallowed by margins
    e(footnote)   program used to implement the footnote display
    e(asbalanced)   factor variables fvset as asbalanced
    e(asobserved)   factor variables fvset as asobserved

Matrices
    e(b)   coefficient vector
    e(outcomes)   outcome values
    e(Cns)   constraints matrix
    e(ilog)   iteration log (up to 20 iterations)
    e(gradient)   gradient vector
    e(V)   variance–covariance matrix of the estimators
    e(V_modelbased)   model-based variance

Functions
    e(sample)   marks estimation sample

Methods and formulas

slogit is implemented as an ado-file.

slogit obtains the maximum likelihood estimates for the stereotype logistic model by using ml; see [R] **ml**. Each set of regression estimates, one set of β_js for each dimension, constitutes one ml model equation. The $d \times (m-1)$ ϕs and the $(m-1)$ θs are ml ancillary parameters.

Without loss of generality, let the base outcome level be the mth level of the dependent variable. Define the row vector $\phi_k = (\phi_{1k}, \ldots, \phi_{dk})$ for $k = 1, \ldots, m-1$, and define the $p \times d$ matrix $\mathbf{B} = (\beta_1, \ldots, \beta_d)$. For observation i, the log odds of outcome level k relative to level m, $k = 1, \ldots, m-1$ is the index

$$\ln\left\{\frac{\Pr(Y_i = k)}{\Pr(Y_i = m)}\right\} = \eta_{ik} = \theta_k - \phi_k\left(\mathbf{x}_i\mathbf{B}\right)'$$
$$= \theta_k - \phi_k\nu_i'$$

The row vector ν_i can be interpreted as a latent variable reducing the p-dimensional vector of covariates to a more interpretable $d < p$ dimension.

The probability of the ith observation having outcome level k is then

$$\Pr(Y_i = k) = p_{ik} = \begin{cases} \dfrac{e^{\eta_{ik}}}{1 + \sum_{j=1}^{m-1} e^{\eta_{ij}}}, & \text{if} \quad k < m \\[3mm] \dfrac{1}{1 + \sum_{j=1}^{m-1} e^{\eta_{ij}}}, & \text{if} \quad k = m \end{cases}$$

from which the log-likelihood function is computed as

$$L = \sum_{i=1}^{n} w_i \sum_{k=1}^{m} I_k(y_i) \ln(p_{ik}) \tag{1}$$

Here w_i is the weight for observation i and

$$I_k(y_i) = \begin{cases} 1, & \text{if observation } y_i \text{ has outcome } k \\ 0, & \text{otherwise} \end{cases}$$

Numeric variables are normalized for numerical stability during optimization where a new double-precision variable $\widetilde{x}_j$ is created from variable x_j, $j = 1, \ldots, p$, such that $\widetilde{x}_j = (x_j - \bar{x}_j)/s_j$. This feature is turned off if you specify nonormalize, or if you use the from() option for initial estimates. Normalization is not performed on byte variables, including the indicator variables generated by [R] xi. The linear equality constraints for regression parameters, if specified, must be scaled also. Assume that a constraint is applied to the regression parameter associated with variable j and dimension i, β_{ji}, and the corresponding element of the constraint matrix (see [P] makecns) is divided by s_j.

After convergence, the parameter estimates for variable j and dimension i—$\widetilde{\beta}_{ji}$, say—are transformed back to their original scale, $\beta_{ji} = \widetilde{\beta}_{ji}/s_j$. For the intercepts, you compute

$$\theta_k = \widetilde{\theta}_k + \sum_{i=1}^{d} \phi_{ik} \sum_{j=1}^{p} \frac{\widetilde{\beta}_{ji}\bar{x}_j}{s_j}$$

Initial values are computed using estimates obtained using mlogit to fit a multinomial logistic model. Let the $p \times (m-1)$ matrix $\widetilde{\mathbf{B}}$ contain the multinomial logistic regression parameters less the $m-1$ intercepts. Each ϕ is initialized with constant values $\min(1/2, 1/d)$, the initialize(constant) option (the default), or, with uniform random numbers, the initialize(random) option. Constraints are then applied to the starting values so that the structure of the $(m-1) \times d$ matrix $\boldsymbol{\Phi}$ is

$$\boldsymbol{\Phi} = \begin{pmatrix} \phi_1 \\ \phi_2 \\ \vdots \\ \phi_{m-1} \end{pmatrix} = \begin{pmatrix} \mathbf{I}_d \\ \widetilde{\boldsymbol{\Phi}} \end{pmatrix}$$

where $\mathbf{I}_d$ is a $d \times d$ identity matrix. Assume that only the corner constraints are used, but any constraints you place on the scale parameters are also applied to the initial scale estimates, so the structure of $\mathbf{\Phi}$ will change accordingly. The ϕ parameters are invariant to the scale of the covariates, so initial estimates in $[\,0, 1\,]$ are reasonable. The constraints guarantee that the rank of $\mathbf{\Phi}$ is at least d, so the initial estimates for the stereotype regression parameters are obtained from $\mathbf{B} = \widetilde{\mathbf{B}}\mathbf{\Phi}(\mathbf{\Phi}'\mathbf{\Phi})^{-1}$.

One other approach for initial estimates is provided: `initialize(svd)`. It starts with the `mlogit` estimates and computes $\widetilde{\mathbf{B}}' = \mathbf{U}\mathbf{D}\mathbf{V}'$, where $\mathbf{U}_{m-1 \times p}$ and $\mathbf{V}_{p \times p}$ are orthonormal matrices and $\mathbf{D}_{p \times p}$ is a diagonal matrix containing the singular values of $\widetilde{\mathbf{B}}$. The estimates for $\mathbf{\Phi}$ and $\mathbf{B}$ are the first d columns of $\mathbf{U}$ and $\mathbf{V}\mathbf{D}$, respectively (Yee and Hastie 2003).

The score for regression coefficients is

$$\mathbf{u}_i(\boldsymbol{\beta}_j) = \frac{\partial L_{ik}}{\partial \boldsymbol{\beta}_j} = \mathbf{x}_i \left(\sum_{l=1}^{m-1} \phi_{jl} p_{il} - \phi_{jk} \right)$$

the score for the scale parameters is

$$u_i(\phi_{jl}) = \frac{\partial L_{ik}}{\partial \phi_{jl}} = \begin{cases} \mathbf{x}_i\boldsymbol{\beta}_j(p_{ik} - 1), & \text{if } l = k \\ \mathbf{x}_i\boldsymbol{\beta}_j p_{il}, & \text{if } l \neq k \end{cases}$$

for $l = 1, \ldots, m-1$; and the score for the intercepts is

$$u_i(\theta_l) = \frac{\partial L_{ik}}{\partial \theta_l} = \begin{cases} 1 - p_{ik}, & \text{if } l = k \\ -p_{il}, & \text{if } l \neq k \end{cases}$$

This command supports the Huber/White/sandwich estimator of the variance and its clustered version using `vce(robust)` and `vce(cluster clustvar)`, respectively. See [P] **_robust**, particularly *Maximum likelihood estimators* and *Methods and formulas*.

`slogit` also supports estimation with survey data. For details on VCEs with survey data, see [SVY] **variance estimation**.

References

Anderson, J. A. 1984. Regression and ordered categorical variables (with discussion). *Journal of the Royal Statistical Society, Series B* 46: 1–30.

Lunt, M. 2001. sg163: Stereotype ordinal regression. *Stata Technical Bulletin* 61: 12–18. Reprinted in *Stata Technical Bulletin Reprints*, vol. 10, pp. 298–307. College Station, TX: Stata Press.

———. 2005. Prediction of ordinal outcomes when the association between predictors and outcome differs between outcome levels. *Statistics in Medicine* 24: 1357–1369.

Tarlov, A. R., J. E. Ware, Jr., S. Greenfield, E. C. Nelson, E. Perrin, and M. Zubkoff. 1989. The medical outcomes study. An application of methods for monitoring the results of medical care. *Journal of the American Medical Association* 262: 925–930.

Wells, K. B., R. D. Hays, M. A. Burnam, W. H. Rogers, S. Greenfield, and J. E. Ware, Jr. 1989. Detection of depressive disorder for patients receiving prepaid or fee-for-service care. Results from the Medical Outcomes Survey. *Journal of the American Medical Association* 262: 3298–3302.

Yee, T. W., and T. J. Hastie. 2003. Reduced-rank vector generalized linear models. *Statistical Modelling* 3: 15–41.

Also see

[R] **slogit postestimation** — Postestimation tools for slogit

[R] **roc** — Receiver operating characteristic (ROC) analysis

[R] **logistic** — Logistic regression, reporting odds ratios

[R] **mlogit** — Multinomial (polytomous) logistic regression

[R] **ologit** — Ordered logistic regression

[R] **oprobit** — Ordered probit regression

[SVY] **svy estimation** — Estimation commands for survey data

[U] **20 Estimation and postestimation commands**

Title

slogit postestimation — Postestimation tools for slogit

Description

The following postestimation commands are available after slogit:

Command	Description
contrast	contrasts and ANOVA-style joint tests of estimates
estat	AIC, BIC, VCE, and estimation sample summary
estat (svy)	postestimation statistics for survey data
estimates	cataloging estimation results
lincom	point estimates, standard errors, testing, and inference for linear combinations of coefficients
lrtest[1]	likelihood-ratio test
margins	marginal means, predictive margins, marginal effects, and average marginal effects
marginsplot	graph the results from margins (profile plots, interaction plots, etc.)
nlcom	point estimates, standard errors, testing, and inference for nonlinear combinations of coefficients
predict	predicted probabilities, estimated index and its approximate standard error
predictnl	point estimates, standard errors, testing, and inference for generalized predictions
pwcompare	pairwise comparisons of estimates
suest	seemingly unrelated estimation
test	Wald tests of simple and composite linear hypotheses
testnl	Wald tests of nonlinear hypotheses

[1] lrtest is not appropriate with svy estimation results.

See the corresponding entries in the *Base Reference Manual* for details, but see [SVY] **estat** for details about estat (svy).

Syntax for predict

predict [*type*] { *stub* | *newvar* | *newvarlist* } [*if*] [*in*] [, *statistic* <u>ou</u>tcome(*outcome*)]

predict [*type*] { *stub* | *newvarlist* } [*if*] [*in*], <u>sc</u>ores

statistic	Description
Main	
<u>pr</u>	probability of one or all of the dependent variable outcomes; the default
xb	index for the kth outcome
stdp	standard error of the index for the kth outcome

If you do not specify outcome(), pr (with one new variable specified), xb, and stdp assume outcome(#1).

You specify one or k new variables with pr, where k is the number of outcomes.

You specify one new variable with xb and stdp.

These statistics are available both in and out of sample; type predict ... if e(sample) ... if wanted only for the estimation sample.

Menu

Statistics > Postestimation > Predictions, residuals, etc.

Options for predict

⌐ Main ⌐

pr, the default, calculates the probability of each of the categories of the dependent variable or the probability of the level specified in outcome(*outcome*). If you specify the outcome(*outcome*) option, you need to specify only one new variable; otherwise, you must specify a new variable for each category of the dependent variable.

xb calculates the index, $\theta_k - \sum_{j=1}^{d} \phi_{jk} \mathbf{x}_i \boldsymbol{\beta}_j$, for outcome level $k \neq$ e(i_base) and dimension $d =$ e(k_dim). It returns a vector of zeros if $k =$ e(i_base). A synonym for xb is index. If outcome() is not specified, outcome(#1) is assumed.

stdp calculates the standard error of the index. A synonym for stdp is seindex. If outcome() is not specified, outcome(#1) is assumed.

outcome(*outcome*) specifies the outcome for which the statistic is to be calculated. equation() is a synonym for outcome(): it does not matter which you use. outcome() or equation() can be specified using

> #1, #2, ..., where #1 means the first category of the dependent variable, #2 means the second category, etc.;
>
> the values of the dependent variable; or
>
> the value labels of the dependent variable if they exist.

scores calculates the equation-level score variables. For models with d dimensions and m levels, $d + (d+1)(m-1)$ new variables are created. Assume $j = 1, \ldots, d$ and $k = 1, \ldots, m$ in the following.

The first d new variables will contain $\partial \ln L / \partial (\mathbf{x}\boldsymbol{\beta}_j)$.

The next $d(m-1)$ new variables will contain $\partial \ln L / \partial \phi_{jk}$.

The last $m-1$ new variables will contain $\partial \ln L / \partial \theta_k$.

Remarks

Once you have fit a stereotype logistic model, you can obtain the predicted probabilities by using the predict command for both the estimation sample and other samples; see [U] **20 Estimation and postestimation commands** and [R] **predict**.

predict without arguments (or with the pr option) calculates the predicted probability of each outcome of the dependent variable. You must therefore give a new variable name for each of the outcomes. To compute the estimated probability of one outcome, you use the outcome(*outcome*) option where *outcome* is the level encoding the outcome. If the dependent variable's levels are labeled, the outcomes can also be identified by the label values (see [D] **label**).

The xb option in conjunction with outcome(*outcome*) specifies that the index be computed for the outcome encoded by level *outcome*. Its approximate standard error is computed if the stdp option is specified. Only one of the pr, xb, or stdp options can be specified with a call to predict.

▷ Example 1

In example 2 of [R] **slogit**, we fit the one-dimensional stereotype model, where the *depvar* is insure with levels $k = 1$ for outcome *Indemnity*, $k = 2$ for *Prepaid*, and $k = 3$ for *Uninsure*. The base outcome for the model is *Indemnity*, so for $k \neq 1$ the vector of indices for the kth level is

$$\eta_k = \theta_k - \phi_k \left(\beta_1 \mathtt{age} + \beta_2 \mathtt{male} + \beta_3 \mathtt{nonwhite} + \beta_4 \mathtt{2.site} + \beta_5 \mathtt{3.site} \right)$$

We estimate the group probabilities by calling predict after slogit.

```
. use http://www.stata-press.com/data/r12/sysdsn1
(Health insurance data)

. slogit insure age male nonwhite i.site, dim(1) base(1) nolog
(output omitted)

. predict pIndemnity pPrepaid pUninsure, p

. list pIndemnity pPrepaid pUninsure insure in 1/10
```

	pIndem~y	pPrepaid	pUnins~e	insure
1.	.5419344	.3754875	.0825782	Indemnity
2.	.4359638	.496328	.0677081	Prepaid
3.	.5111583	.4105107	.0783309	Indemnity
4.	.3941132	.5442234	.0616633	Prepaid
5.	.4655651	.4625064	.0719285	.
6.	.4401779	.4915102	.0683118	Prepaid
7.	.4632122	.4651931	.0715948	Prepaid
8.	.3772302	.5635696	.0592002	.
9.	.4867758	.4383018	.0749225	Uninsure
10.	.5823668	.3295802	.0880531	Prepaid

Observations 5 and 8 are not used to fit the model because insure is missing at these points, but predict estimates the probabilities for these observations since none of the independent variables is missing. You can use if e(sample) in the call to predict to use only those observations that are used to fit the model.

◁

Methods and formulas

All postestimation commands listed above are implemented as ado-files.

predict

Let level b be the base outcome that is used to fit the stereotype logistic regression model of dimension d. The index for observation i and level $k \neq b$ is $\eta_{ik} = \theta_k - \sum_{j=1}^{d} \phi_{jk} \mathbf{x}_i \boldsymbol{\beta}_j$. This is the log odds of outcome encoded as level k relative to that of b so that we define $\eta_{ib} \equiv 0$. The outcome probabilities for this model are defined as $\Pr(Y_i = k) = e^{\eta_{ik}} / \sum_{j=1}^{m} e^{\eta_{ij}}$. Unlike in mlogit, ologit, and oprobit, the index is no longer a linear function of the parameters. The standard error of index η_{ik} is thus computed using the delta method (see also [R] **predictnl**).

The equation-level score for regression coefficients is

$$\frac{\partial \ln L_{ik}}{\partial \mathbf{x}_i \boldsymbol{\beta}_j} = \left(\sum_{l=1}^{m-1} \phi_{jl} p_{il} - \phi_{jk} \right)$$

the equation-level score for the scale parameters is

$$\frac{\partial \ln L_{ik}}{\partial \phi_{jl}} = \begin{cases} \mathbf{x}_i \boldsymbol{\beta}_j (p_{ik} - 1), & \text{if } l = k \\ \mathbf{x}_i \boldsymbol{\beta}_j p_{il}, & \text{if } l \neq k \end{cases}$$

for $l = 1, \ldots, m - 1$; and the equation-level score for the intercepts is

$$\frac{\partial \ln L_{ik}}{\partial \theta_l} = \begin{cases} 1 - p_{ik}, & \text{if } l = k \\ - p_{il}, & \text{if } l \neq k \end{cases}$$

Also see

[R] **slogit** — Stereotype logistic regression

[U] **20 Estimation and postestimation commands**

Title

smooth — Robust nonlinear smoother

Syntax

smooth *smoother* $\left[\right.$, t̲wice $\left.\right]$ *varname* $\left[\right. if \left.\right]$ $\left[\right. in \left.\right]$, g̲enerate(*newvar*)

where *smoother* is specified as $Sm\left[\right. Sm\left[\right. \ldots \left.\right] \left.\right]$ and *Sm* is one of

$$\{\,1\,|\,2\,|\,3\,|\,4\,|\,5\,|\,6\,|\,7\,|\,8\,|\,9\,\}\left[\right. \mathtt{R}\left.\right]$$
$$3\left[\right.\mathtt{R}\left.\right]\mathtt{S}\left[\right.\mathtt{S}\,|\,\mathtt{R}\left.\right]\left[\right.\mathtt{S}\,|\,\mathtt{R}\left.\right]\ldots$$
$$\mathtt{E}$$
$$\mathtt{H}$$

Letters may be specified in lowercase if preferred. Examples of *smoother* $\left[\right.$, twice $\left.\right]$ include

3RSSH	3RSSH,twice	4253H	4253H,twice	43RSR2H,twice
3rssh	3rssh,twice	4253h	4253h,twice	43rsr2h,twice

Menu

Statistics > Nonparametric analysis > Robust nonlinear smoother

Description

smooth applies the specified resistant, nonlinear smoother to *varname* and stores the smoothed series in *newvar*.

Option

generate(*newvar*) is required; it specifies the name of the new variable that will contain the smoothed values.

Remarks

Smoothing is an exploratory data-analysis technique for making the general shape of a series apparent. In this approach (Tukey 1977), the observed data series is assumed to be the sum of an underlying process that evolves smoothly (the smooth) and of an unsystematic noise component (the rough); that is,

$$\text{data} = \text{smooth} + \text{rough}$$

Smoothed values z_t are obtained by taking medians (or some other location estimate) of each point in the original data y_t and a few of the points around it. The number of points used is called the span of the smoother. Thus a span-3 smoother produces z_t by taking the median of y_{t-1}, y_t, and y_{t+1}. smooth provides running median smoothers of spans 1 to 9—indicated by the digit that specifies their span. Median smoothers are resistant to isolated outliers, so they provide robustness to spikes in the data. Because the median is also a nonlinear operator, such smoothers are known as robust (or resistant) nonlinear smoothers.

smooth also provides the Hanning linear, nonrobust smoother, indicated by the letter H. Hanning is a span-3 smoother with binomial weights. Repeated applications of H—HH, HHH, etc.— provide binomial smoothers of span 5, 7, etc. See Cox (1997, 2004) for a graphical application of this fact.

Because one smoother usually cannot adequately separate the smooth from the rough, compound smoothers—multiple smoothers applied in sequence—are used. The smoother 35H, for instance, then smooths the data with a span-3 median smoother, smooths the result with a span-5 median smoother, and finally smooths that result with the Hanning smoother. smooth allows you to specify any number of smoothers in any sequence.

Three refinements can be combined with the running median and Hanning smoothers. First, the endpoints of a smooth can be given special treatment. This is specified by the E operator. Second, smoothing by 3, the span-3 running median, tends to produce flat-topped hills and valleys. The splitting operator, S, "splits" these repeated values, applies the endpoint operator to them, and then "rejoins" the series. Finally, it is sometimes useful to repeat an odd-span median smoother or the splitting operator until the smooth no longer changes. Following a digit or an S with an R specifies this type of repetition.

Even the best smoother may fail to separate the smooth from the rough adequately. To guard against losing any systematic components of the data series, after smoothing, the smoother can be reapplied to the resulting rough, and any recovered signal can be added back to the original smooth. The twice operator specifies this procedure. More generally, an arbitrary smoother can be applied to the rough (using a second smooth command), and the recovered signal can be added back to the smooth. This more general procedure is called reroughing (Tukey 1977).

The details of each of the smoothers and operators are explained in *Methods and formulas* below.

▷ Example 1

smooth is designed to recover the general features of a series that has been contaminated with noise. To demonstrate this, we construct a series, add noise to it, and then smooth the noisy version to recover an estimate of the original data. First, we construct and display the data:

```
. drop _all
. set obs 10
. set seed 123456789
. generate time = _n
. label variable time "Time"
. generate x = _n^3 - 10*_n^2 + 5*_n
. label variable x "Signal"
. generate z = x + 50*rnormal()
. label variable z "Observed series"
```

```
. scatter x z time, c(l .) m(i o) ytitle("")
```

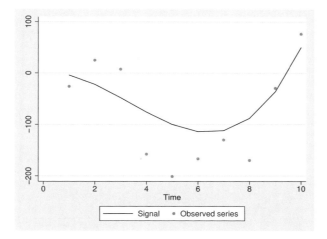

Now we smooth the noisy series, z, assumed to be the only data we would observe:

```
. smooth 4253eh,twice z, gen(sz)
. label variable sz "Smoothed series"
. scatter x z sz time, c(l . l) m(i o i) ytitle("") || scatter sz time,
> c(l . l) m(i o i) ytitle("") clpattern(dash_dot)
```

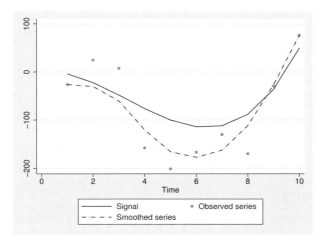

◁

▷ Example 2

Salgado-Ugarte and Curts-García (1993) provide data on the frequencies of observed fish lengths. In this example, the series to be smoothed—the frequencies—is ordered by fish length rather than by time.

```
. use http://www.stata-press.com/data/r12/fishdata, clear
. smooth 4253eh,twice freq, gen(sfreq)
. label var sfreq "4253EH,twice of frequencies"
```

```
. scatter sfreq freq length, c(l .) m(i o)
> title("Smoothed frequencies of fish lengths") ytitle("") xlabel(#4)
```

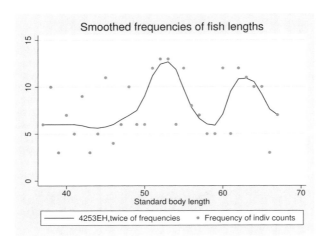

◁

❏ Technical note

   smooth allows missing values at the beginning and end of the series, but missing values in the middle are not allowed. Leading and trailing missing values are ignored. If you wish to ignore missing values in the middle of the series, you must drop the missing observations before using smooth. Doing so, of course, would violate smooth's assumption that observations are equally spaced—each observation represents a year, a quarter, or a month (or a 1-year birth-rate category). In practice, smooth produces good results as long as the spaces between adjacent observations do not vary too much.

   Smoothing is usually applied to time series, but any variable with a natural order can be smoothed. For example, a smoother might be applied to the birth rate recorded by the age of the mothers (birth rate for 17-year-olds, birth rate for 18-year-olds, and so on).

❏

Methods and formulas

   smooth is implemented as an ado-file.

   Methods and formulas are presented under the following headings:

> *Running median smoothers of odd span*
> *Running median smoothers of even span*
> *Repeat operator*
> *Endpoint rule*
> *Splitting operator*
> *Hanning smoother*
> *Twicing*

Running median smoothers of odd span

The smoother 3 defines

$$z_t = \text{median}(y_{t-1}, y_t, y_{t+1})$$

The smoother 5 defines

$$z_t = \text{median}(y_{t-2}, y_{t-1}, y_t, y_{t+1}, y_{t+2})$$

and so on. The smoother 1 defines $z_t = \text{median}(y_t)$, so it does nothing.

Endpoints are handled by using smoothers of shorter, odd span. Thus for 3,

$$z_1 = y_1$$
$$z_2 = \text{median}(y_1, y_2, y_3)$$
$$\vdots$$
$$z_{N-1} = \text{median}(y_{N-2}, y_{N-1}, y_N)$$
$$Z_N = y_N$$

For 5,

$$z_1 = y_1$$
$$z_2 = \text{median}(y_1, y_2, y_3)$$
$$z_3 = \text{median}(y_1, y_2, y_3, y_4, y_5)$$
$$z_4 = \text{median}(y_2, y_3, y_4, y_5, y_6)$$
$$\vdots$$
$$z_{N-2} = \text{median}(y_{N-4}, y_{N-3}, y_{N-2}, y_{N-1}, y_N)$$
$$z_{N-1} = \text{median}(y_{N-2}, y_{N-1}, y_N)$$
$$Z_N = y_N$$

and so on.

Running median smoothers of even span

Define the median() function as returning the linearly interpolated value when given an even number of arguments. Thus the smoother 2 defines

$$z_{t+0.5} = (y_t + y_{t+1})/2$$

The smoother 4 defines $z_{t+0.5}$ as the linearly interpolated median of $(y_{t-1}, y_t, y_{t+1}, y_{t+2})$, and so on. Endpoints are always handled using smoothers of shorter, even span. Thus for 4,

$$z_{0.5} = y_1$$
$$z_{1.5} = \text{median}(y_1, y_2) = (y_1 + y_2)/2$$
$$z_{2.5} = \text{median}(y_1, y_2, y_3, y_4)$$
$$\vdots$$
$$z_{N-2.5} = \text{median}(y_{N-4}, y_{N-3}, y_{N-2}, y_N)$$
$$z_{N-1.5} = \text{median}(y_{N-2}, y_{N-1})$$
$$z_{N-0.5} = \text{median}(y_{N-1}, y_N)$$
$$z_{N+0.5} = y_N$$

As defined above, an even-span smoother increases the length of the series by 1 observation. However, the series can be recentered on the original observation numbers, and the "extra" observation can be eliminated by smoothing the series again with another even-span smoother. For instance, the smooth of 4 illustrated above could be followed by a smooth of 2 to obtain

$$z_1^* = (z_{0.5} + z_{1.5})/2$$
$$z_2^* = (z_{1.5} + z_{2.5})/2$$
$$z_3^* = (z_{2.5} + z_{3.5})/2$$
$$\vdots$$
$$z_{N-2}^* = (z_{N-2.5} + z_{N-1.5})/2$$
$$z_{N-1}^* = (z_{N-1.5} + z_{N-0.5})/2$$
$$z_N^* = (z_{N-0.5} + z_{N+0.5})/2$$

smooth keeps track of the number of even smoothers applied to the data and expands and shrinks the length of the series accordingly. To ensure that the final smooth has the same number of observations as *varname*, smooth requires you to specify an even number of even-span smoothers. However, the pairs of even-span smoothers need not be contiguous; for instance, 4253 and 4523 are both allowed.

Repeat operator

R indicates that a smoother is to be repeated until convergence, that is, until repeated applications of the smoother produce the same series. Thus 3 applies the smoother of running medians of span 3. 33 applies the smoother twice. 3R produces the result of repeating 3 an infinite number of times. R should be used only with odd-span smoothers because even-span smoothers are not guaranteed to converge.

The smoother 453R2 applies a span-4 smoother, followed by a span-5 smoother, followed by repeated applications of a span-3 smoother, followed by a span-2 smoother.

Endpoint rule

The endpoint rule E modifies the values z_1 and z_N according to the following formulas:

$$z_1 = \text{median}(3z_2 - 2z_3, z_1, z_2)$$
$$z_N = \text{median}(3z_{N-2} - 2z_{N-1}, z_N, z_{N-1})$$

When the endpoint rule is not applied, endpoints are typically "copied in"; that is, $z_1 = y_1$ and $z_N = y_N$.

Splitting operator

The smoothers 3 and 3R can produce flat-topped hills and valleys. The split operator attempts to eliminate such hills and valleys by splitting the sequence, applying the endpoint rule E, rejoining the series, and then resmoothing by 3R.

The S operator may be applied only after 3, 3R, or S.

We recommend that the S operator be repeated once (SS) or until no further changes take place (SR).

Hanning smoother

H is the Hanning linear smoother:

$$z_t = (y_{t-1} + 2y_t + y_{t+1})/4$$

Endpoints are copied in: $z_1 = y_1$ and $z_N = y_N$. H should be applied only after all nonlinear smoothers.

Twicing

A smoother divides the data into a smooth and a rough:

$$\text{data} = \text{smooth} + \text{rough}$$

If the smoothing is successful, the rough should exhibit no pattern. Twicing refers to applying the smoother to the observed, calculating the rough, and then applying the smoother to the rough. The resulting "smoothed rough" is then added back to the smooth from the first step.

Acknowledgments

smooth was originally written by William Gould (1992)—at which time it was named nlsm—and was inspired by Salgado-Ugarte and Curts-García (1992). Salgado-Ugarte and Curts-García (1993) subsequently reported anomalies in nlsm's treatment of even-span median smoothers. smooth corrects these problems and incorporates other improvements but otherwise is essentially the same as originally published.

References

Cox, N. J. 1997. gr22: Binomial smoothing plot. *Stata Technical Bulletin* 35: 7–9. Reprinted in *Stata Technical Bulletin Reprints*, vol. 6, pp. 36–38. College Station, TX: Stata Press.

——. 2004. gr22_1: Software update: Binomial smoothing plot. *Stata Journal* 4: 490.

——. 2005. Speaking Stata: Smoothing in various directions. *Stata Journal* 5: 574–593.

Gould, W. W. 1992. sg11.1: Quantile regression with bootstrapped standard errors. *Stata Technical Bulletin* 9: 19–21. Reprinted in *Stata Technical Bulletin Reprints*, vol. 2, pp. 137–139. College Station, TX: Stata Press.

Royston, P., and N. J. Cox. 2005. A multivariable scatterplot smoother. *Stata Journal* 5: 405–412.

Salgado-Ugarte, I. H., and J. Curts-García. 1992. sed7: Resistant smoothing using Stata. *Stata Technical Bulletin* 7: 8–11. Reprinted in *Stata Technical Bulletin Reprints*, vol. 2, pp. 99–103. College Station, TX: Stata Press.

——. 1993. sed7.2: Twice reroughing procedure for resistant nonlinear smoothing. *Stata Technical Bulletin* 11: 14–16. Reprinted in *Stata Technical Bulletin Reprints*, vol. 2, pp. 108–111. College Station, TX: Stata Press.

Sasieni, P. 1998. gr27: An adaptive variable span running line smoother. *Stata Technical Bulletin* 41: 4–7. Reprinted in *Stata Technical Bulletin Reprints*, vol. 7, pp. 63–68. College Station, TX: Stata Press.

Tukey, J. W. 1977. *Exploratory Data Analysis*. Reading, MA: Addison–Wesley.

Velleman, P. F. 1977. Robust nonlinear data smoothers: Definitions and recommendations. *Proceedings of the National Academy of Sciences* 74: 434–436.

——. 1980. Definition and comparison of robust nonlinear data smoothing algorithms. *Journal of the American Statistical Association* 75: 609–615.

Velleman, P. F., and D. C. Hoaglin. 1981. *Applications, Basics, and Computing of Exploratory Data Analysis*. Boston: Duxbury.

Also see

[R] **lowess** — Lowess smoothing

[R] **lpoly** — Kernel-weighted local polynomial smoothing

[TS] **tssmooth** — Smooth and forecast univariate time-series data

Title

> **spearman** — Spearman's and Kendall's correlations

Syntax

Spearman's rank correlation coefficients

> spearman [*varlist*] [*if*] [*in*] [, *spearman_options*]

Kendall's rank correlation coefficients

> ktau [*varlist*] [*if*] [*in*] [, *ktau_options*]

spearman_options	Description
Main	
stats(*spearman_list*)	list of statistics; select up to three statistics; default is stats(rho)
print(#)	significance level for displaying coefficients
star(#)	significance level for displaying with a star
bonferroni	use Bonferroni-adjusted significance level
sidak	use Šidák-adjusted significance level
pw	calculate all pairwise correlation coefficients by using all available data
matrix	display output in matrix form

ktau_options	Description
Main	
stats(*ktau_list*)	list of statistics; select up to six statistics; default is stats(taua)
print(#)	significance level for displaying coefficients
star(#)	significance level for displaying with a star
bonferroni	use Bonferroni-adjusted significance level
sidak	use Šidák-adjusted significance level
pw	calculate all pairwise correlation coefficients by using all available data
matrix	display output in matrix form

by is allowed with spearman and ktau; see [D] **by**.

where the elements of *spearman_list* may be

rho	correlation coefficient
obs	number of observations
p	significance level

and the elements of *ktau_list* may be

taua	correlation coefficient τ_a
taub	correlation coefficient τ_b
score	score
se	standard error of score
obs	number of observations
p	significance level

Menu

spearman

Statistics > Nonparametric analysis > Tests of hypotheses > Spearman's rank correlation

ktau

Statistics > Nonparametric analysis > Tests of hypotheses > Kendall's rank correlation

Description

spearman displays Spearman's rank correlation coefficients for all pairs of variables in *varlist* or, if *varlist* is not specified, for all the variables in the dataset.

ktau displays Kendall's rank correlation coefficients between the variables in *varlist* or, if *varlist* is not specified, for all the variables in the dataset. ktau is intended for use on small- and moderate-sized datasets; it requires considerable computation time for larger datasets.

Options for spearman

⌐ Main ⌐

stats(*spearman_list*) specifies the statistics to be displayed in the matrix of output. stats(rho) is the default. Up to three statistics may be specified; stats(rho obs p) would display the correlation coefficient, number of observations, and significance level. If *varlist* contains only two variables, all statistics are shown in tabular form, and stats(), print(), and star() have no effect unless the matrix option is specified.

print(#) specifies the significance level of correlation coefficients to be printed. Correlation coefficients with larger significance levels are left blank in the matrix. Typing spearman, print(.10) would list only those correlation coefficients that are significant at the 10% level or lower.

star(#) specifies the significance level of correlation coefficients to be marked with a star. Typing spearman, star(.05) would "star" all correlation coefficients significant at the 5% level or lower.

bonferroni makes the Bonferroni adjustment to calculated significance levels. This adjustment affects printed significance levels and the print() and star() options. Thus spearman, print(.05) bonferroni prints coefficients with Bonferroni-adjusted significance levels of 0.05 or less.

sidak makes the Šidák adjustment to calculated significance levels. This adjustment affects printed significance levels and the print() and star() options. Thus spearman, print(.05) sidak prints coefficients with Šidák-adjusted significance levels of 0.05 or less.

pw specifies that correlations be calculated using pairwise deletion of observations with missing values. By default, spearman uses casewise deletion, where observations are ignored if any of the variables in *varlist* are missing.

matrix forces spearman to display the statistics as a matrix, even if *varlist* contains only two variables. matrix is implied if more than two variables are specified.

Options for ktau

> [Main]

stats(*ktau_list*) specifies the statistics to be displayed in the matrix of output. stats(taua) is the default. Up to six statistics may be specified; stats(taua taub score se obs p) would display the correlation coefficients τ_a, τ_b, score, standard error of score, number of observations, and significance level. If *varlist* contains only two variables, all statistics are shown in tabular form and stats(), print(), and star() have no effect unless the matrix option is specified.

print(*#*) specifies the significance level of correlation coefficients to be printed. Correlation coefficients with larger significance levels are left blank in the matrix. Typing ktau, print(.10) would list only those correlation coefficients that are significant at the 10% level or lower.

star(*#*) specifies the significance level of correlation coefficients to be marked with a star. Typing ktau, star(.05) would "star" all correlation coefficients significant at the 5% level or lower.

bonferroni makes the Bonferroni adjustment to calculated significance levels. This adjustment affects printed significance levels and the print() and star() options. Thus ktau, print(.05) bonferroni prints coefficients with Bonferroni-adjusted significance levels of 0.05 or less.

sidak makes the Šidák adjustment to calculated significance levels. This adjustment affects printed significance levels and the print() and star() options. Thus ktau, print(.05) sidak prints coefficients with Šidák-adjusted significance levels of 0.05 or less.

pw specifies that correlations be calculated using pairwise deletion of observations with missing values. By default, ktau uses casewise deletion, where observations are ignored if any of the variables in *varlist* are missing.

matrix forces ktau to display the statistics as a matrix, even if *varlist* contains only two variables. matrix is implied if more than two variables are specified.

Remarks

> Example 1

We wish to calculate the correlation coefficients among marriage rate (mrgrate), divorce rate (divorce_rate), and median age (medage) in state data. We can calculate the standard Pearson correlation coefficients and significance by typing

```
.use http://www.stata-press.com/data/r12/states2
(State data)
. pwcorr mrgrate divorce_rate medage, sig
```

	mrgrate	divorc~e	medage
mrgrate	1.0000		
divorce_rate	0.7895 0.0000	1.0000	
medage	0.0011 0.9941	-0.1526 0.2900	1.0000

We can calculate Spearman's rank correlation coefficients by typing

```
. spearman mrgrate divorce_rate medage, stats(rho p)
(obs=50)
```

```
┌─────────────┐
│ Key         │
├─────────────┤
│ rho         │
│ Sig. level  │
└─────────────┘
```

	mrgrate	divorc~e	medage
mrgrate	1.0000		
divorce_rate	0.6933	1.0000	
	0.0000		
medage	-0.4869	-0.2455	1.0000
	0.0003	0.0857	

The large difference in the results is caused by one observation. Nevada's marriage rate is almost 10 times higher than the state with the next-highest marriage rate. An important feature of the Spearman rank correlation coefficient is its reduced sensitivity to extreme values compared with the Pearson correlation coefficient.

We can calculate Kendall's rank correlations by typing

```
. ktau mrgrate divorce_rate medage, stats(taua taub p)
(obs=50)
```

```
┌─────────────┐
│ Key         │
├─────────────┤
│ tau_a       │
│ tau_b       │
│ Sig. level  │
└─────────────┘
```

	mrgrate	divorc~e	medage
mrgrate	0.9829		
	1.0000		
divorce_rate	0.5110	0.9804	
	0.5206	1.0000	
	0.0000		
medage	-0.3486	-0.1698	0.9845
	-0.3544	-0.1728	1.0000
	0.0004	0.0828	

There are tied values for variables `mrgrate`, `divorce_rate`, and `medage`, so tied ranks are used. As a result, $\tau_a < 1$ on the diagonal (see *Methods and formulas* for the definition of τ_a).

◁

❏ Technical note

According to Conover (1999, 323), "Spearman's ρ tends to be larger than Kendall's τ in absolute value. However, as a test of significance, there is no strong reason to prefer one over the other because both will produce nearly identical results in most cases."

❏

▷ Example 2

We illustrate `spearman` and `ktau` with the auto data, which contains some missing values.

```
.use http://www.stata-press.com/data/r12/auto
(1978 Automobile Data)

. spearman mpg rep78

 Number of obs =        69
Spearman's rho =      0.3098

Test of Ho: mpg and rep78 are independent
     Prob > |t| =      0.0096
```

Because we specified two variables, `spearman` displayed the sample size, correlation, and p-value in tabular form. To obtain just the correlation coefficient displayed in matrix form, we type

```
. spearman mpg rep78, stats(rho) matrix
(obs=69)
```

	mpg	rep78
mpg	1.0000	
rep78	0.3098	1.0000

The `pw` option instructs `spearman` and `ktau` to use all nonmissing observations between a pair of variables when calculating their correlation coefficient. In the output below, some correlations are based on 74 observations, whereas others are based on 69 because 5 observations contain a missing value for `rep78`.

```
. spearman mpg price rep78, pw stats(rho obs p)  star(0.01)
```

Key
rho *Number of obs* *Sig. level*

	mpg	price	rep78
mpg	1.0000 74		
price	-0.5419* 74 0.0000	1.0000 74	
rep78	0.3098* 69 0.0096	0.1028 69 0.4008	1.0000 69

Finally, the `bonferroni` and `sidak` options provide adjusted significance levels:

```
. ktau mpg price rep78, stats(taua taub score se p) bonferroni
(obs=69)
```

```
┌─────────────┐
│ Key         │
├─────────────┤
│ tau_a       │
│ tau_b       │
│ score       │
│ se of score │
│ Sig. level  │
└─────────────┘
```

	mpg	price	rep78
mpg	0.9471		
	1.0000		
	2222.0000		
	191.8600		
price	-0.3973	1.0000	
	-0.4082	1.0000	
	-932.0000	2346.0000	
	192.4561	193.0682	
	0.0000		
rep78	0.2076	0.0648	0.7136
	0.2525	0.0767	1.0000
	487.0000	152.0000	1674.0000
	181.7024	182.2233	172.2161
	0.0224	1.0000	

◁

Charles Edward Spearman (1863–1945) was a British psychologist who made contributions to correlation, factor analysis, test reliability, and psychometrics. After several years' military service, he obtained a PhD in experimental psychology at Leipzig and became a professor at University College London, where he sustained a long program of work on the interpretation of intelligence tests. Ironically, the rank correlation version bearing his name is not the formula he advocated.

Maurice George Kendall (1907–1983) was a British statistician who contributed to rank correlation, time series, multivariate analysis, among other topics, and wrote many statistical texts. Most notably, perhaps, his advanced survey of the theory of statistics went through several editions, later ones with Alan Stuart; the baton has since passed to others. Kendall was employed in turn as a government and business statistician, as a professor at the London School of Economics, as a consultant, and as director of the World Fertility Survey. He was knighted in 1974.

Saved results

spearman saves the following in r():

Scalars
r(N)	number of observations (last variable pair)
r(rho)	ρ (last variable pair)
r(p)	two-sided p-value (last variable pair)

Matrices
r(Nobs)	number of observations
r(Rho)	ρ
r(P)	two-sided p-value

ktau saves the following in r():

Scalars
r(N)	number of observations (last variable pair)
r(tau_a)	τ_a (last variable pair)
r(tau_b)	τ_b (last variable pair)
r(score)	Kendall's score (last variable pair)
r(se_score)	se of score (last variable pair)
r(p)	two-sided p-value (last variable pair)

Matrices
r(Nobs)	number of observations
r(Tau_a)	τ_a
r(Tau_b)	τ_b
r(Score)	Kendall's score
r(Se_Score)	standard error of score
r(P)	two-sided p-value

Methods and formulas

spearman and ktau are implemented as ado-files.

Spearman's (1904) rank correlation is calculated as Pearson's correlation computed on the ranks and average ranks (Conover 1999, 314–315). Ranks are as calculated by egen; see [D] **egen**. The significance is calculated using the approximation

$$p = 2 \times \texttt{ttail}(n - 2, |\widehat{\rho}|\sqrt{n - 2}/\sqrt{1 - \widehat{\rho}^2})$$

For any two pairs of ranks (x_i, y_i) and (x_j, y_j) of one variable pair (varname$_1$, varname$_2$), $1 \le i, j \le n$, where n is the number of observations, define them as concordant if

$$(x_i - x_j)(y_i - y_j) > 0$$

and discordant if this product is less than zero.

Kendall's (1938; also see Kendall and Gibbons [1990] or Bland [2000], 222–225) score S is defined as $C - D$, where C (D) is the number of concordant (discordant) pairs. Let $N = n(n-1)/2$ be the total number of pairs, so τ_a is given by

$$\tau_a = S/N$$

and τ_b is given by

$$\tau_b = \frac{S}{\sqrt{N - U}\sqrt{N - V}}$$

where

$$U = \sum_{i=1}^{N_1} u_i(u_i - 1)/2$$

$$V = \sum_{j=1}^{N_2} v_j(v_j - 1)/2$$

and where N_1 is the number of sets of tied x values, u_i is the number of tied x values in the ith set, N_2 is the number of sets of tied y values, and v_j is the number of tied y values in the jth set. Under the null hypothesis of independence between $varname_1$ and $varname_2$, the variance of S is exactly (Kendall and Gibbons 1990, 66)

$$\text{Var}(S) = \frac{1}{18}\left\{ n(n-1)(2n+5) - \sum_{i=1}^{N_1} u_i(u_i-1)(2u_i+5) - \sum_{j=1}^{N_2} v_j(v_j-1)(2v_j+5) \right\}$$

$$+ \frac{1}{9n(n-1)(n-2)}\left\{\sum_{i=1}^{N_1} u_i(u_i-1)(u_i-2)\right\}\left\{\sum_{j=1}^{N_2} v_j(v_j-1)(v_j-2)\right\}$$

$$+ \frac{1}{2n(n-1)}\left\{\sum_{i=1}^{N_1} u_i(u_i-1)\right\}\left\{\sum_{j=1}^{N_2} v_j(v_j-1)\right\}$$

Using a normal approximation with a continuity correction,

$$z = \frac{|S| - 1}{\sqrt{\text{Var}(S)}}$$

For the hypothesis of independence, the statistics S, τ_a, and τ_b produce equivalent tests and give the same significance.

For Kendall's τ, the normal approximation is surprisingly accurate for sample sizes as small as 8, at least for calculating p-values under the null hypothesis for continuous variables. (See Kendall and Gibbons [1990, chap. 4], who also present some tables for calculating exact p-values for $n < 10$.) For Spearman's ρ, the normal approximation requires larger samples to be valid.

Let v be the number of variables specified so that $k = v(v-1)/2$ correlation coefficients are to be estimated. If bonferroni is specified, the adjusted significance level is $p' = \min(1, kp)$. If sidak is specified, $p' = \min\left\{1, 1 - (1-p)^n\right\}$. See *Methods and formulas* in [R] **oneway** for a more complete description of the logic behind these adjustments.

Early work on rank correlation is surveyed by Kruskal (1958).

Acknowledgment

The original version of ktau was written by Sean Becketti, a past editor of the *Stata Technical Bulletin*.

References

Barnard, G. A. 1997. Kendall, Maurice George. In *Leading Personalities in Statistical Sciences: From the Seventeenth Century to the Present*, ed. N. L. Johnson and S. Kotz, 130–132. New York: Wiley.

Bland, M. 2000. *An Introduction to Medical Statistics*. 3rd ed. Oxford: Oxford University Press.

Conover, W. J. 1999. *Practical Nonparametric Statistics*. 3rd ed. New York: Wiley.

David, H. A., and W. A. Fuller. 2007. Sir Maurice Kendall (1907–1983): A centenary appreciation. *American Statistician* 61: 41–46.

Jeffreys, H. 1961. *Theory of Probability*. 3rd ed. Oxford: Oxford University Press.

Kendall, M. G. 1938. A new measure of rank correlation. *Biometrika* 30: 81–93.

Kendall, M. G., and J. D. Gibbons. 1990. *Rank Correlation Methods*. 5th ed. New York: Oxford University Press.

Kruskal, W. H. 1958. Ordinal measures of association. *Journal of the American Statistical Association* 53: 814–861.

Lovie, P., and A. D. Lovie. 1996. Charles Edward Spearman, F.R.S. (1863–1945). *Notes and Records of the Royal Society of London* 50: 75–88.

Newson, R. 2000a. snp15: somersd—Confidence intervals for nonparametric statistics and their differences. *Stata Technical Bulletin* 55: 47–55. Reprinted in *Stata Technical Bulletin Reprints*, vol. 10, pp. 312–322. College Station, TX: Stata Press.

——. 2000b. snp15.1: Update to somersd. *Stata Technical Bulletin* 57: 35. Reprinted in *Stata Technical Bulletin Reprints*, vol. 10, pp. 322–323. College Station, TX: Stata Press.

——. 2000c. snp15.2: Update to somersd. *Stata Technical Bulletin* 58: 30. Reprinted in *Stata Technical Bulletin Reprints*, vol. 10, p. 323. College Station, TX: Stata Press.

——. 2001. snp15.3: Update to somersd. *Stata Technical Bulletin* 61: 22. Reprinted in *Stata Technical Bulletin Reprints*, vol. 10, p. 324. College Station, TX: Stata Press.

——. 2003. snp15_4: Software update for somersd. *Stata Journal* 3: 325.

——. 2005. snp15_5: Software update for somersd. *Stata Journal* 5: 470.

——. 2006. Confidence intervals for rank statistics: Percentile slopes, differences, and ratios. *Stata Journal* 6: 497–520.

Seed, P. T. 2001. sg159: Confidence intervals for correlations. *Stata Technical Bulletin* 59: 27–28. Reprinted in *Stata Technical Bulletin Reprints*, vol. 10, pp. 267–269. College Station, TX: Stata Press.

Spearman, C. 1904. The proof and measurement of association between two things. *American Journal of Psychology* 15: 72–101.

Wolfe, F. 1997. sg64: pwcorrs: Enhanced correlation display. *Stata Technical Bulletin* 35: 22–25. Reprinted in *Stata Technical Bulletin Reprints*, vol. 6, pp. 163–167. College Station, TX: Stata Press.

——. 1999. sg64.1: Update to pwcorrs. *Stata Technical Bulletin* 49: 17. Reprinted in *Stata Technical Bulletin Reprints*, vol. 9, p. 159. College Station, TX: Stata Press.

Also see

[R] **correlate** — Correlations (covariances) of variables or coefficients

[R] **nptrend** — Test for trend across ordered groups

Title

> **spikeplot** — Spike plots and rootograms

Syntax

> spikeplot *varname* [*if*] [*in*] [*weight*] [, *options*]

options	Description
Main	
<u>r</u>ound(*#*)	round *varname* to nearest multiple of *#* (bin width)
<u>frac</u>tion	make vertical scale the proportion of total values; default is frequencies
root	make vertical scale show square roots of frequencies
Plot	
spike_options	affect rendition of plotted spikes
Add plots	
addplot(*plot*)	add other plots to generated graph
Y axis, X axis, Titles, Legend, Overall, By	
twoway_options	any options documented in [G-3] ***twoway_options***

fweights, aweights, and iweights are allowed; see [U] **11.1.6 weight**.

Menu

Graphics > Distributional graphs > Spike plot and rootogram

Description

spikeplot produces a frequency plot for a variable in which the frequencies are depicted as vertical lines from zero. The frequency may be a count, a fraction, or the square root of the count (Tukey's rootogram, circa 1965). The vertical lines may also originate from a baseline other than zero at the user's option.

Options

Main

round(*#*) rounds the values of *varname* to the nearest multiple of *#*. This action effectively specifies the bin width.

fraction specifies that the vertical scale be the proportion of total values (percentage) rather than the count.

root specifies that the vertical scale show square roots. This option may not be specified if fraction is specified.

Plot

spike_options affect the rendition of the plotted spikes; see [G-2] **graph twoway spike**.

2042

⌐‾‾‾‾⌐ Add plots ⌐‾‾

addplot(*plot*) provides a way to add other plots to the generated graph. See [G-3] ***addplot_option***.

⌐‾‾‾‾⌐ Y axis, X axis, Titles, Legend, Overall, By ⌐‾‾

twoway_options are any of the options documented in [G-3] ***twoway_options***. These include options for titling the graph (see [G-3] ***title_options***), options for saving the graph to disk (see [G-3] ***saving_option***), and the by() option (see [G-3] ***by_option***).

Remarks

▷ Example 1

Cox and Brady (1997a) present an illustrative example using the age structure of the population of Ghana from the 1960 census (rounded to the nearest 1,000). The dataset has ages from 0 (less than 1 year) to 90. To view the distribution of ages, we would like to use each integer from 0 to 90 as the bins for the dataset.

```
. use http://www.stata-press.com/data/r12/ghanaage
. spikeplot age [fw=pop], ytitle("Population in 1000s") xlab(0(10)90)
> xmtick(5(10)85)
```

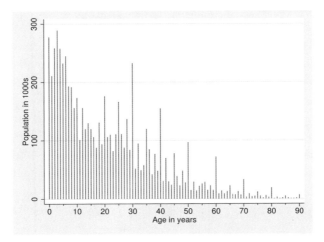

The resulting graph shows a "heaping" of ages at the multiples of 5. Also, ages ending in even numbers are more frequent than ages ending in odd numbers (except for 5). This preference for reporting ages is well known in demography and other social sciences.

Note also that we used the ytitle() option to override the default title of "Frequency" and that we used the xlab() and xmtick() options with *numlist*s to further customize the resulting graph. See [U] **11.1.8 numlist** for details on specifying *numlist*s.

◁

▷ Example 2

The rootogram is a plot of the square-root transformation of the frequency counts. The square root of a normal distribution is a multiple of another normal distribution.

```
. clear
. set seed 1234567
. set obs 5000
obs was 0, now 5000
. generate normal = rnormal()
. label variable normal "Gaussian(0,1) random numbers"
. spikeplot normal, round(.10) xlab(-4(1)4)
```

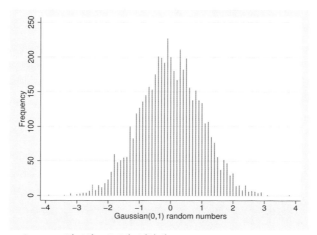

```
. spikeplot normal, round(.10) xlab(-4(1)4) root
```

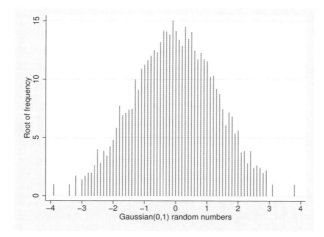

Interpreting a histogram in terms of normality is thus similar to interpreting the rootogram for normality.

This example also shows how the round() option is used to bin the values for a spike plot of a continuous variable.

◁

▷ Example 3

spikeplot can also be used to produce time-series plots. *varname* should be the time variable, and weights should be specified as the values for those times. To get a plot of daily rainfalls, we type

    . spikeplot day [w=rain] if rain, ytitle("Daily rainfall in mm")

The base() option of graph twoway spike may be used to set a different baseline, such as when we want to show variations relative to an average or to some other measure of level.   ◁

Methods and formulas

spikeplot is implemented as an ado-file.

Acknowledgments

The original version of spikeplot was written by Nicholas J. Cox of Durham University and Anthony R. Brady of the Imperial College School of Medicine (1997a, 1997b).

References

Cox, N. J., and A. R. Brady. 1997a. gr25: Spike plots for histograms, rootograms, and time-series plots. *Stata Technical Bulletin* 36: 8–11. Reprinted in *Stata Technical Bulletin Reprints*, vol. 6, pp. 50–54. College Station, TX: Stata Press.

——. 1997b. gr25.1: Spike plots for histograms, rootograms, and time-series plots: Update. *Stata Technical Bulletin* 40: 12. Reprinted in *Stata Technical Bulletin Reprints*, vol. 7, p. 58. College Station, TX: Stata Press.

Tukey, J. W. 1965. The future of processes of data analysis. In *The Collected Works of John W. Tukey, Volume IV: Philosophy and Principles of Data Analysis: 1965–1986*, ed. L. V. Jones, 123–126. Monterey, CA: Wadsworth & Brooks/Cole.

Also see

[R] **histogram** — Histograms for continuous and categorical variables

Title

> **ssc** — Install and uninstall packages from SSC

Syntax

Summary of packages most recently added or updated at SSC

> ssc new [, <u>sav</u>ing(*filename*[, replace]) type]

Summary of most popular packages at SSC

> ssc hot [, n(*#*) <u>auth</u>or(*name*)]

Describe a specified package at SSC

> ssc <u>d</u>escribe { *pkgname* | *letter* } [, saving(*filename*[, replace])]

Install a specified package from SSC

> ssc <u>inst</u>all *pkgname* [, all replace]

Uninstall from your computer a previously installed package from SSC

> ssc uninstall *pkgname*

Type a specific file stored at SSC

> ssc type *filename* [, asis]

Copy a specific file from SSC to your computer

> ssc copy *filename* [, <u>p</u>lus <u>p</u>ersonal replace <u>pub</u>lic <u>bin</u>ary]

where *letter* in ssc describe is a–z or _ .

Description

ssc works with packages (and files) from the Statistical Software Components (SSC) archive, which is often called the Boston College Archive and is provided by http://repec.org.

The SSC has become the premier Stata download site for user-written software on the web. ssc provides a convenient interface to the resources available there. For example, on Statalist (see http://www.stata.com/statalist/), users will often write

> The program can be found by typing ssc install newprogramname.

Typing that would load everything associated with newprogramname, including the help files.

If you are searching for what is available, type ssc new and ssc hot, and see [R] **search**. search searches the SSC and other places, too. search provides a GUI interface from which programs can be installed, including the programs at the SSC archive.

You can uninstall particular packages by using ssc uninstall. For the packages that you keep, see [R] **adoupdate** for an automated way of keeping those packages up to date.

Command overview

ssc new summarizes the packages made available or updated recently. Output is presented in the Stata Viewer, and from there you may click to find out more about individual packages or to install them.

ssc hot lists the most popular packages—popular based on a moving average of the number of downloads in the past three months. By default, 10 packages are listed.

ssc describe *pkgname* describes, but does not install, the specified package. Use search to find packages; see [R] **search**. If you know the package name but do not know the exact spelling, type ssc describe followed by one letter, a–z or _ (underscore), to list all the packages starting with that letter.

ssc install *pkgname* installs the specified package. You do not have to describe a package before installing it. (You may also install a package by using net install; see [R] **net**.)

ssc uninstall *pkgname* removes the previously installed package from your computer. It does not matter how the package was installed. (ssc uninstall is a synonym for ado uninstall, so either may be used to installed any package.)

ssc type *filename* types a specific file stored at SSC. ssc cat is a synonym for ssc type, which may appeal to those familiar with Unix.

ssc copy *filename* copies a specific file stored at SSC to your computer. By default, the file is copied to the current directory, but you can use options to change this. ssc copy is a rarely used alternative to ssc install ... , all. ssc cp is a synonym for ssc copy.

Options for use with ssc new

saving(*filename*[, replace]) specifies that the "what's new" summary be saved in *filename*. If *filename* is specified without a suffix, *filename*.smcl is assumed. If saving() is not specified, saving(ssc_results.smcl) is assumed.

type specifies that the "what's new" results be displayed in the Results window rather than in the Viewer.

Options for use with ssc hot

n(#) specifies the number of packages to list; n(10) is the default. Specify n(.) to list all packages in order of popularity.

author(*name*) lists the 10 most popular packages by the specified author. If n(#) is also specified, the top # packages are listed.

Option for use with ssc describe

saving(*filename*[, replace]) specifies that, in addition to the descriptions being displayed on your screen, it be saved in the specified file.

If *filename* is specified without an extension, .smcl will be assumed, and the file will be saved as a SMCL file.

If *filename* is specified with an extension, no default extension is added. If the extension is .log, the file will be stored as a text file.

If replace is specified, *filename* is replaced if it already exists.

Options for use with ssc install

all specifies that any ancillary files associated with the package be downloaded to your current directory, in addition to the program and help files being installed. Ancillary files are files that do not end in .ado or .sthlp and typically contain datasets or examples of the use of the new command.

You can find out which files are associated with the package by typing ssc describe *pkgname* before or after installing. If you install without using the all option and then want the ancillary files, you can ssc install again.

replace specifies that any files being downloaded that already exist on your computer be replaced by the downloaded files. If replace is not specified and any files already exist, none of the files from the package is downloaded or installed.

It is better not to specify the replace option and wait to see if there is a problem. If there is a problem, it is usually better to uninstall the old package by using ssc uninstall or ado uninstall (which are, in fact, the same command).

Option for use with ssc type

asis affects how files with the suffixes .smcl and .sthlp are displayed. The default is to interpret SMCL directives the file might contain. asis specifies that the file be displayed in raw, uninterpreted form.

Options for use with ssc copy

plus specifies that the file be copied to the PLUS directory, the directory where user-written additions are installed. Typing sysdir will display the identity of the PLUS directory on your computer; see [P] **sysdir**.

personal specifies that the file be copied to your PERSONAL directory as reported by sysdir; see [P] **sysdir**.

If neither plus nor personal is specified, the default is to copy the file to the current directory.

replace specifies that, if the file already exists on your computer, the new file replace it.

public specifies that the new file be made readable by everyone; otherwise, the file will be created according to the default permission you have set with your operating system.

binary specifies that the file being copied is a binary file and that it is to be copied as is. The default is to assume that the file is a text file and change the end-of-line characters to those appropriate for your computer/operating system.

Remarks

Users can add new features to Stata, and some users choose to make new features that they have written available to others via the web. The files that comprise a new feature are called a package, and a package usually consists of one or more ado-files and help files. The net command (see [R] **net**) makes it reasonably easy to install and uninstall packages regardless of where they are on the web. One site, the SSC, has become particularly popular as a repository for additions to Stata. Command ssc is an easier to use version of net designed especially for the SSC.

Many packages are available at the SSC. Packages have names, such as oaxaca, estout, or egenmore. At SSC, capitalization is not significant, so Oaxaca, ESTOUT, and EGENmore are ways of writing the same package names.

When you type

```
. ssc install oaxaca
```

the files associated with the package are downloaded and installed on your computer. Package names usually correspond to the names of the command being added to Stata, so one would expect that installing the package oaxaca will add command oaxaca to Stata on your computer, and expect that typing help oaxaca will provide the documentation. That is the situation here, but that is not always so. Before or after installing a package, type ssc describe *pkgname* to obtain the details.

▷ Example 1

ssc new summarizes the packages most recently made available or updated. Output is presented in the Viewer, from which you may click on a package name to find out more or install it. For example,

```
. ssc new
(contacting http://repec.org)

    (output omitted )

KHB
  module to decompose total effects into direct and indirect via KHB-method
  Authors: Ulrich Kohler  Kristian Karlson       Req: Stata version 11
  Revised: 2011-05-02

    (output omitted )

TODUMMY
  module to create dummy variables
  Authors: Daniel Klein          Req: Stata version 11
  Revised: 2011-05-07

    (output omitted )
```

```
End of recent additions and updates
```

ssc hot provides a list of the most popular packages at SSC.

```
. ssc hot
```

Top 10 packages at SSC

Rank	Apr2011 # hits	Package	Author(s)
1	5267.5	outreg2	Roy Wada
2	4226.8	estout	Ben Jann
3	2102.7	psmatch2	Barbara Sianesi, Edwin Leuven
4	2006.2	ivreg2	Mark E Schaffer, Christopher F Baum Steven Stillman
5	1224.0	ranktest	Mark E Schaffer, Frank Kleibergen
6	1126.5	gllamm	Sophia Rabe-Hesketh
7	1067.8	xtabond2	David Roodman
8	928.3	tabout	Ian Watson
9	895.5	xtivreg2	Mark E Schaffer
10	789.3	outreg	John Luke Gallup

```
(Click on package name for description)
```

Use the n(#) option to change the number of packages listed:

. ssc hot, n(20)

Top 20 packages at SSC

Rank	Apr2011 # hits	Package	Author(s)
1	5267.5	outreg2	Roy Wada
2	4226.8	estout	Ben Jann
3	2102.7	psmatch2	Barbara Sianesi, Edwin Leuven
4	2006.2	ivreg2	Mark E Schaffer, Christopher F Baum, Steven Stillman
5	1224.0	ranktest	Mark E Schaffer, Frank Kleibergen
6	1126.5	gllamm	Sophia Rabe-Hesketh
7	1067.8	xtabond2	David Roodman
8	928.3	tabout	Ian Watson
9	895.5	xtivreg2	Mark E Schaffer
10	789.3	outreg	John Luke Gallup
11	782.8	usespss	Sergiy Radyakin
12	736.5	winsor	Nicholas J. Cox
13	734.0	hprescott	Christopher F Baum
14	571.5	overid	Vince Wiggins, Steven Stillman, Mark E Schaffer, Christopher F Baum
15	565.5	fre	Ben Jann
16	477.0	whitetst	Nicholas J. Cox, Christopher F Baum
17	420.1	spmap	Maurizio Pisati
18	419.5	shp2dta	Kevin Crow
19	416.3	egenmore	Nicholas J. Cox
20	405.2	ice	Patrick Royston

(Click on package name for description)

The author(*name*) option allows you to list the most popular packages by a specific person:

. ssc hot, author(baum)

Top 10 packages at SSC by author Baum

Rank	Apr2011 # hits	Package	Author(s)
4	2006.2	ivreg2	Mark E Schaffer, Christopher F Baum, Steven Stillman
13	734.0	hprescott	Christopher F Baum
14	571.5	overid	Vince Wiggins, Steven Stillman, Mark E Schaffer, Christopher F Baum
16	477.0	whitetst	Nicholas J. Cox, Christopher F Baum
31	360.5	xttest3	Christopher F Baum
35	342.0	ivendog	Mark E Schaffer, Christopher F Baum, Steven Stillman
44	289.5	xttest2	Christopher F Baum
49	275.0	tsmktim	Christopher F Baum, Vince Wiggins
62	223.8	outtable	Joao Pedro Azevedo, Christopher F Baum
65	217.0	ipshin	Fabian Bornhorst, Christopher F Baum

(Click on package name for description)

ssc describe *pkgname* describes, but does not install, the specified package. You must already know the name of the package. See [R] **search** for assistance in searching for packages. Sometimes you know the package name, but you do not know the exact spelling. Then you can type ssc describe followed by one letter, a–z or _, to list all the packages starting with that letter; even so, using search is better.

```
. ssc describe khb
```

---

package **khb** from http://fmwww.bc.edu/repec/bocode/k

TITLE
      'KHB': module to decompose total effects into direct and indirect via KHB
> -method
DESCRIPTION/AUTHOR(S)
       decomposes the total effect of a variable into direct and
       indirect effects using the KHB-method developed by Karlson, Holm,
       and Breen (2011). The method is developed for binary and logit
       probit models, but this command also includes other nonlinear
       probability models (ordered and multinomial) and linear
       regression. Contrary to other decomposition methods, the
       KHB-method gives unbiased decompositions, decomposes effects of
       both discrete and continuous variables, and provides analytically
       derived statistical tests for many models of the GLM family.
       KW: decomposition
       KW: effects
       KW: probit
       KW: nonlinear probability model
       Requires: Stata version 11
       Distribution-Date: 20110502
       Author: Ulrich Kohler, WZB
       Support: email kohler@wz-berlin.de
       Author: Kristian Karlson
       Support: email kbk@dpu.dk
INSTALLATION FILES   (type **net install khb**)
       khb.ado
       khb.sthlp

(type -**ssc install khb**- to install)

The default setting for the saving() option is for the output to be saved with the .smcl extension. You could also save the file with a log extension, and in this case, the file would be stored as a text file.

```
. ssc describe k, saving(k.index)
  (output omitted )
. ssc describe khb, saving(khb.log)
  (output omitted )
```

ssc install *pkgname* installs the specified package. You do not have to describe a package before installing it. There are ways of installing packages other than ssc install, such as net; see [R] **net**. It does not matter how a package is installed. For instance, a package can be installed using net and still be uninstalled using ssc.

```
. ssc install khb
checking khb consistency and verifying not already installed...
installing into C:\ado\plus\...
installation complete.
```

ssc uninstall *pkgname* removes the specified, previously installed package from your computer. You can uninstall immediately after installation or at any time in the future. (Technical note: ssc uninstall is a synonym for ado uninstall, so it can uninstall any installed package, not just packages obtained from the SSC.)

```
. ssc uninstall khb
```
package **khb** from http://fmwww.bc.edu/repec/bocode/k
     'KHB': module to decompose total effects into direct and indirect via KHB-method
(package uninstalled)

`ssc type` *filename* types a specific file stored at the SSC. Although not shown in the syntax diagram, `ssc cat` is a synonym for `ssc type`, which may appeal to those familiar with Unix. To view only the `khb` help file for the `khb` package, you would type

> **Title**
>   **khb** Decomposition of total effects into direct and indirect effects using
>   the KHB-method
> (*output omitted*)

`ssc copy` *filename* copies a specific file stored at the SSC to your computer. By default, the file is copied to the current directory, but you can use options to change this. `ssc copy` is a rarely used alternative to `ssc install ... , all`. `ssc cp` is a synonym for `ssc copy`.

```
. ssc copy khb.ado
(file khb.ado copied to current directory)
```
◁

For more details on the SSC archive and for information on how to submit your own programs to the SSC, see http://repec.org/bocode/s/sscsubmit.html.

Methods and formulas

`ssc` is implemented as an ado-file.

Acknowledgments

`ssc` is based on `archutil` by Nicholas J. Cox, Durham University, and Christopher Baum, Boston College. The reworking of the original was done with their blessing and their participation.

Christopher Baum maintains the Stata-related files stored at the SSC archive. We thank him for this contribution to the Stata community.

References

Baum, C. F., and N. J. Cox. 1999. ip29: Metadata for user-written contributions to the Stata programming language. *Stata Technical Bulletin* 52: 10–12. Reprinted in *Stata Technical Bulletin Reprints*, vol. 9, pp. 121–124. College Station, TX: Stata Press.

Cox, N. J., and C. F. Baum. 2000. ip29.1: Metadata for user-written contributions to the Stata programming language. *Stata Technical Bulletin* 54: 21–22. Reprinted in *Stata Technical Bulletin Reprints*, vol. 9, pp. 124–126. College Station, TX: Stata Press.

Also see

[R] **adoupdate** — Update user-written ado-files

[R] **net** — Install and manage user-written additions from the Internet

[R] **search** — Search Stata documentation

[R] **sj** — Stata Journal and STB installation instructions

[P] **sysdir** — Query and set system directories

Title

> **stem** — Stem-and-leaf displays

Syntax

stem *varname* $\left[\textit{if} \right]$ $\left[\textit{in} \right]$ $\left[, \textit{options} \right]$

options	Description
Main	
<u>p</u>rune	do not print stems that have no leaves
<u>r</u>ound(#)	round data to this value; default is round(1)
<u>d</u>igits(#)	digits per leaf; default is digits(1)
<u>l</u>ines(#)	number of stems per interval of 10^{digits}
<u>w</u>idth(#)	stem width; equal to $10^{\text{digits}}/\text{width}$

by is allowed; see [D] **by**.

Menu

Statistics > Summaries, tables, and tests > Distributional plots and tests > Stem-and-leaf display

Description

stem displays stem-and-leaf plots.

Options

┌─ Main └───

prune prevents printing any stems that have no leaves.

round(#) rounds the data to this value and displays the plot in these units. If round() is not specified, noninteger data will be rounded automatically.

digits(#) sets the number of digits per leaf. The default is 1.

lines(#) sets the number of stems per every data interval of 10^{digits}. The value of lines() must divide 10^{digits}; that is, if digits(1) is specified, then lines() must divide 10. If digits(2) is specified, then lines() must divide 100, etc. Only one of lines() or width() may be specified. If neither is specified, an appropriate value will be set automatically.

width(#) sets the width of a stem. lines() is equal to $10^{\text{digits}}/\text{width}$, and this option is merely an alternative way of setting lines(). The value of width() must divide 10^{digits}. Only one of width() or lines() may be specified. If neither is specified, an appropriate value will be set automatically.

Note: If lines() or width() is not specified, digits() may be decreased in some circumstances to make a better-looking plot. If lines() or width() is set, the user-specified value of digits() will not be altered.

Remarks

▷ Example 1

Stem-and-leaf displays are a compact way to present considerable information about a batch of data. For instance, using our automobile data (described in [U] **1.2.2 Example datasets**):

```
. use http://www.stata-press.com/data/r12/auto
(1978 Automobile Data)
. stem mpg
Stem-and-leaf plot for mpg (Mileage (mpg))
  1t │ 22
  1f │ 44444455
  1s │ 66667777
  1. │ 88888888899999999
  2* │ 00011111
  2t │ 22222333
  2f │ 444455555
  2s │ 666
  2. │ 8889
  3* │ 001
  3t │
  3f │ 455
  3s │
  3. │
  4* │ 1
```

The stem-and-leaf display provides a way to list our data. The expression to the left of the vertical bar is called the stem; the digits to the right are called the leaves. All the stems that begin with the same digit and the corresponding leaves, written beside each other, reconstruct an observation of the data. Thus, if we look at the four stems that begin with the digit 1 and their corresponding leaves, we see that we have two cars rated at 12 mpg, 6 cars at 14, 2 at 15, and so on. The car with the highest mileage rating in our data is rated at 41 mpg.

The above plot is a five-line plot with lines() equal to 5 (five lines per interval of 10) and width() equal to 2 (two leaves per stem).

Instead, we could specify lines(2):

```
. stem mpg, lines(2)
Stem-and-leaf plot for mpg (Mileage (mpg))
  1* │ 22444444
  1. │ 55666677778888888888899999999
  2* │ 00011111222223334444
  2. │ 555556668889
  3* │ 0014
  3. │ 55
  4* │ 1
```

stem mpg, width(5) would produce the same plot as above.

The stem-and-leaf display provides a crude histogram of our data, one not so pretty as that produced by histogram (see [R] **histogram**), but one that is nonetheless informative.

◁

▷ Example 2

The miles per gallon rating fits easily into a stem-and-leaf display because, in our data, it has two digits. However, `stem` does not require two digits.

```
. stem price, lines(1) digits(3)

Stem-and-leaf plot for price (Price)
     3*** | 291,299,667,748,798,799,829,895,955,984,995
     4*** | 010,060,082,099,172,181,187,195,296,389,424,425,453,482,499, ... (26)
     5*** | 079,104,172,189,222,379,397,705,719,788,798,799,886,899
     6*** | 165,229,295,303,342,486,850
     7*** | 140,827
     8*** | 129,814
     9*** | 690,735
    10*** | 371,372
    11*** | 385,497,995
    12*** | 990
    13*** | 466,594
    14*** | 500
    15*** | 906
```

The (26) at the right of the second stem shows that there were 26 leaves on this stem—too many to display on one line.

We can make a more compact stem-and-leaf plot by rounding. To display `stem` in units of 100, we could type

```
. stem price, round(100)

Stem-and-leaf plot for price (Price)

price rounded to nearest multiple of 100
plot in units of 100
     3* | 33778889
     4* | 000011122223444455555667777899
     5* | 11222447788899
     6* | 2233359
     7* | 18
     8* | 18
     9* | 77
    10* | 44
    11* | 45
    12* | 0
    13* | 056
    14* | 5
    15* | 9
```

`price`, in our data, has four or five digits. `stem` presented the display in terms of units of 100, so a car that cost \$3,291 was treated for display purposes as \$3,300.

◁

❑ Technical note

Stem-and-leaf diagrams have been used in Japanese railway timetables, as shown in Tufte (1990, 46–47).

❑

Saved results

stem saves the following in r():

Scalars
    r(width)   width of a stem
    r(digits)   number of digits per leaf; default is 1

Macros
    r(round)   number specified in round()

Methods and formulas

stem is implemented as an ado-file.

References

Cox, N. J. 2007. Speaking Stata: Turning over a new leaf. *Stata Journal* 7: 413–433.

Emerson, J. D., and D. C. Hoaglin. 1983. Stem-and-leaf displays. In *Understanding Robust and Exploratory Data Analysis*, ed. D. C. Hoaglin, F. Mosteller, and J. W. Tukey, 7–32. New York: Wiley.

Tufte, E. R. 1990. *Envisioning Information.* Cheshire, CT: Graphics Press.

Tukey, J. W. 1972. Some graphic and semigraphic displays. In *Statistical Papers in Honor of George W. Snedecor*, ed. T. A. Bancroft and S. A. Brown, 293–316. Ames, IA: Iowa State University Press.

——. 1977. *Exploratory Data Analysis.* Reading, MA: Addison–Wesley.

Also see

[R] **histogram** — Histograms for continuous and categorical variables

[R] **lv** — Letter-value displays

Title

stepwise — Stepwise estimation

Syntax

stepwise [, *options*] : *command*

options	Description
Model	
*pr(#)	significance level for removal from the model
*pe(#)	significance level for addition to the model
Model2	
forward	perform forward-stepwise selection
hierarchical	perform hierarchical selection
lockterm1	keep the first term
lr	perform likelihood-ratio test instead of Wald test
Reporting	
display_options	control column formats and line width

* At least one of pr(#) or pe(#) must be specified.

by and xi are allowed; see [U] **11.1.10 Prefix commands**.

Weights are allowed if *command* allows them; see [U] **11.1.6 weight**.

All postestimation commands behave as they would after *command* without the stepwise prefix; see the postestimation manual entry for *command*.

See [U] **20 Estimation and postestimation commands** for more capabilities of estimation commands.

Menu

Statistics > Other > Stepwise estimation

Description

stepwise performs stepwise estimation. Typing

. stepwise, pr(#): *command*

performs backward-selection estimation for *command*. The stepwise selection method is determined by the following option combinations:

options	Description
pr(#)	backward selection
pr(#) hierarchical	backward hierarchical selection
pr(#) pe(#)	backward stepwise
pe(#)	forward selection
pe(#) hierarchical	forward hierarchical selection
pr(#) pe(#) forward	forward stepwise

command defines the estimation command to be executed. The following Stata commands are supported by `stepwise`:

clogit	nbreg	regress
cloglog	ologit	scobit
glm	oprobit	stcox
intreg	poisson	stcrreg
logistic	probit	streg
logit	qreg	tobit

`stepwise` expects *command* to have the following form:

command_name [*depvar*] *term* [*term* ...] [*if*] [*in*] [*weight*] [, *command_options*]

where *term* is either *varname* or (*varlist*) (a *varlist* in parentheses indicates that this group of variables is to be included or excluded together). *depvar* is not present when *command_name* is `stcox`, `stcrreg`, or `streg`; otherwise, *depvar* is assumed to be present. For `intreg`, *depvar* is actually two dependent variable names (*depvar*$_1$ and *depvar*$_2$).

`sw` is a synonym for `stepwise`.

Options

    ┌─── Model ┐

`pr(#)` specifies the significance level for removal from the model; terms with $p \geq$ `pr()` are eligible for removal.

`pe(#)` specifies the significance level for addition to the model; terms with $p <$ `pe()` are eligible for addition.

    ┌─── Model 2 ┐

`forward` specifies the forward-stepwise method and may be specified only when both `pr()` and `pe()` are also specified. Specifying both `pr()` and `pe()` without `forward` results in backward-stepwise selection. Specifying only `pr()` results in backward selection, and specifying only `pe()` results in forward selection.

`hierarchical` specifies hierarchical selection.

`lockterm1` specifies that the first term be included in the model and not be subjected to the selection criteria.

`lr` specifies that the test of term significance be the likelihood-ratio test. The default is the less computationally expensive Wald test; that is, the test is based on the estimated variance–covariance matrix of the estimators.

    ┌─── Reporting ┐

display_options: `cformat(%fmt)`, `pformat(%fmt)`, `sformat(%fmt)`, and `nolstretch`; see [R] **estimation options**.

Remarks

Remarks are presented under the following headings:

Introduction
Search logic for a step
Full search logic
Examples
Estimation sample considerations
Messages
Programming for stepwise

Introduction

Typing

```
. stepwise, pr(.10): regress y1 x1 x2 d1 d2 d3 x4 x5
```

performs a backward-selection search for the regression model y1 on x1, x2, d1, d2, d3, x4, and x5. In this search, each explanatory variable is said to be a term. Typing

```
. stepwise, pr(.10): regress y1 x1 x2 (d1 d2 d3) (x4 x5)
```

performs a similar backward-selection search, but the variables d1, d2, and d3 are treated as one term, as are x4 and x5. That is, d1, d2, and d3 may or may not appear in the final model, but they appear or do not appear together.

▷ Example 1

Using the automobile dataset, we fit a backward-selection model of mpg:

```
. use http://www.stata-press.com/data/r12/auto
. gen weight2 = weight*weight
. stepwise, pr(.2): regress mpg weight weight2 displ gear turn headroom
> foreign price
                              begin with full model
p = 0.7116 >= 0.2000   removing headroom
p = 0.6138 >= 0.2000   removing displacement
p = 0.3278 >= 0.2000   removing price
```

Source	SS	df	MS		
Model	1736.31455	5	347.262911		
Residual	707.144906	68	10.3991898		
Total	2443.45946	73	33.4720474		

Number of obs =   74
F(5,   68) =   33.39
Prob > F   = 0.0000
R-squared   = 0.7106
Adj R-squared = 0.6893
Root MSE   = 3.2248

mpg	Coef.	Std. Err.	t	P>|t|	[95% Conf. Interval]
weight	-.0158002	.0039169	-4.03	0.000	-.0236162 -.0079842
weight2	1.77e-06	6.20e-07	2.86	0.006	5.37e-07 3.01e-06
foreign	-3.615107	1.260844	-2.87	0.006	-6.131082 -1.099131
gear_ratio	2.011674	1.468831	1.37	0.175	-.9193321 4.94268
turn	-.3087038	.1763099	-1.75	0.084	-.6605248 .0431172
_cons	59.02133	9.3903	6.29	0.000	40.28327 77.75938

This estimation treated each variable as its own term and thus considered each one separately. The engine displacement and gear ratio should really be considered together:

```
. stepwise, pr(.2): regress mpg weight weight2 (displ gear) turn headroom
> foreign price
                        begin with full model
p = 0.7116 >= 0.2000  removing headroom
p = 0.3944 >= 0.2000  removing displacement gear_ratio
p = 0.2798 >= 0.2000  removing price
```

Source	SS	df	MS		
Model	1716.80842	4	429.202105		
Residual	726.651041	69	10.5311745		
Total	2443.45946	73	33.4720474		

```
                                              Number of obs =      74
                                              F(  4,    69) =   40.76
                                              Prob > F      =  0.0000
                                              R-squared     =  0.7026
                                              Adj R-squared =  0.6854
                                              Root MSE      =  3.2452
```

mpg	Coef.	Std. Err.	t	P>|t|	[95% Conf. Interval]
weight	-.0160341	.0039379	-4.07	0.000	-.0238901 -.0081782
weight2	1.70e-06	6.21e-07	2.73	0.008	4.58e-07 2.94e-06
foreign	-2.758668	1.101772	-2.50	0.015	-4.956643 -.5606925
turn	-.2862724	.176658	-1.62	0.110	-.6386955 .0661508
_cons	65.39216	8.208778	7.97	0.000	49.0161 81.76823

◁

Search logic for a step

Before discussing the complete search logic, consider the logic for a step—the first step—in detail. The other steps follow the same logic. If you type

```
. stepwise, pr(.20): regress y1 x1 x2 (d1 d2 d3) (x4 x5)
```

the logic is

1. Fit the model y on x1 x2 d1 d2 d3 x4 x5.
2. Consider dropping x1.
3. Consider dropping x2.
4. Consider dropping d1 d2 d3.
5. Consider dropping x4 x5.
6. Find the term above that is least significant. If its significance level is ≥ 0.20, remove that term.

If you type

```
. stepwise, pr(.20) hierarchical: regress y1 x1 x2 (d1 d2 d3) (x4 x5)
```

the logic would be different because the hierarchical option states that the terms are ordered. The initial logic would become

1. Fit the model y on x1 x2 d1 d2 d3 x4 x5.
2. Consider dropping x4 x5—the last term.
3. If the significance of this last term is ≥ 0.20, remove the term.

The process would then stop or continue. It would stop if x4 x5 were not dropped, and otherwise, stepwise would continue to consider the significance of the next-to-last term, d1 d2 d3.

Specifying pe() rather than pr() switches to forward estimation. If you type

```
. stepwise, pe(.20): regress y1 x1 x2 (d1 d2 d3) (x4 x5)
```

`stepwise` performs forward-selection search. The logic for the first step is

      1. Fit a model of y on nothing (meaning a constant).
      2. Consider adding x1.
      3. Consider adding x2.
      4. Consider adding d1 d2 d3.
      5. Consider adding x4 x5.
      6. Find the term above that is most significant. If its significance level is < 0.20, add that term.

As with backward estimation, if you specify `hierarchical`,

    . stepwise, pe(.20) hierarchical: regress y1 x1 x2 (d1 d2 d3) (x4 x5)

the search for the most significant term is restricted to the next term:

      1. Fit a model of y on nothing (meaning a constant).
      2. Consider adding x1—the first term.
      3. If the significance is < 0.20, add the term.

If x1 were added, `stepwise` would next consider x2; otherwise, the search process would stop.

    `stepwise` can also use a stepwise selection logic that alternates between adding and removing terms. The full logic for all the possibilities is given below.

Full search logic

Option	Logic
pr() (backward selection)	Fit the full model on all explanatory variables. While the least-significant term is "insignificant", remove it and reestimate.
pr() hierarchical (backward hierarchical selection)	Fit full model on all explanatory variables. While the last term is "insignificant", remove it and reestimate.

pr() pe()
(backward stepwise)
	Fit full model on all explanatory variables.
	If the least-significant term is "insignificant", remove it and
	    reestimate; otherwise, stop.
	Do that again: if the least-significant term is "insignificant",
	    remove it and reestimate; otherwise, stop.
	Repeatedly,
	        if the most-significant excluded term is "significant", add
	            it and reestimate;
	        if the least-significant included term is "insignificant",
	            remove it and reestimate;
	until neither is possible.

pe() (forward selection)	Fit "empty" model. While the most-significant excluded term is "significant", add it and reestimate.
pe() hierarchical (forward hierarchical selection)	Fit "empty" model. While the next term is "significant", add it and reestimate.

pr() pe() forward
(forward stepwise)
	Fit "empty" model.
	If the most-significant excluded term is "significant",
	    add it and reestimate; otherwise, stop.
	Do that again: if the most-significant excluded term is
	    "significant", add it and reestimate; otherwise, stop.
	Repeatedly,
	        if the least-significant included term is "insignificant",
	            remove it and reestimate;
	        if the most-significant excluded term is "significant",
	            add it and reestimate;
	until neither is possible.

Examples

The following two statements are equivalent; both include solely single-variable terms:

```
. stepwise, pr(.2): regress price mpg weight displ
. stepwise, pr(.2): regress price (mpg) (weight) (displ)
```

The following two statements are equivalent; the last term in each is r1, ..., r4:

```
. stepwise, pr(.2) hierarchical: regress price mpg weight displ (r1-r4)
. stepwise, pr(.2) hierarchical: regress price (mpg) (weight) (displ) (r1-r4)
```

To group variables weight and displ into one term, type

```
. stepwise, pr(.2) hierarchical: regress price mpg (weight displ) (r1-r4)
```

stepwise can be used with commands other than regress; for instance,

```
. stepwise, pr(.2): logit outcome (sex weight) treated1 treated2
. stepwise, pr(.2): logistic outcome (sex weight) treated1 treated2
```

Either statement would fit the same model because logistic and logit both perform logistic regression; they differ only in how they report results; see [R] **logit** and [R] **logistic**.

We use the lockterm1 option to force the first term to be included in the model. To keep treated1 and treated2 in the model no matter what, we type

```
. stepwise, pr(.2) lockterm1: logistic outcome (treated1 treated2) ...
```

After stepwise estimation, we can type stepwise without arguments to redisplay results,

```
. stepwise
 (output from logistic appears )
```

or type the underlying estimation command:

```
. logistic
 (output from logistic appears )
```

At estimation time, we can specify options unique to the command being stepped:

```
. stepwise, pr(.2): logit outcome (sex weight) treated1 treated2, or
```

or is logit's option to report odds ratios rather than coefficients; see [R] **logit**.

Estimation sample considerations

Whether you use backward or forward estimation, stepwise forms an estimation sample by taking observations with nonmissing values of all the variables specified (except for $depvar_1$ and $depvar_2$ for intreg). The estimation sample is held constant throughout the stepping. Thus if you type

```
. stepwise, pr(.2) hierarchical: regress amount sk edul sval
```

and variable sval is missing in half the data, that half of the data will not be used in the reported model, even if sval is not included in the final model.

The function e(sample) identifies the sample that was used. e(sample) contains 1 for observations used and 0 otherwise. For instance, if you type

```
. stepwise, pr(.2) pe(.10): logistic outcome x1 x2 (x3 x4) (x5 x6 x7)
```

and the final model is outcome on x1, x5, x6, and x7, you could re-create the final regression by typing

    . logistic outcome x1 x5 x6 x7 if e(sample)

You could obtain summary statistics within the estimation sample of the independent variables by typing

    . summarize x1 x5 x6 x7 if e(sample)

If you fit another model, e(sample) will automatically be redefined. Typing

    . stepwise, lock pr(.2): logistic outcome (x1 x2) (x3 x4) (x5 x6 x7)

would automatically drop e(sample) and re-create it.

Messages

note: _____ dropped because of collinearity

Each term is checked for collinearity, and variables within the term are dropped if collinearity is found. For instance, say that you type

    . stepwise, pr(.2): regress y x1 x2 (r1-r4) (x3 x4)

and assume that variables r1 through r4 are mutually exclusive and exhaustive dummy variables—perhaps r1, ..., r4 indicate in which of four regions the subject resides. One of the r1, ..., r4 variables will be automatically dropped to identify the model.

This message should cause you no concern.

Error message: between-term collinearity, variable _____

After removing any within-term collinearity, if stepwise still finds collinearity between terms, it refuses to continue. For instance, assume that you type

    . stepwise, pr(.2): regress y1 x1 x2 (d1-d8) (r1-r4)

Assume that r1, ..., r4 identify in which of four regions the subject resides, and that d1, ..., d8 identify the same sort of information, but more finely. r1, say, amounts to d1 and d2; r2 to d3, d4, and d5; r3 to d6 and d7; and r4 to d8. You can estimate the d* variables or the r* variables, but not both.

It is your responsibility to specify noncollinear terms.

note: _____ dropped because of estimability
note: _____ obs. dropped because of estimability

You probably received this message in fitting a logistic or probit model. Regardless of estimation strategy, stepwise checks that the full model can be fit. The indicated variable had a 0 or infinite standard error.

For logistic, logit, and probit, this message is typically caused by one-way causation. Assume that you type

    . stepwise, pr(.2): logistic outcome (x1 x2 x3) d1

and assume that variable d1 is an indicator (dummy) variable. Further assume that whenever d1 = 1, outcome = 1 in the data. Then the coefficient on d1 is infinite. One (conservative) solution to this problem is to drop the d1 variable and the d1==1 observations. The underlying estimation commands probit, logit, and logistic report the details of the difficulty and solution; stepwise simply accumulates such problems and reports the above summary messages. Thus if you see this message, you could type

```
. logistic outcome x1 x2 x3 d1
```

to see the details. Although you should think carefully about such situations, Stata's solution of dropping the offending variables and observations is, in general, appropriate.

Programming for stepwise

stepwise requires that *command_name* follow standard Stata syntax and allow the if qualifier; see [U] **11 Language syntax**. Furthermore, *command_name* must have sw or swml as a program property; see [P] **program properties**. If *command_name* has swml as a property, *command_name* must save the log-likelihood value in e(ll) and model degrees of freedom in e(df_m).

Saved results

stepwise saves whatever is saved by the underlying estimation command.

Also, stepwise saves stepwise in e(stepwise).

Methods and formulas

stepwise is implemented as an ado-file.

Some statisticians do not recommend stepwise procedures; see Sribney (1998) for a summary.

References

Beale, E. M. L. 1970. Note on procedures for variable selection in multiple regression. *Technometrics* 12: 909–914.

Bendel, R. B., and A. A. Afifi. 1977. Comparison of stopping rules in forward "stepwise" regression. *Journal of the American Statistical Association* 72: 46–53.

Berk, K. N. 1978. Comparing subset regression procedures. *Technometrics* 20: 1–6.

Draper, N., and H. Smith. 1998. *Applied Regression Analysis*. 3rd ed. New York: Wiley.

Efroymson, M. A. 1960. Multiple regression analysis. In *Mathematical Methods for Digital Computers*, ed. A. Ralston and H. S. Wilf, 191–203. New York: Wiley.

Gorman, J. W., and R. J. Toman. 1966. Selection of variables for fitting equations to data. *Technometrics* 8: 27–51.

Hocking, R. R. 1976. The analysis and selection of variables in linear regression. *Biometrics* 32: 1–49.

Hosmer, D. W., Jr., and S. Lemeshow. 2000. *Applied Logistic Regression*. 2nd ed. New York: Wiley.

Kennedy, W. J., Jr., and T. A. Bancroft. 1971. Model-building for prediction in regression based on repeated significance tests. *Annals of Mathematical Statistics* 42: 1273–1284.

Lindsey, C., and S. J. Sheather. 2010. Variable selection in linear regression. *Stata Journal* 10: 650–669.

Mantel, N. 1970. Why stepdown procedures in variable selection. *Technometrics* 12: 621–625.

———. 1971. More on variable selection and an alternative approach (letter to the editor). *Technometrics* 13: 455–457.

Sribney, W. M. 1998. FAQ: What are some of the problems with stepwise regression? http://www.stata.com/support/faqs/stat/stepwise.html.

Wang, Z. 2000. sg134: Model selection using the Akaike information criterion. *Stata Technical Bulletin* 54: 47–49. Reprinted in *Stata Technical Bulletin Reprints*, vol. 9, pp. 335–337. College Station, TX: Stata Press.

Williams, R. 2007. Stata tip 46: Step we gaily, on we go. *Stata Journal* 7: 272–274.

Also see

[R] **nestreg** — Nested model statistics

Title

> **suest** — Seemingly unrelated estimation

Syntax

> suest *namelist* [, *options*]

where *namelist* is a list of one or more names under which estimation results were saved via estimates store; see [R] **estimates store**. Wildcards may be used. * and _all refer to all stored results. A period (.) may be used to refer to the last estimation results, even if they have not (yet) been stored.

options	Description
SE/Robust	
svy	survey data estimation
vce(*vcetype*)	*vcetype* may be <u>r</u>obust or <u>cl</u>uster *clustvar*
Reporting	
<u>l</u>evel(#)	set confidence level; default is level(95)
<u>dir</u>	display a table describing the models
<u>ef</u>orm(*string*)	report exponentiated coefficients and label as *string*
display_options	control column formats, row spacing, line width, and display of omitted variables and base and empty cells
<u>coefl</u>egend	display legend instead of statistics

coeflegend does not appear in the dialog box.

Menu

Statistics > Postestimation > Tests > Seemingly unrelated estimation

Description

suest is a postestimation command; see [U] **20 Estimation and postestimation commands**.

suest combines the estimation results—parameter estimates and associated (co)variance matrices—stored under *namelist* into one parameter vector and *simultaneous* (co)variance matrix of the sandwich/robust type. This (co)variance matrix is appropriate even if the estimates were obtained on the same or on overlapping data.

Typical applications of suest are tests for intramodel and cross-model hypotheses using test or testnl, for example, a generalized Hausman specification test. lincom and nlcom may be used after suest to estimate linear combinations and nonlinear functions of coefficients. suest may also be used to adjust a standard VCE for clustering or survey design effects.

Different estimators are allowed, for example, a regress model and a probit model; the only requirement is that predict produce equation-level scores with the score option after an estimation command. The models may be estimated on different samples, due either to explicit if or in selection or to missing values. If weights are applied, the same weights (type and values) should be applied to all models in *namelist*. The estimators should be estimated without vce(robust) or vce(cluster *clustvar*) options. suest returns the robust VCE, allows the vce(cluster *clustvar*) option, and automatically works with results from the svy prefix command (only for vce(linearized)). See example 7 in [SVY] **svy postestimation** for an example using suest with svy: ologit.

Because suest posts its results like a proper estimation command, its results can be stored via estimates store. Moreover, like other estimation commands, suest typed without arguments replays the results.

Options

──────┐ SE/Robust └──

svy specifies that estimation results should be modified to reflect the survey design effects according to the svyset specifications, see [SVY] **svyset**.

The svy option is implied when suest encounters survey estimation results from the svy prefix; see [SVY] **svy**. Poststratification is allowed only with survey estimation results from the svy prefix.

vce(*vcetype*) specifies the type of standard error reported, which includes types that are robust to some kinds of misspecification and that allow for intragroup correlation; see [R] **vce_option**.

The vce() option may not be combined with the svy option or estimation results from the svy prefix.

──────┐ Reporting └──

level(#) specifies the confidence level, as a percentage, for confidence intervals of the coefficients; see [R] **level**.

dir displays a table describing the models in *namelist* just like estimates dir *namelist*.

eform(*string*) displays the coefficient table in exponentiated form: for each coefficient, $\exp(b)$ rather than b is displayed, and standard errors and confidence intervals are transformed. *string* is the table header that will be displayed above the transformed coefficients and must be 11 characters or fewer, for example, eform("Odds ratio").

display_options: noomitted, vsquish, noemptycells, baselevels, allbaselevels, cformat(% *fmt*), pformat(% *fmt*), sformat(% *fmt*), and nolstretch; see [R] **estimation options**.

The following option is available with suest but is not shown in the dialog box:

coeflegend; see [R] **estimation options**.

Remarks

Remarks are presented under the following headings:

    Using suest
    Remarks on regress
    Testing the assumption of the independence of irrelevant alternatives
    Testing proportionality
    Testing cross-model hypotheses

Using suest

If you plan to use suest, you must take precautions when fitting the original models. These restrictions are relaxed when using svy commands; see [SVY] **svy postestimation**.

1. suest works with estimation commands that allow predict to generate equation-level score variables when supplied with the score (or scores) option. For example, equation-level score variables are generated after running mlogit by typing

   . predict sc*, scores

2. Estimation should take place *without* the vce(robust) or vce(cluster *clustvar*) option. suest always computes the robust estimator of the (co)variance, and suest has a vce(cluster *clustvar*) option.

   The within-model covariance matrices computed by suest are identical to those obtained by specifying a vce(robust) or vce(cluster *clustvar*) option during estimation. suest, however, also estimates the between-model covariances of parameter estimates.

3. Finally, the estimation results to be combined should be stored by estimates store; see [R] **estimates store**.

After estimating and storing a series of estimation results, you are ready to combine the estimation results with suest,

   . suest *name1* $\big[$ *name2* ... $\big]$ $\big[$, vce(cluster *clustvar*) $\big]$

and you can subsequently use postestimation commands, such as test, to test hypotheses. Here an important issue is how suest assigns names to the equations. If you specify one model *name*, the original equation names are left unchanged; otherwise, suest constructs new equation names. The coefficients of a single-equation model (such as logit and poisson) that was estimate stored under name *X* are collected under equation *X*. With a multiequation model stored under name *X*, suest prefixes *X_* to an original equation name *eq*, forming equation name, *X_eq*.

❑ Technical note

Earlier we said that standard errors from suest are identical to those obtained by specifying the vce(robust) option with each command individually. Thus if you fit a logistic model using logit with the vce(robust) option, you will get the same standard errors when you type

   . suest .

directly after logit using the same data without the vce(robust) option.

This is not true for multiple estimation results when the estimation samples are not all the same. The standard errors from suest will be slightly smaller than those from individual model fits using the vce(robust) option because suest uses a larger number of observations to estimate the simultaneous (co)variance matrix.

❑

❑ Technical note

In rare circumstances, suest may have to truncate equation names to 32 characters. When equation names are not unique because of truncation, suest numbers the equations within models, using equations named *X_#*.

❑

Remarks on regress

regress (see [R] **regress**) does not include its ancillary parameter, the residual variance, in its coefficient vector and (co)variance matrix. Moreover, while the score option is allowed with predict after regress, a score variable is generated for the mean but not for the variance parameter. suest contains special code that assigns the equation name mean to the coefficients for the mean, adds the equation lnvar for the log variance, and computes the appropriate two score variables itself.

Testing the assumption of the independence of irrelevant alternatives

The multinomial logit model and the closely related conditional logit model satisfy a probabilistic version of the assumption of the independence of irrelevant alternatives (IIA), implying that the ratio of the probabilities for two alternatives does not depend on what other alternatives are available. Hausman and McFadden (1984) proposed a test for this assumption that is implemented in the hausman command. The standard Hausman test has several limitations. First, the test statistic may be undefined because the estimated VCE does not satisfy the required asymptotic properties of the test. Second, the classic Hausman test applies only to the test of the equality of two estimators. Third, the test requires access to a fully efficient estimator; such an estimator may not be available, for example, if you are analyzing complex survey data. Using suest can overcome these three limitations.

▷ Example 1

In our first example, we follow the analysis of the type of health insurance reported in [R] **mlogit** and demonstrate the hausman command with the suest/test combination. We fit the full multinomial logit model for all three alternatives and two restricted multinomial models in which one alternative is excluded. After fitting each of these models, we store the results by using the store subcommand of estimates. title() simply documents the models.

```
. use http://www.stata-press.com/data/r12/sysdsn4
(Health insurance data)

. mlogit insure age male

Iteration 0:   log likelihood = -555.85446
Iteration 1:   log likelihood = -551.32973
Iteration 2:   log likelihood = -551.32802
Iteration 3:   log likelihood = -551.32802
```

Multinomial logistic regression	Number of obs	=	615
	LR chi2(4)	=	9.05
	Prob > chi2	=	0.0598
Log likelihood = -551.32802	Pseudo R2	=	0.0081

insure	Coef.	Std. Err.	z	P>|z|	[95% Conf. Interval]
Indemnity	(base outcome)					
Prepaid						
age	-.0100251	.0060181	-1.67	0.096	-.0218204	.0017702
male	.5095747	.1977893	2.58	0.010	.1219147	.8972346
_cons	.2633838	.2787575	0.94	0.345	-.2829708	.8097383
Uninsure						
age	-.0051925	.0113821	-0.46	0.648	-.0275011	.0171161
male	.4748547	.3618462	1.31	0.189	-.2343508	1.18406
_cons	-1.756843	.5309602	-3.31	0.001	-2.797506	-.7161803

```
. estimates store m1, title(all three insurance forms)
```

```
. quietly mlogit insure age male if insure != "Uninsure":insure
. estimates store m2, title(insure != "Uninsure":insure)
. quietly mlogit insure age male if insure != "Prepaid":insure
. estimates store m3, title(insure != "Prepaid":insure)
```

Having performed the three estimations, we inspect the results. `estimates dir` provides short descriptions of the models that were stored using `estimates store`. Typing `estimates table` lists the coefficients, displaying blanks for a coefficient not contained in a model.

```
. estimates dir
```

name	command	depvar	npar	title
m1	mlogit	insure	9	all three insurance forms
m2	mlogit	insure	6	insure != Uninsure :insure
m3	mlogit	insure	6	insure != Prepaid :insure

```
. estimates table m1 m2 m3, star stats(N ll) keep(Prepaid: Uninsure:)
```

Variable	m1	m2	m3
Prepaid			
age	-.01002511	-.01015205	
male	.50957468**	.51440033**	
_cons	.26338378	.26780432	
Uninsure			
age	-.00519249		-.00410547
male	.47485472		.45910738
_cons	-1.7568431***		-1.8017743***
Statistics			
N	615	570	338
ll	-551.32802	-390.48643	-131.76807

legend: * p<0.05; ** p<0.01; *** p<0.001

Comparing the coefficients between models does not suggest substantial differences. We can formally test that coefficients are the same for the full model m1 and the restricted models m2 and m3 by using the `hausman` command. `hausman` expects the models to be specified in the order "always consistent" first and "efficient under H_0" second.

```
. hausman m2 m1, alleqs constant
```

	—— Coefficients —— (b) m2	(B) m1	(b-B) Difference	sqrt(diag(V_b-V_B)) S.E.
age	-.0101521	-.0100251	-.0001269	.
male	.5144003	.5095747	.0048256	.0123338
_cons	.2678043	.2633838	.0044205	.

```
                  b = consistent under Ho and Ha; obtained from mlogit
          B = inconsistent under Ha, efficient under Ho; obtained from mlogit
    Test:  Ho:  difference in coefficients not systematic
              chi2(3) = (b-B)'[(V_b-V_B)^(-1)](b-B)
                      =        0.08
          Prob>chi2 =        0.9944
          (V_b-V_B is not positive definite)
```

```
. hausman m3 m1, alleqs constant
```

	── Coefficients ── (b) m3	(B) m1	(b-B) Difference	sqrt(diag(V_b-V_B)) S.E.
age	-.0041055	-.0051925	.001087	.0021355
male	.4591074	.4748547	-.0157473	.
_cons	-1.801774	-1.756843	-.0449311	.1333421

```
                    b = consistent under Ho and Ha; obtained from mlogit
          B = inconsistent under Ha, efficient under Ho; obtained from mlogit
    Test:  Ho:  difference in coefficients not systematic

        chi2(3) = (b-B)'[(V_b-V_B)^(-1)](b-B)
                =    -0.18    chi2<0 ==> model fitted on these
                              data fails to meet the asymptotic
                              assumptions of the Hausman test;
                              see suest for a generalized test
```

According to the test of m1 against m2, we cannot reject the hypothesis that the coefficients of m1 and m2 are the same. The second Hausman test is not well defined—something that happens fairly often. The problem is due to the estimator of the variance $V(b-B)$ as $V(b)-V(B)$, which is a feasible estimator only asymptotically. Here it simply is not a proper variance matrix, and the Hausman test becomes undefined.

suest m1 m2 estimates the simultaneous (co)variance of the coefficients of models m1 and m2. Although suest is technically a postestimation command, it acts like an estimation command in that it stores the simultaneous coefficients in e(b) and the full (co)variance matrix in e(V). We could have used the estat vce command to display the full (co)variance matrix to show that the cross-model covariances were indeed estimated. Typically, we would not have a direct interest in e(V).

```
. suest m1 m2, noomitted
Simultaneous results for m1, m2
```

Number of obs   =   615

	Coef.	Robust Std. Err.	z	P>|z|	[95% Conf. Interval]
m1_Indemnity					
m1_Prepaid					
age	-.0100251	.0059403	-1.69	0.091	-.0216679 .0016176
male	.5095747	.1988159	2.56	0.010	.1199027 .8992467
_cons	.2633838	.277307	0.95	0.342	-.280128 .8068956
m1_Uninsure					
age	-.0051925	.0109005	-0.48	0.634	-.0265571 .0161721
male	.4748547	.3677326	1.29	0.197	-.2458879 1.195597
_cons	-1.756843	.4971383	-3.53	0.000	-2.731216 -.78247
m2_Indemnity					
m2_Prepaid					
age	-.0101521	.0058988	-1.72	0.085	-.0217135 .0014094
male	.5144003	.1996133	2.58	0.010	.1231654 .9056352
_cons	.2678043	.2744019	0.98	0.329	-.2700134 .8056221

suest created equation names by combining the name under which we stored the results using estimates store with the original equation names. Thus, in the simultaneous estimation result, equation Prepaid originating in model m1 is named m1_Prepaid. According to the McFadden–Hausman specification of a test for IIA, the coefficients of the equations m1_PrePaid and m2_PrePaid should be equal. This equality can be tested easily with the test command. The cons option specifies that the intercept _cons be included in the test.

```
. test [m1_Prepaid = m2_Prepaid], cons

 ( 1)  [m1_Prepaid]age - [m2_Prepaid]age = 0
 ( 2)  [m1_Prepaid]male - [m2_Prepaid]male = 0
 ( 3)  [m1_Prepaid]_cons - [m2_Prepaid]_cons = 0

           chi2(  3) =     0.89
         Prob > chi2 =     0.8266
```

The Hausman test via suest is comparable to that computed by hausman, but they use different estimators of the variance of the difference of the estimates. The hausman command estimates $V(b-B)$ by $V(b) - V(B)$, whereas suest estimates $V(b-B)$ by $V(b) - \text{cov}(b, B) - \text{cov}(B, b) + V(B)$. One advantage of the second estimator is that it is always admissible, so the resulting test is always well defined. This quality is illustrated in the Hausman-type test of IIA comparing models m1 and m3.

```
. suest m1 m3, noomitted
Simultaneous results for m1, m3
                                                   Number of obs    =       615
```

	Coef.	Robust Std. Err.	z	P>|z|	[95% Conf.	Interval]
m1_Indemnity						
m1_Prepaid						
age	-.0100251	.0059403	-1.69	0.091	-.0216679	.0016176
male	.5095747	.1988159	2.56	0.010	.1199027	.8992467
_cons	.2633838	.277307	0.95	0.342	-.280128	.8068956
m1_Uninsure						
age	-.0051925	.0109005	-0.48	0.634	-.0265571	.0161721
male	.4748547	.3677326	1.29	0.197	-.2458879	1.195597
_cons	-1.756843	.4971383	-3.53	0.000	-2.731216	-.78247
m3_Indemnity						
m3_Uninsure						
age	-.0041055	.0111185	-0.37	0.712	-.0258974	.0176865
male	.4591074	.3601307	1.27	0.202	-.2467357	1.164951
_cons	-1.801774	.5226351	-3.45	0.001	-2.82612	-.7774283

```
. test [m1_Uninsure = m3_Uninsure], cons

 ( 1)  [m1_Uninsure]age - [m3_Uninsure]age = 0
 ( 2)  [m1_Uninsure]male - [m3_Uninsure]male = 0
 ( 3)  [m1_Uninsure]_cons - [m3_Uninsure]_cons = 0

           chi2(  3) =     1.49
         Prob > chi2 =     0.6845
```

Although the classic Hausman test computed by hausman is not defined here, the suest-based test is just fine. We cannot reject the equality of the common coefficients across m1 and m3.

A second advantage of the suest approach is that we can estimate the (co)variance matrix of the multivariate normal distribution of the estimators of the three models m1, m2, and m3 and test that the common coefficients are equal.

```
. suest m*, noomitted

Simultaneous results for m1, m2, m3
```

	Coef.	Robust Std. Err.	z	P>|z|	[95% Conf. Interval]	
m1_Indemnity						
m1_Prepaid						
age	-.0100251	.0059403	-1.69	0.091	-.0216679	.0016176
male	.5095747	.1988159	2.56	0.010	.1199027	.8992467
_cons	.2633838	.277307	0.95	0.342	-.280128	.8068956
m1_Uninsure						
age	-.0051925	.0109005	-0.48	0.634	-.0265571	.0161721
male	.4748547	.3677326	1.29	0.197	-.2458879	1.195597
_cons	-1.756843	.4971383	-3.53	0.000	-2.731216	-.78247
m2_Indemnity						
m2_Prepaid						
age	-.0101521	.0058988	-1.72	0.085	-.0217135	.0014094
male	.5144003	.1996133	2.58	0.010	.1231654	.9056352
_cons	.2678043	.2744019	0.98	0.329	-.2700134	.8056221
m3_Indemnity						
m3_Uninsure						
age	-.0041055	.0111185	-0.37	0.712	-.0258974	.0176865
male	.4591074	.3601307	1.27	0.202	-.2467357	1.164951
_cons	-1.801774	.5226351	-3.45	0.001	-2.82612	-.7774283

Number of obs = 615

```
. test [m1_Prepaid = m2_Prepaid], cons notest

 ( 1)  [m1_Prepaid]age - [m2_Prepaid]age = 0
 ( 2)  [m1_Prepaid]male - [m2_Prepaid]male = 0
 ( 3)  [m1_Prepaid]_cons - [m2_Prepaid]_cons = 0

. test [m1_Uninsure = m3_Uninsure], cons acc

 ( 1)  [m1_Prepaid]age - [m2_Prepaid]age = 0
 ( 2)  [m1_Prepaid]male - [m2_Prepaid]male = 0
 ( 3)  [m1_Prepaid]_cons - [m2_Prepaid]_cons = 0
 ( 4)  [m1_Uninsure]age - [m3_Uninsure]age = 0
 ( 5)  [m1_Uninsure]male - [m3_Uninsure]male = 0
 ( 6)  [m1_Uninsure]_cons - [m3_Uninsure]_cons = 0

           chi2(  6) =     1.95
         Prob > chi2 =    0.9240
```

Again we do not find evidence against the correct specification of the multinomial logit for type of insurance. The classic Hausman test assumes that one of the estimators (named B in hausman) is efficient, that is, it has minimal (asymptotic) variance. This assumption ensures that $V(b) - V(B)$ is an admissible, viable estimator for $V(b - B)$. The assumption that we have an efficient estimator is a restrictive one. It is violated, for instance, if our data are clustered. We want to adjust for clustering via a vce(cluster *clustvar*) option by requesting the cluster-adjusted sandwich estimator of variance. Consequently, in such a case, hausman cannot be used. This problem does not exist with the suest version of the Hausman test. To illustrate this feature, we suppose that the data are clustered by city—we constructed an imaginary variable cityid for this illustration. If we plan to apply suest, we would not specify the vce(cluster *clustvar*) option at the time of estimation.

suest has a vce(cluster *clustvar*) option. Thus we do not need to refit the models; we can call suest and test right away.

```
. suest m1 m2, vce(cluster cityid) noomitted
Simultaneous results for m1, m2

                                          Number of obs    =       615
                              (Std. Err. adjusted for 260 clusters in cityid)
```

	Coef.	Robust Std. Err.	z	P>|z|	[95% Conf. Interval]	
m1_Indemnity						
m1_Prepaid						
age	-.0100251	.005729	-1.75	0.080	-.0212538	.0012035
male	.5095747	.1910496	2.67	0.008	.1351244	.884025
_cons	.2633838	.2698797	0.98	0.329	-.2655708	.7923384
m1_Uninsure						
age	-.0051925	.0104374	-0.50	0.619	-.0256495	.0152645
male	.4748547	.3774021	1.26	0.208	-.2648399	1.214549
_cons	-1.756843	.4916613	-3.57	0.000	-2.720481	-.7932048
m2_Indemnity						
m2_Prepaid						
age	-.0101521	.0057164	-1.78	0.076	-.0213559	.0010518
male	.5144003	.1921385	2.68	0.007	.1378158	.8909848
_cons	.2678043	.2682193	1.00	0.318	-.2578959	.7935045

```
. test [m1_Prepaid = m2_Prepaid], cons
 ( 1)  [m1_Prepaid]age - [m2_Prepaid]age = 0
 ( 2)  [m1_Prepaid]male - [m2_Prepaid]male = 0
 ( 3)  [m1_Prepaid]_cons - [m2_Prepaid]_cons = 0
           chi2(  3) =      0.79
         Prob > chi2 =      0.8529
```

suest provides some descriptive information about the clustering on cityid. Like any other estimation command, suest informs us that the standard errors are adjusted for clustering. The Hausman-type test obtained from the test command uses a simultaneous (co)variance of m1 and m2 appropriately adjusted for clustering. In this example, we still do not have reason to conclude that the multinomial logit model in this application is misspecified, that is, that IIA is violated.

◁

The multinomial logistic regression model is a special case of the conditional logistic regression model; see [R] **clogit**. Like the multinomial logistic regression model, the conditional logistic regression model also makes the IIA assumption. Consider an example, introduced in [R] **asclogit**, in which the demand for American, Japanese, and European cars is modeled in terms of the number of local dealers of the respective brands and of some individual attributes incorporated in interaction with the nationality of cars. We want to perform a Hausman-type test for IIA comparing the decision between all nationalities with the decision between non-American cars. The following code fragment demonstrates how to conduct a Hausman test for IIA via suest in this case.

```
. clogit choice japan europe maleJap maleEur incJap incEur dealer, group(id)
. estimates store allcars
. clogit choice japan maleJap incJap dealer if car!=1 , group(id)
```

```
. estimates store foreign
. suest allcars foreign
. test [allcars_choice=foreign_choice], common
```

Testing proportionality

The applications of suest that we have discussed so far concern Hausman-type tests for misspecification. To test such a hypothesis, we compared two estimators that have the same probability limit if the hypothesis holds true, but otherwise have different limits. We may also want to compare the coefficients of models (estimators) for other substantive reasons. Although we most often want to test whether coefficients differ between models or estimators, we may occasionally want to test other constraints (see Hausman and Ruud [1987]).

▷ Example 2

In this example, using simulated labor market data for siblings, we consider two dependent variables, income (inc) and whether a person was promoted in the last year (promo). We apply familiar economic arguments regarding human capital, according to which employees have a higher income and a higher probability of being promoted, by having more human capital. Human capital is acquired through formal education (edu) and on-the-job training experience (exp). We study whether income and promotion are "two sides of the same coin", that is, whether they reflect a common latent variable, "human capital". Accordingly, we want to compare the effects of different aspects of human capital on different outcome variables.

We estimate fairly simple labor market equations. The income model is estimated with regress, and the estimation results are stored under the name Inc.

```
. use http://www.stata-press.com/data/r12/income
. regress inc edu exp male
```

Source	SS	df	MS		Number of obs =	277
					F(3, 273) =	42.34
Model	2058.44672	3	686.148908		Prob > F =	0.0000
Residual	4424.05183	273	16.2053181		R-squared =	0.3175
					Adj R-squared =	0.3100
Total	6482.49855	276	23.4873136		Root MSE =	4.0256

inc	Coef.	Std. Err.	t	P>|t|	[95% Conf. Interval]	
edu	2.213707	.243247	9.10	0.000	1.734828	2.692585
exp	1.47293	.231044	6.38	0.000	1.018076	1.927785
male	.5381153	.4949466	1.09	0.278	-.436282	1.512513
_cons	1.255497	.3115808	4.03	0.000	.642091	1.868904

```
. est store Inc
```

Being sibling data, the observations are clustered on family of origin, famid. In the estimation of the regression parameters, we did not specify a vce(cluster famid) option to adjust standard errors for clustering on family (famid). Thus the standard errors reported by regress are potentially flawed. This problem will, however, be corrected by specifying a vce(cluster *clustvar*) option with suest.

Next we estimate the promotion equation with probit and again store the results under an appropriate name.

```
. probit promo edu exp male, nolog
```

Probit regression   Number of obs   =   277
                                                     LR chi2(3)   =   49.76
                                                     Prob > chi2   =   0.0000
Log likelihood = -158.43888   Pseudo R2   =   0.1357

promo	Coef.	Std. Err.	z	P>|z|	[95% Conf. Interval]	
edu	.4593002	.0898537	5.11	0.000	.2831901	.6354102
exp	.3593023	.0805774	4.46	0.000	.2013735	.5172312
male	.2079983	.1656413	1.26	0.209	-.1166527	.5326494
_cons	-.464622	.1088166	-4.27	0.000	-.6778985	-.2513454

```
. est store Promo
```

The coefficients in the income and promotion equations definitely seem to be different. However, because the scales of the two variables are different, we would not expect the coefficients to be equal. The correct hypothesis here is that the proportionality of the coefficients of the two models, apart from the constant, are equal. This formulation would still reflect that the relative effects of the different aspects of human capital do not differ between the dependent variables. We can obtain a nonlinear Wald test for the hypothesis of proportionality by using the testnl command on the combined estimation results of the two estimators. Thus we first have to form the combined estimation results. At this point, we specify the vce(cluster famid) option to adjust for the clustering of observations on famid.

```
. suest Inc Promo, vce(cluster famid)
```

Simultaneous results for Inc, Promo

                                                     Number of obs   =   277

                                      (Std. Err. adjusted for 135 clusters in famid)

	Coef.	Robust Std. Err.	z	P>|z|	[95% Conf. Interval]	
Inc_mean						
edu	2.213707	.2483907	8.91	0.000	1.72687	2.700543
exp	1.47293	.1890583	7.79	0.000	1.102383	1.843478
male	.5381153	.4979227	1.08	0.280	-.4377952	1.514026
_cons	1.255497	.3374977	3.72	0.000	.594014	1.916981
Inc_lnvar						
_cons	2.785339	.079597	34.99	0.000	2.629332	2.941347
Promo_promo						
edu	.4593002	.0886982	5.18	0.000	.2854549	.6331454
exp	.3593023	.079772	4.50	0.000	.2029522	.5156525
male	.2079983	.1691053	1.23	0.219	-.1234419	.5394386
_cons	-.464622	.1042169	-4.46	0.000	-.6688833	-.2603607

The standard errors reported by suest are identical to those reported by the respective estimation commands when invoked with the vce(cluster famid) option. We are now ready to test for proportionality:

$$H_0 : \frac{\beta_{\text{edu}}^{\text{Income}}}{\beta_{\text{edu}}^{\text{Promotion}}} = \frac{\beta_{\text{exp}}^{\text{Income}}}{\beta_{\text{exp}}^{\text{Promotion}}} = \frac{\beta_{\text{male}}^{\text{Income}}}{\beta_{\text{male}}^{\text{Promotion}}}$$

It is straightforward to translate this into syntax suitable for testnl, recalling that the coefficient of variable v in equation eq is denoted by $[eq]v$.

```
. testnl [Inc_mean]edu/[Promo_promo]edu =
>         [Inc_mean]exp/[Promo_promo]exp =
>         [Inc_mean]male/[Promo_promo]male

  (1)   [Inc_mean]edu/[Promo_promo]edu = [Inc_mean]exp/[Promo_promo]exp
  (2)   [Inc_mean]edu/[Promo_promo]edu = [Inc_mean]male/[Promo_promo]male

            chi2(2) =        0.61
        Prob > chi2 =        0.7385
```

From the evidence, we fail to reject the hypotheses that the coefficients of the income and promotion equations are proportional. Thus it is not unreasonable to assume that income and promotion can be explained by the same latent variable, "labor market success".

A disadvantage of the nonlinear Wald test is that it is not invariant with respect to representation: a Wald test for a mathematically equivalent formulation of the nonlinear constraint usually leads to a different test result. An equivalent formulation of the proportionality hypothesis is

$$H_0: \quad \beta_{\text{edu}}^{\text{Income}}\beta_{\text{exp}}^{\text{Promotion}} = \beta_{\text{edu}}^{\text{Promotion}}\beta_{\text{exp}}^{\text{Income}} \quad \text{and}$$
$$\beta_{\text{edu}}^{\text{Income}}\beta_{\text{male}}^{\text{Promotion}} = \beta_{\text{edu}}^{\text{Promotion}}\beta_{\text{male}}^{\text{Income}}$$

This formulation is "more linear" in the coefficients. The asymptotic χ^2 distribution of the nonlinear Wald statistic can be expected to be more accurate for this representation.

```
. testnl ([Inc_mean]edu*[Promo_promo]exp = [Inc_mean]exp*[Promo_promo]edu)
>        ([Inc_mean]edu*[Promo_promo]male = [Inc_mean]male*[Promo_promo]edu)

  (1)   [Inc_mean]edu*[Promo_promo]exp = [Inc_mean]exp*[Promo_promo]edu
  (2)   [Inc_mean]edu*[Promo_promo]male = [Inc_mean]male*[Promo_promo]edu

            chi2(2) =        0.46
        Prob > chi2 =        0.7936
```

Here the two representations lead to similar test statistics and p-values. As before, we fail to reject the hypothesis of proportionality of the coefficients of the two models.

◁

Testing cross-model hypotheses

▷ Example 3

In this example, we demonstrate how some cross-model hypotheses can be tested using the facilities already available in most estimation commands. This demonstration will explain the intricate relationship between the cluster adjustment of the robust estimator of variance and the suest command. It will also be made clear that a new facility is required to perform more general cross-model testing.

We want to test whether the effect of x_1 on the binary variable y_1 is the same as the effect of x_2 on the binary y_2; see Clogg, Petkova, and Haritou (1995). In this setting, x_1 may equal x_2, and y_1 may equal y_2. We assume that logistic regression models can be used to model the responses, and for simplicity, we ignore further predictor variables in these models. If the two logit models are fit on independent samples so that the estimators are (stochastically) independent, a Wald test for `_b[x1]` = `_b[x2]` rejects the null hypothesis if

$$\frac{\widehat{b}(x_1) - \widehat{b}(x_2)}{\left[\widehat{\sigma}^2\left\{\widehat{b}(x_1)\right\} + \widehat{\sigma}^2\left\{\widehat{b}(x_2)\right\}\right]^{1/2}}$$

is larger than the appropriate χ_1^2 threshold. If the models are fit on the *same* sample (or on dependent samples), so that the estimators are stochastically dependent, the above test that ignores the covariance between the estimators is not appropriate.

It is instructive to see how this problem can be tackled by "stacking" data. In the stacked format, we doubled the number of observations. The dependent variable is y_1 in the first half of the data and is y_2 in the second half of the data. The predictor variable z_1 is set to x_1 in the first half of the expanded data and to 0 in the rest. Similarly, z_2 is 0 in the first half and x_2 in the second half. The following diagram illustrates the transformation, in the terminology of the `reshape` command, from wide to long format.

$$
\begin{pmatrix}
\text{id} & y_1 & y_2 & x_1 & x_2 \\
\hline
1 & y_{11} & y_{21} & x_{11} & x_{21} \\
2 & y_{12} & y_{22} & x_{12} & x_{22} \\
3 & y_{13} & y_{23} & x_{13} & x_{23}
\end{pmatrix}
\implies
\begin{pmatrix}
\text{id} & y & z_1 & z_2 & \text{model} \\
\hline
1 & y_{11} & x_{11} & 0 & 1 \\
2 & y_{12} & x_{12} & 0 & 1 \\
3 & y_{13} & x_{13} & 0 & 1 \\
1 & y_{21} & 0 & x_{21} & 2 \\
2 & y_{22} & 0 & x_{22} & 2 \\
3 & y_{23} & 0 & x_{23} & 2
\end{pmatrix}
$$

The observations in the long format data organization are *clustered* on the original subjects and are identified with the identifier `id`. The clustering on `id` has to be accounted for when fitting a simultaneous model. The simplest way to deal with clustering is to use the cluster adjustment of the robust or sandwich estimator; see [P] **_robust**. The data manipulation can be accomplished easily with the `stack` command; see [D] **stack**. Subsequently, we fit a simultaneous logit model and perform a Wald test for the hypothesis that the coefficients of `z1` and `z2` are the same. A full setup to obtain the cross-model Wald test could then be as follows:

```
. generate zero = 0          // a variable that is always 0
. generate one  = 1          // a variable that is always 1
. generate two  = 2          // a variable that is always 2
. stack  id y1 x1 zero one   id y2 zero x2 two, into(id y z1 z2 model)
. generate model2 = (model==2)
. logit y model2 z1 z2, vce(cluster id)
. test _b[z1] = _b[z2]
```

The coefficient of `z1` represents the effect of `x1` on `y1`, and similarly, `z2` for the effect of `x2` on `y2`. The variable `model2` is a dummy for the "second model", which is included to allow the intercept in the second model to differ from that in the first model. The estimates of the coefficient of `z1` and its standard error in the combined model are the same as the estimates of the coefficient of `z1` and its standard error if we fit the model on the unstacked data.

```
. logit y1 x1, vce(robust)
```

The vce(cluster *clustvar*) option specified with the logit command for the stacked data ensures that the covariances of _b[z1] and _b[z2] are indeed estimated. This estimation ensures that the Wald test for the equality of the coefficients is correct. If we had not specified the vce(cluster *clustvar*) option, the (co)variance matrix of the coefficients would have been block-diagonal; that is, the covariances of _b[z1] and _b[z2] would have been 0. Then test would have effectively used the invalid formula for the Wald test for two independent samples.

In this example, the two logit models were fit on the same data. The same setup would apply, without modification, when the two logit models were fit on overlapping data that resulted, for instance, if the y or x variables were missing in some observations.

The suest command allows us to obtain the above Wald test more efficiently by avoiding the data manipulation, obviating the need to fit a model with twice the number of coefficients. The test statistic produced by the above code fragment is *identical* to that obtained via suest on the original (unstacked) data:

```
. logit y1 x1
. estimates store M1
. logit y2 x2
. estimates store M2
. suest M1 M2
. test [M1]x1=[M2]x2
```

The stacking method can be applied not only to the testing of cross-model hypotheses for logit models but also to any estimation command that supports the vce(cluster *clustvar*) option. The stacking approach clearly generalizes to stacking more than two logit or other models, testing more general linear hypotheses, and testing nonlinear cross-model hypotheses (see [R] **testnl**). In all these cases, suest would yield identical statistical results but at smaller costs in terms of data management, computer storage, and computer time.

Is suest nothing but a convenience command? No, there are two disadvantages to the stacking method, both of which are resolved via suest. First, if the models include ancillary parameters (in a regression model, the residual variance; in an ordinal response model, the cutpoints; and in lognormal survival-time regression, the time scale parameter), these parameters are constrained to be equal between the stacked models. In suest, this constraint is relaxed. Second, the stacking method does not generalize to compare different statistical models, such as a probit model and a regression model. As demonstrated in the previous section, suest can deal with this situation.

◁

Saved results

suest saves the following in e():

Scalars
e(N)	number of observations
e(N_clust)	number of clusters
e(rank)	rank of e(V)

Macros
e(cmd)	suest
e(eqnames#)	original names of equations of model #
e(names)	list of model names
e(wtype)	weight type
e(wexp)	weight expression
e(clustvar)	name of cluster variable
e(vce)	*vcetype* specified in vce()
e(vcetype)	title used to label Std. Err.
e(properties)	b V

Matrices
e(b)	stacked coefficient vector of the models
e(V)	variance–covariance matrix of the estimators

Functions
e(sample)	marks estimation sample

Methods and formulas

suest is implemented as an ado-file.

The estimation of the simultaneous (co)variance of a series of k estimators is a nonstandard application of the sandwich estimator, as implemented by the command [P] **_robust**. You may want to read this manual entry before reading further.

The starting point is that we have fit k different models on the *same* data—partially overlapping or nonoverlapping data are special cases. We want to derive the *simultaneous* distribution of these k estimators, for instance, to test a cross-estimator hypothesis H_0. As in the framework of Hausman testing, H_0 will often be of the form that different estimators have the same probability limit under some hypothesis, while the estimators have different limits if the hypothesis is violated.

We consider (vector) estimators $\widehat{\beta}_i$ to be defined as "the" solution of the estimation equations $\mathbf{G}_i$,

$$\mathbf{G}_i(\mathbf{b}_i) = \sum_j w_{ij}\mathbf{u}_{ij}(\mathbf{b}_i) = \mathbf{0}, \qquad i = 1, \ldots, k$$

We refer to the $\mathbf{u}_{ij}$ as the "scores". Specifying some weights $w_{ij} = 0$ trivially accommodates for partially overlapping or even disjoint data. Under "suitable regularity conditions" (see White [1982; 1996] for details), the $\widehat{\beta}_i$ are asymptotically normally distributed, with the variance estimated consistently by the sandwich estimator

$$V_i = \mathrm{Var}(\widehat{\beta}_i) = \mathbf{D}_i^{-1} \sum_j w_{ij}\mathbf{u}_{ij}\mathbf{u}_{ij}' \, \mathbf{D}_i^{-1}$$

where $\mathbf{D}_i$ is the Jacobian of $\mathbf{G}_i$ evaluated at $\widehat{\beta}_i$. In the context of maximum likelihood estimation, $\mathbf{D}_i$ can be estimated consistently by (minus) the Hessian of the log likelihood or by the Fisher information matrix. If the model is also well specified, the sandwiched term $(\sum_j w_{ij} \mathbf{u}_{ij} \mathbf{u}'_{ij})$ converges in probability to $\mathbf{D}_i$, so V_i may be consistently estimated by $\mathbf{D}_i^{-1}$.

To derive the simultaneous distribution of the estimators, we consider the "stacked" estimation equation,

$$\mathbf{G}(\widehat{\beta}) = \left\{ \mathbf{G}_1(\widehat{\beta}_1)' \quad \mathbf{G}_1(\widehat{\beta}_2)' \quad \ldots \quad \mathbf{G}_k(\widehat{\beta}_k)' \right\}' = \mathbf{0}$$

Under "suitable regularity conditions" (see White [1996] for details), $\widehat{\beta}$ is asymptotically *jointly* normally distributed. The Jacobian and scores of the simultaneous equation are easily expressed in the Jacobian and scores of the separate equations. The Jacobian of $\mathbf{G}$,

$$\mathbf{D}(\widehat{\beta}) = \left. \frac{d\mathbf{G}(\beta)}{d\beta} \right|_{\beta=\widehat{\beta}}$$

is block diagonal with blocks $\mathbf{D}_1, \ldots, \mathbf{D}_k$. The inverse of $\mathbf{D}(\widehat{\beta})$ is again block diagonal, with the inverses of $\mathbf{D}_i$ on the diagonal. The scores $\mathbf{u}$ of $\mathbf{G}$ are simply obtained as the *concatenated* scores of the separate equations:

$$\mathbf{u}_j = (\mathbf{u}'_{1j} \quad \mathbf{u}'_{2j} \quad \ldots \quad \mathbf{u}'_{kj})'$$

Out-of-sample (that is, where $w_{ij} = 0$) values of the score variables are defined as 0 (thus we drop the i subscript from the common weight variable). The sandwich estimator for the asymptotic variance of $\widehat{\beta}$ reads

$$V = \mathrm{Var}(\widehat{\beta}) = \mathbf{D}(\widehat{\beta})^{-1} \left(\sum_j w_j \mathbf{u}_j \mathbf{u}'_j \right) \mathbf{D}(\widehat{\beta})^{-1}$$

Taking a "partitioned" look at this expression, we see that $V(\widehat{\beta}_i)$ is estimated by

$$\mathbf{D}_i^{-1} \left(\sum_j w_j \mathbf{u}_{ij} \mathbf{u}'_{ij} \right) \mathbf{D}_i^{-1}$$

which is, yet again, the familiar sandwich estimator for $\widehat{\beta}_i$ based on the separate estimation equation $\mathbf{G}_i$. Thus considering several estimators simultaneously in this way does not affect the estimators of the asymptotic variances of these estimators. However, as a bonus of stacking, we obtained a sandwich-type estimate of the *covariance* V_{ih} of estimators $\widehat{\beta}_i$ and $\widehat{\beta}_h$,

$$V_{ih} = \mathrm{Cov}(\widehat{\beta}_i, \widehat{\beta}_h) = \mathbf{D}_i^{-1} \left(\sum_j w_j \mathbf{u}_{ij} \mathbf{u}'_{ih} \right) \mathbf{D}_h^{-1}$$

which is also obtained by White (1982).

This estimator for the covariance of estimators is an application of the cluster modification of the sandwich estimator proposed by Rogers (1993). Consider the stacked data format as discussed in the logit example, and assume that Stata would be able to estimate a "stacked model" in which different models apply to different observations, for example, a probit model for the first half, a regression model for the second half, and a one-to-one cluster relation between the first and second half. If there are no common parameters to both models, the score statistics of parameters for the stacked models are zero in the half of the data in which they do not occur. In Rogers' method, we have to sum the score statistics over the observations within a cluster. This step boils down to concatenating the score statistics at the level of the cluster.

We compare the sandwich estimator of the (co)variance V_{12} of two estimators with the estimator of variance $\widetilde{V}_{12}$ applied in the classic Hausman test. Hausman (1978) showed that if $\widehat{\beta}_1$ is consistent under H_0 and $\widehat{\beta}_2$ is efficient under H_0, then asymptotically

$$\mathrm{Cov}(\widehat{\beta}_1, \widehat{\beta}_2) = \mathrm{Var}(\widehat{\beta}_2)$$

and so $\mathrm{var}(\widehat{\beta}_1 - \widehat{\beta}_2)$ is consistently estimated by $V_1 - V_2$.

Acknowledgment

suest was written by Jeroen Weesie, Department of Sociology, Utrecht University, The Netherlands. This research is supported by grant PGS 50-370 by The Netherlands Organization for Scientific Research.

An earlier version of suest was published in the *Stata Technical Bulletin* (1999). The current version of suest is not backward compatible with the STB version because of the introduction of new ways to manage estimation results via the estimates command.

References

Arminger, G. 1990. Testing against misspecification in parametric rate models. In *Event History Analysis in Life Course Research*, ed. K. U. Mayer and N. B. Tuma, 146–158. Madison: University of Wisconsin Press.

Clogg, C. C., E. Petkova, and A. Haritou. 1995. Statistical methods for comparing regression coefficients between models. *American Journal of Sociology* 100: 1261–1312. (With comments by P. D. Allison and a reply by C. C. Clogg, E. Petkova, and T. Cheng).

Gourieroux, C., and A. Monfort. 1997. *Time Series and Dynamic Models*. Trans. ed. G. M. Gallo. Cambridge: Cambridge University Press.

Hausman, J. A. 1978. Specification tests in econometrics. *Econometrica* 46: 1251–1271.

Hausman, J. A., and D. L. McFadden. 1984. Specification tests for the multinomial logit model. *Econometrica* 52: 1219–1240.

Hausman, J. A., and P. A. Ruud. 1987. Specifying and testing econometric models for rank-ordered data. *Journal of Econometrics* 34: 83–104.

Huber, P. J. 1967. The behavior of maximum likelihood estimates under nonstandard conditions. In Vol. 1 of *Proceedings of the Fifth Berkeley Symposium on Mathematical Statistics and Probability*, 221–233. Berkeley: University of California Press.

Rogers, W. H. 1993. sg16.4: Comparison of nbreg and glm for negative binomial. *Stata Technical Bulletin* 16: 7. Reprinted in *Stata Technical Bulletin Reprints*, vol. 3, pp. 82–84. College Station, TX: Stata Press.

Weesie, J. 1999. sg121: Seemingly unrelated estimation and the cluster-adjusted sandwich estimator. *Stata Technical Bulletin* 52: 34–47. Reprinted in *Stata Technical Bulletin Reprints*, vol. 9, pp. 231–248. College Station, TX: Stata Press.

White, H. 1982. Maximum likelihood estimation of misspecified models. *Econometrica* 50: 1–25.

——. 1996. *Estimation, Inference and Specification Analysis*. Cambridge: Cambridge University Press.

Also see

[R] **estimates** — Save and manipulate estimation results

[R] **hausman** — Hausman specification test

[R] **lincom** — Linear combinations of estimators

[R] **nlcom** — Nonlinear combinations of estimators

[R] **test** — Test linear hypotheses after estimation

[R] **testnl** — Test nonlinear hypotheses after estimation

[P] **_robust** — Robust variance estimates

Title

> **summarize** — Summary statistics

Syntax

> summarize [*varlist*] [*if*] [*in*] [*weight*] [, *options*]

options	Description
Main	
detail	display additional statistics
meanonly	suppress the display; calculate only the mean; programmer's option
format	use variable's display format
separator(#)	draw separator line after every # variables; default is separator(5)
display_options	control spacing and base and empty cells

varlist may contain factor variables; see [U] **11.4.3 Factor variables**.

varlist may contain time-series operators; see [U] **11.4.4 Time-series varlists**.

by is allowed; see [D] **by**.

aweights, fweights, and iweights are allowed. However, iweights may not be used with the detail
   option; see [U] **11.1.6 weight**.

Menu

Statistics > Summaries, tables, and tests > Summary and descriptive statistics > Summary statistics

Description

summarize calculates and displays a variety of univariate summary statistics. If no *varlist* is
specified, summary statistics are calculated for all the variables in the dataset.

Also see [R] **ci** for calculating the standard error and confidence intervals of the mean.

Options

> ⌐ Main ⌐

detail produces additional statistics, including skewness, kurtosis, the four smallest and largest
   values, and various percentiles.

meanonly, which is allowed only when detail is not specified, suppresses the display of results
   and calculation of the variance. Ado-file writers will find this useful for fast calls.

format requests that the summary statistics be displayed using the display formats associated with
   the variables rather than the default g display format; see [U] **12.5 Formats: Controlling how
   data are displayed**.

separator(#) specifies how often to insert separation lines into the output. The default is sepa-
   rator(5), meaning that a line is drawn after every five variables. separator(10) would draw
   a line after every 10 variables. separator(0) suppresses the separation line.

display_options: vsquish, noemptycells, baselevels, allbaselevels; see [R] **estimation options**.

Remarks

summarize can produce two different sets of summary statistics. Without the detail option, the number of nonmissing observations, the mean and standard deviation, and the minimum and maximum values are presented. With detail, the same information is presented along with the variance, skewness, and kurtosis; the four smallest and four largest values; and the 1st, 5th, 10th, 25th, 50th (median), 75th, 90th, 95th, and 99th percentiles.

▷ Example 1: summarize with the separator() option

We have data containing information on various automobiles, among which is the variable mpg, the mileage rating. We can obtain a quick summary of the mpg variable by typing

```
. use http://www.stata-press.com/data/r12/auto
(1978 Automobile Data)

. summarize mpg
```

Variable	Obs	Mean	Std. Dev.	Min	Max
mpg	74	21.2973	5.785503	12	41

We see that we have 74 observations. The mean of mpg is 21.3 miles per gallon, and the standard deviation is 5.79. The minimum is 12, and the maximum is 41.

If we had not specified the variable (or variables) we wanted to summarize, we would have obtained summary statistics on all the variables in the dataset:

```
. summarize, separator(4)
```

Variable	Obs	Mean	Std. Dev.	Min	Max
make	0				
price	74	6165.257	2949.496	3291	15906
mpg	74	21.2973	5.785503	12	41
rep78	69	3.405797	.9899323	1	5
headroom	74	2.993243	.8459948	1.5	5
trunk	74	13.75676	4.277404	5	23
weight	74	3019.459	777.1936	1760	4840
length	74	187.9324	22.26634	142	233
turn	74	39.64865	4.399354	31	51
displacement	74	197.2973	91.83722	79	425
gear_ratio	74	3.014865	.4562871	2.19	3.89
foreign	74	.2972973	.4601885	0	1

There are only 69 observations on rep78, so some of the observations are missing. There are no observations on make because it is a string variable.

◁

The idea of the mean is quite old (Plackett 1958), but its extension to a scheme of moment-based measures was not done until the end of the 19th century. Between 1893 and 1905, Pearson discussed and named the standard deviation, skewness, and kurtosis, but he was not the first to use any of these. Thiele (1889), in contrast, had earlier firmly grasped the notion that the m_r provide a systematic basis for discussing distributions. However, even earlier anticipations can also be found. For example, Euler in 1778 used m_2 and m_3 in passing in a treatment of estimation (Hald 1998, 87), but seemingly did not build on that.

Similarly, the idea of the median is quite old. The history of the interquartile range is tangled up with that of the probable error, a long-popular measure. Extending this in various ways to a more general approach based on quantiles (to use a later term) occurred to several people in the nineteenth century. Galton (1875) is a nice example, particularly because he seems so close to the key idea of the quantiles as a function, which took another century to reemerge strongly.

Thorvald Nicolai Thiele (1838–1910) was a Danish scientist who worked in astronomy, mathematics, actuarial science, and statistics. He made many pioneering contributions to statistics, several of which were overlooked until recently. Thiele advocated graphical analysis of residuals checking for trends, symmetry of distributions, and changes of sign, and he even warned against overinterpreting such graphs.

▷ Example 2: summarize with the detail option

The detail option provides all the information of a normal summarize and more. The format of the output also differs, as shown here:

```
. summarize mpg, detail

                         Mileage (mpg)

          Percentiles      Smallest
 1%           12              12
 5%           14              12
10%           14              14         Obs                  74
25%           18              14         Sum of Wgt.          74

50%           20                         Mean            21.2973
                            Largest      Std. Dev.      5.785503
75%           25              34
90%           29              35         Variance       33.47205
95%           34              35         Skewness       .9487176
99%           41              41         Kurtosis       3.975005
```

As in the previous example, we see that the mean of mpg is 21.3 miles per gallon and that the standard deviation is 5.79. We also see the various percentiles. The median of mpg (the 50th percentile) is 20 miles per gallon. The 25th percentile is 18, and the 75th percentile is 25.

When we performed summarize, we learned that the minimum and maximum were 12 and 41, respectively. We now see that the four smallest values in our dataset are 12, 12, 14, and 14. The four largest values are 34, 35, 35, and 41. The skewness of the distribution is 0.95, and the kurtosis is 3.98. (A normal distribution would have a skewness of 0 and a kurtosis of 3.)

Skewness is a measure of the lack of symmetry of a distribution. If the distribution is symmetric, the coefficient of skewness is 0. If the coefficient is negative, the median is usually greater than the mean and the distribution is said to be skewed left. If the coefficient is positive, the median is usually less than the mean and the distribution is said to be skewed right. *Kurtosis* (from the Greek *kyrtosis*, meaning curvature) is a measure of peakedness of a distribution. The smaller the coefficient of kurtosis, the flatter the distribution. The normal distribution has a coefficient of kurtosis of 3 and provides a convenient benchmark.   ◁

▷ Example 3: summarize with the by prefix

summarize can usefully be combined with the by *varlist*: prefix. In our dataset, we have a variable, foreign, that distinguishes foreign and domestic cars. We can obtain summaries of mpg and weight within each subgroup by typing

```
. by foreign: summarize mpg weight
```

-> foreign = Domestic

Variable	Obs	Mean	Std. Dev.	Min	Max
mpg	52	19.82692	4.743297	12	34
weight	52	3317.115	695.3637	1800	4840

-> foreign = Foreign

Variable	Obs	Mean	Std. Dev.	Min	Max
mpg	22	24.77273	6.611187	14	41
weight	22	2315.909	433.0035	1760	3420

Domestic cars in our dataset average 19.8 miles per gallon, whereas foreign cars average 24.8.

Because by *varlist*: can be combined with summarize, it can also be combined with summarize, detail:

```
. by foreign: summarize mpg, detail
```

-> foreign = Domestic

Mileage (mpg)

	Percentiles	Smallest		
1%	12	12		
5%	14	12		
10%	14	14	Obs	52
25%	16.5	14	Sum of Wgt.	52
50%	19		Mean	19.82692
		Largest	Std. Dev.	4.743297
75%	22	28		
90%	26	29	Variance	22.49887
95%	29	30	Skewness	.7712432
99%	34	34	Kurtosis	3.441459

-> foreign = Foreign

Mileage (mpg)

	Percentiles	Smallest		
1%	14	14		
5%	17	17		
10%	17	17	Obs	22
25%	21	18	Sum of Wgt.	22
50%	24.5		Mean	24.77273
		Largest	Std. Dev.	6.611187
75%	28	31		
90%	35	35	Variance	43.70779
95%	35	35	Skewness	.657329
99%	41	41	Kurtosis	3.10734

◁

❑ Technical note

summarize respects display formats if we specify the format option. When we type summarize price weight, we obtain

```
. summarize price weight
```

Variable	Obs	Mean	Std. Dev.	Min	Max
price	74	6165.257	2949.496	3291	15906
weight	74	3019.459	777.1936	1760	4840

The display is accurate but is not as aesthetically pleasing as we may wish, particularly if we plan to use the output directly in published work. By placing formats on the variables, we can control how the table appears:

```
. format price weight %9.2fc
. summarize price weight, format
```

Variable	Obs	Mean	Std. Dev.	Min	Max
price	74	6,165.26	2,949.50	3,291.00	15,906.00
weight	74	3,019.46	777.19	1,760.00	4,840.00

❑

If you specify a weight (see [U] **11.1.6 weight**), each observation is multiplied by the value of the weighting expression before the summary statistics are calculated so that the weighting expression is interpreted as the discrete density of each observation.

▷ Example 4: summarize with factor variables

You can also use summarize to obtain summary statistics for factor variables. For example, if you type

```
. summarize i.rep78
```

Variable	Obs	Mean	Std. Dev.	Min	Max
rep78					
2	69	.115942	.3225009	0	1
3	69	.4347826	.4993602	0	1
4	69	.2608696	.4423259	0	1
5	69	.1594203	.3687494	0	1

you obtain the sample proportions for four of the five levels of the rep78 variable. For example, 11.6% of the 69 cars with nonmissing values of rep78 fall into repair category two. When you use factor-variable notation, the base category is suppressed by default. If you type

```
. summarize bn.rep78
```

Variable	Obs	Mean	Std. Dev.	Min	Max
rep78					
1	69	.0289855	.1689948	0	1
2	69	.115942	.3225009	0	1
3	69	.4347826	.4993602	0	1
4	69	.2608696	.4423259	0	1
5	69	.1594203	.3687494	0	1

the notation bn.rep78 indicates that Stata should not suppress the base category so that we see the proportions for all five levels.

We could have used `tabulate oneway rep78` to obtain the sample proportions along with the cumulative proportions. Alternatively, we could have used `proportions rep78` to obtain the sample proportions along with the standard errors of the proportions instead of the standard deviations of the proportions.

◁

▷ Example 5: summarize with weights

We have 1980 census data on each of the 50 states. Included in our variables is `medage`, the median age of the population of each state. If we type `summarize medage`, we obtain unweighted statistics:

```
. use http://www.stata-press.com/data/r12/census
(1980 Census data by state)
. summarize medage
```

Variable	Obs	Mean	Std. Dev.	Min	Max
medage	50	29.54	1.693445	24.2	34.7

Also among our variables is `pop`, the population in each state. Typing `summarize medage [w=pop]` produces population-weighted statistics:

```
. summarize medage [w=pop]
(analytic weights assumed)
```

Variable	Obs	Weight	Mean	Std. Dev.	Min	Max
medage	50	225907472	30.11047	1.66933	24.2	34.7

The number listed under `Weight` is the sum of the weighting variable, `pop`, indicating that there are roughly 226 million people in the United States. The pop-weighted mean of `medage` is 30.11 (compared with 29.54 for the unweighted statistic), and the weighted standard deviation is 1.67 (compared with 1.69).

◁

▷ Example 6: summarize with weights and the detail option

We can obtain detailed summaries of weighted data as well. When we do this, *all* the statistics are weighted, including the percentiles.

```
. summarize medage [w=pop], detail
(analytic weights assumed)
```

<pre>
                             Median age

              Percentiles   Smallest
 1%   27.1   24.2
 5%   27.7   26.1
10%   28.2   27.1   Obs   50
25%   29.2   27.4   Sum of Wgt.   225907472

50%   29.9   Mean   30.11047
                                Largest   Std. Dev.   1.66933
75%   30.9   32
90%   32.1   32.1   Variance   2.786661
95%   32.2   32.2   Skewness   .5281972
99%   34.7   34.7   Kurtosis   4.494223
</pre>

❏ Technical note

If you are writing a program and need to access the mean of a variable, the meanonly option provides for fast calls. For example, suppose that your program reads as follows:

```
program mean
        summarize '1', meanonly
        display "   mean = " r(mean)
end
```

The result of executing this is

```
. mean price
  mean = 6165.2568
```

❏

Saved results

summarize saves the following in r():

Scalars

r(N)	number of observations	r(p50)	50th percentile (detail only)
r(mean)	mean	r(p75)	75th percentile (detail only)
r(skewness)	skewness (detail only)	r(p90)	90th percentile (detail only)
r(min)	minimum	r(p95)	95th percentile (detail only)
r(max)	maximum	r(p99)	99th percentile (detail only)
r(sum_w)	sum of the weights	r(Var)	variance
r(p1)	1st percentile (detail only)	r(kurtosis)	kurtosis (detail only)
r(p5)	5th percentile (detail only)	r(sum)	sum of variable
r(p10)	10th percentile (detail only)	r(sd)	standard deviation
r(p25)	25th percentile (detail only)		

Methods and formulas

Let x denote the variable on which we want to calculate summary statistics, and let x_i, $i = 1, \ldots, n$, denote an individual observation on x. Let v_i be the weight, and if no weight is specified, define $v_i = 1$ for all i.

Define V as the *sum of the weight*:

$$V = \sum_{i=1}^{n} v_i$$

Define w_i to be v_i normalized to sum to n, $w_i = v_i(n/V)$.

The *mean*, $\overline{x}$, is defined as

$$\overline{x} = \frac{1}{n}\sum_{i=1}^{n} w_i x_i$$

The *variance*, s^2, is defined as

$$s^2 = \frac{1}{n-1}\sum_{i=1}^{n} w_i(x_i - \overline{x})^2$$

The *standard deviation*, s, is defined as $\sqrt{s^2}$.

Define m_r as the rth moment about the mean $\overline{x}$:

$$m_r = \frac{1}{n}\sum_{i=1}^{n} w_i(x_i - \overline{x})^r$$

The *coefficient of skewness* is then defined as $m_3 m_2^{-3/2}$. The *coefficient of kurtosis* is defined as $m_4 m_2^{-2}$.

Let $x_{(i)}$ refer to the x in ascending order, and let $w_{(i)}$ refer to the corresponding weights of $x_{(i)}$. The four smallest values are $x_{(1)}$, $x_{(2)}$, $x_{(3)}$, and $x_{(4)}$. The four largest values are $x_{(n)}$, $x_{(n-1)}$, $x_{(n-2)}$, and $x_{(n-3)}$.

To obtain the pth *percentile*, which we will denote as $x_{[p]}$, let $P = np/100$. Let

$$W_{(i)} = \sum_{j=1}^{i} w_{(j)}$$

Find the first index i such that $W_{(i)} > P$. The pth percentile is then

$$x_{[p]} = \begin{cases} \dfrac{x_{(i-1)} + x_{(i)}}{2} & \text{if } W_{(i-1)} = P \\ x_{(i)} & \text{otherwise} \end{cases}$$

References

Cox, N. J. 2010. Speaking Stata: The limits of sample skewness and kurtosis. *Stata Journal* 10: 482–495.

David, H. A. 2001. First (?) occurrence of common terms in statistics and probability. In *Annotated Readings in the History of Statistics*, ed. H. A. David and A. W. F. Edwards, 209–246. New York: Springer.

Galton, F. 1875. Statistics by intercomparison, with remarks on the law of frequency of error. *Philosophical Magazine* 49: 33–46.

Gleason, J. R. 1997. sg67: Univariate summaries with boxplots. *Stata Technical Bulletin* 36: 23–25. Reprinted in *Stata Technical Bulletin Reprints*, vol. 6, pp. 179–183. College Station, TX: Stata Press.

——. 1999. sg67.1: Update to univar. *Stata Technical Bulletin* 51: 27–28. Reprinted in *Stata Technical Bulletin Reprints*, vol. 9, pp. 159–161. College Station, TX: Stata Press.

Hald, A. 1998. *A History of Mathematical Statistics from 1750 to 1930*. New York: Wiley.

Hamilton, L. C. 1996. *Data Analysis for Social Scientists*. Belmont, CA: Duxbury.

——. 2009. *Statistics with Stata (Updated for Version 10)*. Belmont, CA: Brooks/Cole.

Kirkwood, B. R., and J. A. C. Sterne. 2003. *Essential Medical Statistics*. 2nd ed. Malden, MA: Blackwell.

Lauritzen, S. L. 2002. *Thiele: Pioneer in Statistics*. Oxford: Oxford University Press.

Plackett, R. L. 1958. Studies in the history of probability and statistics: VII. The principle of the arithmetic mean. *Biometrika* 45: 130–135.

Stuart, A., and J. K. Ord. 1994. *Kendall's Advanced Theory of Statistics: Distribution Theory, Vol I*. 6th ed. London: Arnold.

Thiele, T. N. 1889. *Forelæsringer over Almindelig Iagttagelseslære: Sandsynlighedsregning og mindste Kvadraters Methode*. Kjøbenhavn: C.A. Reitzel. (English translation included in Lauritzen 2002).

Weisberg, H. F. 1992. *Central Tendency and Variability*. Newbury Park, CA: Sage.

Also see

[R] **ameans** — Arithmetic, geometric, and harmonic means

[R] **centile** — Report centile and confidence interval

[R] **mean** — Estimate means

[R] **proportion** — Estimate proportions

[R] **ratio** — Estimate ratios

[R] **table** — Tables of summary statistics

[R] **tabstat** — Display table of summary statistics

[R] **tabulate, summarize()** — One- and two-way tables of summary statistics

[R] **total** — Estimate totals

[D] **codebook** — Describe data contents

[D] **describe** — Describe data in memory or in file

[D] **inspect** — Display simple summary of data's attributes

[ST] **stsum** — Summarize survival-time data

[SVY] **svy estimation** — Estimation commands for survey data

[XT] **xtsum** — Summarize xt data

Title

sunflower — Density-distribution sunflower plots

Syntax

sunflower *yvar xvar* [*if*] [*in*] [*weight*] [, *options*]

options	Description
Main	
nograph	do not show graph
notable	do not show summary table; implied when by() is specified
marker_options	affect rendition of markers drawn at the plotted points
Bins/Petals	
binwidth(#)	width of the hexagonal bins
binar(#)	aspect ratio of the hexagonal bins
bin_options	affect rendition of hexagonal bins
light(#)	minimum observations for a light sunflower; default is light(3)
dark(#)	minimum observations for a dark sunflower; default is dark(13)
xcenter(#)	x-coordinate of the reference bin
ycenter(#)	y-coordinate of the reference bin
petalweight(#)	observations in a dark sunflower petal
petallength(#)	length of sunflower petal as a percentage
petal_options	affect rendition of sunflower petals
flowersonly	show petals only; do not render bins
nosinglepetal	suppress single petals
Add plots	
addplot(*plot*)	add other plots to generated graph
Y axis, X axis, Titles, Legend, Overall, By	
twoway_options	any options documented in [G-3] **twoway_options**

bin_options	Description
$\boxed{\underline{\text{l}}|\underline{\text{d}}}$ bstyle(*areastyle*)	overall look of hexagonal bins
$\boxed{\underline{\text{l}}|\underline{\text{d}}}$ bcolor(*colorstyle*)	outline and fill color
$\boxed{\underline{\text{l}}|\underline{\text{d}}}$ bfcolor(*colorstyle*)	fill color
$\boxed{\underline{\text{l}}|\underline{\text{d}}}$ blstyle(*linestyle*)	overall look of outline
$\boxed{\underline{\text{l}}|\underline{\text{d}}}$ blcolor(*colorstyle*)	outline color
$\boxed{\underline{\text{l}}|\underline{\text{d}}}$ blwidth(*linewidthstyle*)	thickness of outline

petal_options	Description
$\boxed{\underline{\text{l}}|\underline{\text{d}}}$ flstyle(*linestyle*)	overall style of sunflower petals
$\boxed{\underline{\text{l}}|\underline{\text{d}}}$ flcolor(*colorstyle*)	color of sunflower petals
$\boxed{\underline{\text{l}}|\underline{\text{d}}}$ flwidth(*linewidthstyle*)	thickness of sunflower petals

All options are *rightmost*; see [G-4] **concept: repeated options**.

fweights are allowed; see [U] **11.1.6 weight**.

Menu

Graphics > Smoothing and densities > Density-distribution sunflower plot

Description

sunflower draws density-distribution sunflower plots (Plummer and Dupont 2003). These plots are useful for displaying bivariate data whose density is too great for conventional scatterplots to be effective.

A sunflower is several line segments of equal length, called petals, that radiate from a central point. There are two varieties of sunflowers: light and dark. Each petal of a light sunflower represents 1 observation. Each petal of a dark sunflower represents several observations. Dark and light sunflowers represent high- and medium-density regions of the data, and marker symbols represent individual observations in low-density regions.

The plane defined by the variables *yvar* and *xvar* is divided into contiguous hexagonal bins. The number of observations contained within a bin determines how the bin will be represented.

- When there are fewer than light(*#*) observations in a bin, each point is plotted using the usual marker symbols in a scatterplot.

- Bins with at least light(*#*) but fewer than dark(*#*) observations are represented by a light sunflower.

- Bins with at least dark(*#*) observations are represented by a dark sunflower.

Options

┌─ Main ┐

nograph prevents the graph from being generated.

`notable` prevents the summary table from being displayed. This option is implied when the `by()` option is specified.

marker_options affect the rendition of markers drawn at the plotted points, including their shape, size, color, and outline; see [G-3] *marker_options*.

⌐───────── Bins/Petals ⌐──

`binwidth(#)` specifies the horizontal width of the hexagonal bins in the same units as *xvar*. By default,

$$binwidth = \text{max}(\text{rbw}, \text{nbw})$$

where

$$\text{rbw} = \text{range of } xvar/40$$

$$\text{nbw} = \text{range of } xvar/\text{max}(1,\text{nb})$$

and

$$\text{nb} = \text{int}(\text{min}(\text{sqrt}(n),10*\text{log10}(n)))$$

where

$$n = \text{the number of observations in the dataset}$$

`binar(#)` specifies the aspect ratio for the hexagonal bins. The height of the bins is given by

$$binheight = binwidth \times \# \times 2/\sqrt{3}$$

where *binheight* and *binwidth* are specified in the units of *yvar* and *xvar*, respectively. The default is `binar(r)`, where *r* results in the rendering of regular hexagons.

bin_options affect how the hexagonal bins are rendered.

`lbstyle(`*areastyle*`)` and `dbstyle(`*areastyle*`)` specify the look of the light and dark hexagonal bins, respectively. The options listed below allow you to change each attribute, but `lbstyle()` and `dbstyle()` provide the starting points. See [G-4] *areastyle* for a list of available area styles.

`lbcolor(`*colorstyle*`)` and `dbcolor(`*colorstyle*`)` specify one color to be used both to outline the shape and to fill the interior of the light and dark hexagonal bins, respectively. See [G-4] *colorstyle* for a list of color choices.

`lbfcolor(`*colorstyle*`)` and `dbfcolor(`*colorstyle*`)` specify the color to be used to fill the interior of the light and dark hexagonal bins, respectively. See [G-4] *colorstyle* for a list of color choices.

`lblstyle(`*linestyle*`)` and `dblstyle(`*linestyle*`)` specify the overall style of the line used to outline the area, which includes its pattern (solid, dashed, etc.), thickness, and color. The other options listed below allow you to change the line's attributes, but `lblstyle()` and `dblstyle()` are the starting points. See [G-4] *linestyle* for a list of choices.

`lblcolor(`*colorstyle*`)` and `dblcolor(`*colorstyle*`)` specify the color to be used to outline the light and dark hexagonal bins, respectively. See [G-4] *colorstyle* for a list of color choices.

`lblwidth(`*linewidthstyle*`)` and `dblwidth(`*linewidthstyle*`)` specify the thickness of the line to be used to outline the light and dark hexagonal bins, respectively. See [G-4] *linewidthstyle* for a list of choices.

light(#) specifies the minimum number of observations needed for a bin to be represented by a light sunflower. The default is light(3).

dark(#) specifies the minimum number of observations needed for a bin to be represented by a dark sunflower. The default is dark(13).

xcenter(#) and ycenter(#) specify the center of the reference bin. The default values are the median values of *xvar* and *yvar*, respectively. The centers of the other bins are implicitly defined by the location of the reference bin together with the common bin width and height.

petalweight(#) specifies the number of observations represented by each petal of a dark sunflower. The default value is chosen so that the maximum number of petals on a dark sunflower is 14.

petallength(#) specifies the length of petals in the sunflowers. The value specified is interpreted as a percentage of half the bin width. The default is 100%.

petal_options affect how the sunflower petals are rendered.

lflstyle(*linestyle*) and dflstyle(*linestyle*) specify the overall style of the light and dark sunflower petals, respectively.

lflcolor(*colorstyle*) and dflcolor(*colorstyle*) specify the color of the light and dark sunflower petals, respectively.

lflwidth(*linewidthstyle*) and dflwidth(*linewidthstyle*) specify the width of the light and dark sunflower petals, respectively.

flowersonly suppresses rendering of the bins. This option is equivalent to specifying lbcolor(none) and dbcolor(none).

nosinglepetal suppresses flowers from being drawn in light bins that contain only 1 observation and dark bins that contain as many observations as the petal weight (see the petalweight() option).

⌐ Add plots ⌐

addplot(*plot*) provides a way to add other plots to the generated graph; see [G-3] *addplot_option*.

⌐ Y axis, X axis, Titles, Legend, Overall, By ⌐

twoway_options are any of the options documented in [G-3] *twoway_options*. These include options for titling the graph (see [G-3] *title_options*), options for saving the graph to disk (see [G-3] *saving_option*), and the by() option (see [G-3] *by_option*).

Remarks

See Dupont (2009, 87–92) for a discussion of sunflower plots and how to create them using Stata.

▷ Example 1

Using the auto dataset, we want to examine the relationship between weight and mpg. To do that, we type

```
. use http://www.stata-press.com/data/r12/auto
(1978 Automobile Data)

. sunflower mpg weight, binwid(500) petalw(2) dark(8)
Bin width         =        500
Bin height        =    8.38703
Bin aspect ratio  =   .0145268
Max obs in a bin  =         15
Light             =          3
Dark              =          8
X-center          =       3190
Y-center          =         20
Petal weight      =          2
```

flower type	petal weight	No. of petals	No. of flowers	estimated obs.	actual obs.
none				10	10
light	1	3	1	3	3
light	1	4	2	8	8
light	1	7	3	21	21
dark	2	4	1	8	8
dark	2	5	1	10	9
dark	2	8	1	16	15
				76	74

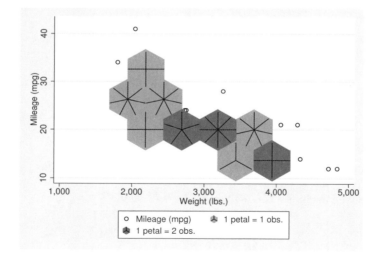

The three darkly shaded sunflowers immediately catch our eyes, indicating a group of eight cars that are heavy (nearly 4,000 pounds) and fuel inefficient and two groups of cars that get about 20 miles per gallon and weight in the neighborhood of 3,000 pounds, one with 10 cars and one with 8 cars. The lighter sunflowers with seven petals each indicate groups of seven cars that share similar weight and fuel economy characteristics. To obtain the number of cars in each group, we counted the number of petals in each flower and consulted the graph legend to see how many observations each petal represents.

◁

Methods and formulas

sunflower is implemented as an ado-file.

Acknowledgments

We thank William D. Dupont and W. Dale Plummer Jr. (Vanderbilt University), authors of the original sunflower command, for their assistance in producing this version.

References

Cleveland, W. S., and R. McGill. 1984. The many faces of a scatterplot. *Journal of the American Statistical Association* 79: 807–822.

Dupont, W. D. 2009. *Statistical Modeling for Biomedical Researchers: A Simple Introduction to the Analysis of Complex Data.* 2nd ed. Cambridge: Cambridge University Press.

Dupont, W. D., and W. D. Plummer, Jr. 2005. Using density-distribution sunflower plots to explore bivariate relationships in dense data. *Stata Journal* 5: 371–384.

Huang, C., J. A. McDonald, and W. Stuetzle. 1997. Variable resolution bivariate plots. *Journal of Computational and Graphical Statistics* 6: 383–396.

Levy, D. 1999. *50 years of discovery: Medical milestones from the National Heart, Lung, and Blood Institute's Framingham Heart Study.* Hoboken, NJ: Center for Bio-Medical Communication.

Plummer, W. D., Jr., and W. D. Dupont. 2003. Density distribution sunflower plots. *Journal of Statistical Software* 8: 1–11.

Steichen, T. J., and N. J. Cox. 1999. flower: Stata module to draw sunflower plots. Boston College Department of Economics, Statistical Software Components S393001. http://ideas.repec.org/c/boc/bocode/s393001.html.

Title

> **sureg** — Zellner's seemingly unrelated regression

Syntax

Basic syntax

> sureg (*depvar*$_1$ *varlist*$_1$) (*depvar*$_2$ *varlist*$_2$) ... (*depvar*$_N$ *varlist*$_N$)
>
> $\big[$ *if* $\big]$ $\big[$ *in* $\big]$ $\big[$ *weight* $\big]$

Full syntax

> sureg ($\big[$*eqname*$_1$:$\big]$*depvar*$_{1a}$ $\big[$*depvar*$_{1b}$ ... = $\big]$*varlist*$_1$ $\big[$, <u>noc</u>onstant $\big]$)
>
> ($\big[$*eqname*$_2$:$\big]$*depvar*$_{2a}$ $\big[$*depvar*$_{2b}$ ... = $\big]$*varlist*$_2$ $\big[$, <u>noc</u>onstant $\big]$)
>
> ...
>
> ($\big[$*eqname*$_N$:$\big]$*depvar*$_{Na}$ $\big[$*depvar*$_{Nb}$ ... = $\big]$*varlist*$_N$ $\big[$, <u>noc</u>onstant $\big]$)
>
> $\big[$ *if* $\big]$ $\big[$ *in* $\big]$ $\big[$ *weight* $\big]$ $\big[$, *options* $\big]$

Explicit equation naming (*eqname*:) cannot be combined with multiple dependent variables in an equation specification.

options	Description
Model	
<u>i</u>sure	iterate until estimates converge
<u>c</u>onstraints(*constraints*)	apply specified linear constraints
df adj.	
<u>sm</u>all	report small-sample statistics
dfk	use small-sample adjustment
dfk2	use alternate adjustment
Reporting	
<u>l</u>evel(*#*)	set confidence level; default is level(95)
<u>corr</u>	perform Breusch–Pagan test
<u>nocns</u>report	do not display constraints
display_options	control column formats, row spacing, line width, and display of omitted variables and base and empty cells
Optimization	
optimization_options	control the optimization process; seldom used
<u>noh</u>eader	suppress header table from above coefficient table
<u>not</u>able	suppress coefficient table
<u>coefl</u>egend	display legend instead of statistics

varlist$_1$, ..., *varlist*$_N$ may contain factor variables; see [U] **11.4.3 Factor variables**. You must have the same levels of factor variables in all equations that have factor variables.

depvars and the *varlist*s may contain time-series operators; see [U] **11.4.4 Time-series varlists**.

bootstrap, by, jackknife, rolling, and statsby are allowed; see [U] **11.1.10 Prefix commands**.

Weights are not allowed with the bootstrap prefix; see [R] **bootstrap**.

aweights are not allowed with the jackknife prefix; see [R] **jackknife**.

aweights and fweights are allowed; see [U] **11.1.6 weight**.

noheader, notable, and coeflegend do not appear in the dialog box.

See [U] **20 Estimation and postestimation commands** for more capabilities of estimation commands.

Menu

Statistics > Linear models and related > Multiple-equation models > Seemingly unrelated regression

Description

sureg fits seemingly unrelated regression models (Zellner 1962; Zellner and Huang 1962; Zellner 1963). The acronyms SURE and SUR are often used for the estimator.

Options

> ⌐ Model ⌐

isure specifies that sureg iterate over the estimated disturbance covariance matrix and parameter estimates until the parameter estimates converge. Under seemingly unrelated regression, this iteration converges to the maximum likelihood results. If this option is not specified, sureg produces two-step estimates.

constraints(*constraints*); see [R] **estimation options**.

> ⌐ df adj. ⌐

small specifies that small-sample statistics be computed. It shifts the test statistics from chi-squared and z statistics to F statistics and t statistics. Although the standard errors from each equation are computed using the degrees of freedom for the equation, the degrees of freedom for the t statistics are all taken to be those for the first equation.

dfk specifies the use of an alternate divisor in computing the covariance matrix for the equation residuals. As an asymptotically justified estimator, sureg by default uses the number of sample observations (n) as a divisor. When the dfk option is set, a small-sample adjustment is made and the divisor is taken to be $\sqrt{(n - k_i)(n - k_j)}$, where k_i and k_j are the numbers of parameters in equations i and j, respectively.

dfk2 specifies the use of an alternate divisor in computing the covariance matrix for the equation residuals. When the dfk2 option is set, the divisor is taken to be the mean of the residual degrees of freedom from the individual equations.

> ⌐ Reporting ⌐

level(*#*); see [R] **estimation options**.

corr displays the correlation matrix of the residuals between equations and performs a Breusch–Pagan test for independent equations; that is, the disturbance covariance matrix is diagonal.

nocnsreport; see [R] **estimation options**.

display_options: <u>noomit</u>ted, vsquish, <u>noempty</u>cells, <u>base</u>levels, <u>allbase</u>levels,
    cformat(%*fmt*), pformat(%*fmt*), sformat(%*fmt*), and nolstretch; see [R] **estimation options**.

⌐ Optimization ⌐

optimization_options control the iterative process that minimizes the sum of squared errors when
    isure is specified. These options are seldom used.

<u>iter</u>ate(*#*) specifies the maximum number of iterations. When the number of iterations equals *#*,
    the optimizer stops and presents the current results, even if the convergence tolerance has not been
    reached. The default value of iterate() is the current value of set maxiter (see [R] **maximize**),
    which is iterate(16000) if maxiter has not been changed.

<u>trace</u> adds to the iteration log a display of the current parameter vector

nolog suppresses the display of the iteration log.

<u>tol</u>erance(*#*) specifies the tolerance for the coefficient vector. When the relative change in the
    coefficient vector from one iteration to the next is less than or equal to *#*, the optimization process
    is stopped. tolerance(1e-6) is the default.

The following options are available with sureg but are not shown in the dialog box:

noheader suppresses display of the table reporting F statistics, R-squared, and root mean squared
    error above the coefficient table.

notable suppresses display of the coefficient table.

coeflegend; see [R] **estimation options**.

Remarks

Seemingly unrelated regression models are so called because they appear to be joint estimates
from several regression models, each with its own error term. The regressions are related because the
(contemporaneous) errors associated with the dependent variables may be correlated. Chapter 5 of
Cameron and Trivedi (2010) contains a discussion of the seemingly unrelated regression model and
the feasible generalized least-squares estimator underlying it.

▷ Example 1

When we fit models with the same set of right-hand-side variables, the seemingly unrelated
regression results (in terms of coefficients and standard errors) are the same as fitting the models
separately (using, say, regress). The same is true when the models are nested. Even in such cases,
sureg is useful when we want to perform joint tests. For instance, let us assume that we think

$$\texttt{price} = \beta_0 + \beta_1 \texttt{foreign} + \beta_2 \texttt{length} + u_1$$
$$\texttt{weight} = \gamma_0 + \gamma_1 \texttt{foreign} + \gamma_2 \texttt{length} + u_2$$

Because the models have the same set of explanatory variables, we could estimate the two equations
separately. Yet, we might still choose to estimate them with sureg because we want to perform the
joint test $\beta_1 = \gamma_1 = 0$.

We use the small and dfk options to obtain small-sample statistics comparable with regress or
mvreg.

```
. use http://www.stata-press.com/data/r12/auto
(1978 Automobile Data)

. sureg (price foreign length) (weight foreign length), small dfk
Seemingly unrelated regression
```

Equation	Obs	Parms	RMSE	"R-sq"	F-Stat	P
price	74	2	2474.593	0.3154	16.35	0.0000
weight	74	2	250.2515	0.8992	316.54	0.0000

	Coef.	Std. Err.	t	P>|t|	[95% Conf. Interval]	
price						
foreign	2801.143	766.117	3.66	0.000	1286.674	4315.611
length	90.21239	15.83368	5.70	0.000	58.91219	121.5126
_cons	-11621.35	3124.436	-3.72	0.000	-17797.77	-5444.93
weight						
foreign	-133.6775	77.47615	-1.73	0.087	-286.8332	19.4782
length	31.44455	1.601234	19.64	0.000	28.27921	34.60989
_cons	-2850.25	315.9691	-9.02	0.000	-3474.861	-2225.639

These two equations have a common set of regressors, and we could have used a shorthand syntax to specify the equations:

```
. sureg (price weight = foreign length), small dfk
```

Here the results presented by sureg are the same as if we had estimated the equations separately:

```
. regress price foreign length
(output omitted )
. regress weight foreign length
(output omitted )
```

There is, however, a difference. We have allowed u_1 and u_2 to be correlated and have estimated the full variance–covariance matrix of the coefficients. sureg has estimated the correlations, but it does not report them unless we specify the corr option. We did not remember to specify corr when we fit the model, but we can redisplay the results:

```
. sureg, notable noheader corr

Correlation matrix of residuals:

          price   weight
 price   1.0000
weight   0.5840   1.0000

Breusch-Pagan test of independence: chi2(1) =    25.237, Pr = 0.0000
```

The notable and noheader options prevented sureg from redisplaying the header and coefficient tables. We find that, for the same cars, the correlation of the residuals in the price and weight equations is 0.5840 and that we can reject the hypothesis that this correlation is zero.

We can test that the coefficients on foreign are jointly zero in both equations—as we set out to do—by typing test foreign; see [R] **test**. When we type a variable without specifying the equation, that variable is tested for zero in all equations in which it appears:

```
. test foreign
 ( 1)  [price]foreign = 0
 ( 2)  [weight]foreign = 0
       F(  2,   142) =   17.99
            Prob > F =    0.0000
```

◁

▷ Example 2

When the models do not have the same set of explanatory variables and are not nested, sureg may lead to more efficient estimates than running the models separately as well as allowing joint tests. This time, let us assume that we believe

$$\text{price} = \beta_0 + \beta_1\text{foreign} + \beta_2\text{mpg} + \beta_3\text{displ} + u_1$$
$$\text{weight} = \gamma_0 + \gamma_1\text{foreign} + \gamma_2\text{length} + u_2$$

To fit this model, we type

```
. sureg (price foreign mpg displ) (weight foreign length), corr
Seemingly unrelated regression
```

Equation	Obs	Parms	RMSE	"R-sq"	chi2	P
price	74	3	2165.321	0.4537	49.64	0.0000
weight	74	2	245.2916	0.8990	661.84	0.0000

	Coef.	Std. Err.	z	P>|z|	[95% Conf. Interval]	
price						
foreign	3058.25	685.7357	4.46	0.000	1714.233	4402.267
mpg	-104.9591	58.47209	-1.80	0.073	-219.5623	9.644042
displacement	18.18098	4.286372	4.24	0.000	9.779842	26.58211
_cons	3904.336	1966.521	1.99	0.047	50.0263	7758.645
weight						
foreign	-147.3481	75.44314	-1.95	0.051	-295.2139	.517755
length	30.94905	1.539895	20.10	0.000	27.93091	33.96718
_cons	-2753.064	303.9336	-9.06	0.000	-3348.763	-2157.365

```
Correlation matrix of residuals:

        price  weight
 price  1.0000
weight  0.3285  1.0000

Breusch-Pagan test of independence: chi2(1) =    7.984, Pr = 0.0047
```

In comparison, if we had fit the price model separately,

```
. regress price foreign mpg displ
```

Source	SS	df	MS
Model	294104790	3	98034929.9
Residual	340960606	70	4870865.81
Total	635065396	73	8699525.97

Number of obs =	74
F(3, 70) =	20.13
Prob > F =	0.0000
R-squared =	0.4631
Adj R-squared =	0.4401
Root MSE =	2207

price	Coef.	Std. Err.	t	P>|t|	[95% Conf. Interval]	
foreign	3545.484	712.7763	4.97	0.000	2123.897	4967.072
mpg	-98.88559	63.17063	-1.57	0.122	-224.8754	27.10426
displacement	22.40416	4.634239	4.83	0.000	13.16146	31.64686
_cons	2796.91	2137.873	1.31	0.195	-1466.943	7060.763

The coefficients are slightly different, but the standard errors are uniformly larger. This would still be true if we specified the dfk option to make a small-sample adjustment to the estimated covariance of the disturbances.   ◁

❑ Technical note

Constraints can be applied to SURE models using Stata's standard syntax for constraints. For a general discussion of constraints, see [R] **constraint**; for examples similar to seemingly unrelated regression models, see [R] **reg3**.

❑

Arnold Zellner (1927–2010) was born in New York. He studied physics at Harvard and economics at Berkeley, and then he taught economics at the Universities of Washington and Wisconsin before settling in Chicago in 1966. Among his many major contributions to econometrics and statistics are his work on seemingly unrelated regression, three-stage least squares, and Bayesian econometrics.

Saved results

sureg saves the following in e():

Scalars
e(N)	number of observations
e(k)	number of parameters
e(k_eq)	number of equations in e(b)
e(mss_#)	model sum of squares for equation #
e(df_m#)	model degrees of freedom for equation #
e(rss_#)	residual sum of squares for equation #
e(df_r)	residual degrees of freedom
e(r2_#)	R-squared for equation #
e(F_#)	F statistic for equation # (small only)
e(rmse_#)	root mean squared error for equation #
e(dfk2_adj)	divisor used with VCE when dfk2 specified
e(ll)	log likelihood
e(chi2_#)	χ^2 for equation #
e(p_#)	significance for equation #
e(cons_#)	1 if equation # has a constant, 0 otherwise
e(chi2_bp)	Breusch–Pagan χ^2
e(df_bp)	degrees of freedom for Breusch–Pagan χ^2 test
e(cons_#)	1 when equation # has a constant; 0, otherwise
e(rank)	rank of e(V)
e(ic)	number of iterations

Macros
e(cmd)	sureg
e(cmdline)	command as typed
e(method)	sure or isure
e(depvar)	names of dependent variables
e(exog)	names of exogenous variables
e(eqnames)	names of equations
e(wtype)	weight type
e(wexp)	weight expression
e(corr)	correlation structure
e(small)	small
e(dfk)	alternate divisor (dfk or dfk2 only)
e(properties)	b V
e(predict)	program used to implement predict
e(marginsok)	predictions allowed by margins
e(marginsnotok)	predictions disallowed by margins
e(asbalanced)	factor variables fvset as asbalanced
e(asobserved)	factor variables fvset as asobserved

Matrices
e(b)	coefficient vector
e(Cns)	constraints matrix
e(Sigma)	$\widehat{\Sigma}$ matrix
e(V)	variance–covariance matrix of the estimators

Functions
e(sample)	marks estimation sample

Methods and formulas

sureg is implemented as an ado-file.

sureg uses the asymptotically efficient, feasible, generalized least-squares algorithm described in Greene (2012, 292–304). The computing formulas are given on page 293–294.

The R-squared reported is the percent of variance explained by the predictors. It may be used for descriptive purposes, but R-squared is not a well-defined concept when GLS is used.

`sureg` will refuse to compute the estimators if the same equation is named more than once or the covariance matrix of the residuals is singular.

The Breusch and Pagan (1980) χ^2 statistic—a Lagrange multiplier statistic—is given by

$$\lambda = T \sum_{m=1}^{M} \sum_{n=1}^{m-1} r_{mn}^2$$

where r_{mn} is the estimated correlation between the residuals of the M equations and T is the number of observations. It is distributed as χ^2 with $M(M-1)/2$ degrees of freedom.

References

Breusch, T. S., and A. R. Pagan. 1980. The Lagrange multiplier test and its applications to model specification in econometrics. *Review of Economic Studies* 47: 239–253.

Cameron, A. C., and P. K. Trivedi. 2010. *Microeconometrics Using Stata*. Rev. ed. College Station, TX: Stata Press.

Greene, W. H. 2012. *Econometric Analysis*. 7th ed. Upper Saddle River, NJ: Prentice Hall.

McDowell, A. W. 2004. From the help desk: Seemingly unrelated regression with unbalanced equations. *Stata Journal* 4: 442–448.

Rossi, P. E. 1989. The ET interview: Professor Arnold Zellner. *Econometric Theory* 5: 287–317.

Weesie, J. 1999. sg121: Seemingly unrelated estimation and the cluster-adjusted sandwich estimator. *Stata Technical Bulletin* 52: 34–47. Reprinted in *Stata Technical Bulletin Reprints*, vol. 9, pp. 231–248. College Station, TX: Stata Press.

Zellner, A. 1962. An efficient method of estimating seemingly unrelated regressions and tests for aggregation bias. *Journal of the American Statistical Association* 57: 348–368.

——. 1963. Estimators for seemingly unrelated regression equations: Some exact finite sample results. *Journal of the American Statistical Association* 58: 977–992.

Zellner, A., and D. S. Huang. 1962. Further properties of efficient estimators for seemingly unrelated regression equations. *International Economic Review* 3: 300–313.

Also see

[R] **sureg postestimation** — Postestimation tools for sureg

[R] **mvreg** — Multivariate regression

[R] **nlsur** — Estimation of nonlinear systems of equations

[R] **reg3** — Three-stage estimation for systems of simultaneous equations

[R] **regress** — Linear regression

[TS] **dfactor** — Dynamic-factor models

Stata Structural Equation Modeling Reference Manual

[U] **20 Estimation and postestimation commands**

Title

sureg postestimation — Postestimation tools for sureg

Description

The following postestimation commands are available after `sureg`:

Command	Description
contrast	contrasts and ANOVA-style joint tests of estimates
estat	AIC, BIC, VCE, and estimation sample summary
estimates	cataloging estimation results
lincom	point estimates, standard errors, testing, and inference for linear combinations of coefficients
margins	marginal means, predictive margins, marginal effects, and average marginal effects
marginsplot	graph the results from margins (profile plots, interaction plots, etc.)
nlcom	point estimates, standard errors, testing, and inference for nonlinear combinations of coefficients
predict	predictions, residuals, influence statistics, and other diagnostic measures
predictnl	point estimates, standard errors, testing, and inference for generalized predictions
pwcompare	pairwise comparisons of estimates
test	Wald tests of simple and composite linear hypotheses
testnl	Wald tests of nonlinear hypotheses

See the corresponding entries in the *Base Reference Manual* for details.

Syntax for predict

predict [*type*] *newvar* [*if*] [*in*] [, <u>equation</u>(*eqno*[,*eqno*]) *statistic*]

statistic	Description
Main	
xb	linear prediction; the default
stdp	standard error of the linear prediction
<u>residuals</u>	residuals
<u>difference</u>	difference between the linear predictions of two equations
stddp	standard error of the difference in linear predictions

These statistics are available both in and out of sample; type `predict ... if e(sample) ...` if wanted only for the estimation sample.

Menu

Statistics > Postestimation > Predictions, residuals, etc.

Options for predict

      ⌐ Main ⌐

equation(*eqno*[,*eqno*]) specifies to which equation(s) you are referring.

> equation() is filled in with one *eqno* for the xb, stdp, and residuals options. equation(#1) would mean that the calculation is to be made for the first equation, equation(#2) would mean the second, and so on. You could also refer to the equations by their names. equation(income) would refer to the equation named income and equation(hours) to the equation named hours.

> If you do not specify equation(), the results are the same as if you specified equation(#1).

> difference and stddp refer to between-equation concepts. To use these options, you must specify two equations, for example, equation(#1,#2) or equation(income,hours). When two equations must be specified, equation() is required.

xb, the default, calculates the linear prediction (fitted values)—the prediction of $x_j b$ for the specified equation.

stdp calculates the standard error of the prediction for the specified equation. It can be thought of as the standard error of the predicted expected value or mean for the observation's covariate pattern. The standard error of the prediction is also referred to as the standard error of the fitted value.

residuals calculates the residuals.

difference calculates the difference between the linear predictions of two equations in the system. With equation(#1,#2), difference computes the prediction of equation(#1) minus the prediction of equation(#2).

stddp is allowed only after you have previously fit a multiple-equation model. The standard error of the difference in linear predictions $(x_{1j} b - x_{2j} b)$ between equations 1 and 2 is calculated.

For more information on using predict after multiple-equation estimation commands, see [R] **predict**.

Remarks

For an example of cross-equation testing of parameters using the test command, see example 1 in [R] **sureg**.

▷ Example 1

In example 1 of [R] **sureg**, we fit a seemingly unrelated regressions model of price and weight. Here we obtain the fitted values.

```
. use http://www.stata-press.com/data/r12/auto
(1978 Automobile Data)
. sureg (price foreign length) (weight foreign length), small dfk
 (output omitted )
. predict phat, equation(price)
(option xb assumed; fitted values)
. predict what, equation(weight)
(option xb assumed; fitted values)
```

```
. summarize price phat weight what
```

Variable	Obs	Mean	Std. Dev.	Min	Max
price	74	6165.257	2949.496	3291	15906
phat	74	6165.257	1656.407	1639.872	9398.138
weight	74	3019.459	777.1936	1760	4840
what	74	3019.459	736.9666	1481.199	4476.331

Just as in single-equation OLS regression, in a SURE model the sample mean of the fitted values for an equation equals the sample mean of the dependent variable.

◁

▷ Example 2

Suppose that for whatever reason we were interested in the difference between the predicted values of price and weight. predict has an option to compute this difference in one step:

```
. predict diff, equation(price, weight) difference
```

diff is the same as phat - what:

```
. generate mydiff = phat - what
. summarize diff mydiff
```

Variable	Obs	Mean	Std. Dev.	Min	Max
diff	74	3145.797	1233.26	-132.2275	5505.914
mydiff	74	3145.797	1233.26	-132.2275	5505.914

◁

Methods and formulas

All postestimation commands listed above are implemented as ado-files.

Also see

[R] **sureg** — Zellner's seemingly unrelated regression

[U] **20 Estimation and postestimation commands**

Title

> **swilk** — Shapiro–Wilk and Shapiro–Francia tests for normality

Syntax

Shapiro–Wilk normality test

    swilk *varlist* [*if*] [*in*] [, *swilk_options*]

Shapiro–Francia normality test

    sfrancia *varlist* [*if*] [*in*][, *sfrancia_options*]

swilk_options	Description
Main	
generate(*newvar*)	create *newvar* containing W test coefficients
lnnormal	test for three-parameter lognormality
noties	do not use average ranks for tied values

sfrancia_options	Description
Main	
boxcox	use the Box–Cox transformation for W'; the default is to use the log transformation
noties	do not use average ranks for tied values

by is allowed with swilk and sfrancia; see [D] **by**.

Menu

swilk

Statistics > Summaries, tables, and tests > Distributional plots and tests > Shapiro-Wilk normality test

sfrancia

Statistics > Summaries, tables, and tests > Distributional plots and tests > Shapiro-Francia normality test

Description

    swilk performs the Shapiro–Wilk W test for normality, and sfrancia performs the Shapiro–Francia W' test for normality. swilk can be used with $4 \leq n \leq 2000$ observations, and sfrancia can be used with $5 \leq n \leq 5000$ observations; see [R] **sktest** for a test allowing more observations. See [MV] **mvtest normality** for multivariate tests of normality.

2111

Options for swilk

````
  ┌─ Main ────────────────────────────────────────────────────────────────────
````

generate(*newvar*) creates new variable *newvar* containing the W test coefficients.

lnnormal specifies that the test be for three-parameter lognormality, meaning that $\ln(X - k)$ is tested for normality, where k is calculated from the data as the value that makes the skewness coefficient zero. When simply testing $\ln(X)$ for normality, do not specify this option. See [R] **lnskew0** for estimation of k.

noties suppresses use of averaged ranks for tied values when calculating the W test coefficients.

Options for sfrancia

````
  ┌─ Main ────────────────────────────────────────────────────────────────────
````

boxcox specifies that the Box–Cox transformation of Royston (1983) for calculating W' test coefficients be used instead of the default log transformation (Royston 1993a). Under the Box–Cox transformation, the normal approximation to the sampling distribution of W', used by sfrancia, is valid for $5 \le n \le 1000$. Under the log transformation, it is valid for $10 \le n \le 5000$.

noties suppresses use of averaged ranks for tied values when calculating the W' test coefficients.

Remarks

▷ Example 1

Using our automobile dataset, we will test whether the variables mpg and trunk are normally distributed:

```
. use http://www.stata-press.com/data/r12/auto
(1978 Automobile Data)
. swilk mpg trunk
                   Shapiro-Wilk W test for normal data
    Variable │     Obs        W          V          z       Prob>z
─────────────┼──────────────────────────────────────────────────────
         mpg │      74     0.94821      3.335      2.627     0.00430
       trunk │      74     0.97921      1.339      0.637     0.26215
. sfrancia mpg trunk
                   Shapiro-Francia W' test for normal data
    Variable │     Obs        W'         V'         z       Prob>z
─────────────┼──────────────────────────────────────────────────────
         mpg │      74     0.94872      3.650      2.510     0.00604
       trunk │      74     0.98446      1.106      0.195     0.42271
```

We can reject the hypothesis that mpg is normally distributed, but we cannot reject that trunk is normally distributed.

The values reported under W and W' are the Shapiro–Wilk and Shapiro–Francia test statistics. The tests also report V and V', which are more appealing indexes for departure from normality. The median values of V and V' are 1 for samples from normal populations. Large values indicate nonnormality. The 95% critical values of V (V'), which depend on the sample size, are between 1.2 and 2.4 (2.0 and 2.8); see Royston (1991b). There is no more information in V (V') than in W (W')—one is just the transform of the other.

◁

▷ Example 2

We have data on a variable called studytime, which we suspect is distributed lognormally:

```
. use http://www.stata-press.com/data/r12/cancer
(Patient Survival in Drug Trial)
. generate lnstudytime = ln(studytime)
. swilk lnstudytime
                  Shapiro-Wilk W test for normal data
    Variable |    Obs       W         V         z       Prob>z

 lnstudytime |     48    0.92731    3.311    2.547     0.00543
```

We can reject the lognormal assumption. We do *not* specify the lnnormal option when testing for lognormality. The lnnormal option is for three-parameter lognormality.

◁

▷ Example 3

Having discovered that $\ln(\text{studytime})$ is not distributed normally, we now test that $\ln(\text{studytime} - k)$ is normally distributed, where k is chosen so that the resulting skewness is zero. We obtain the estimate for k from lnskew0; see [R] **lnskew0**:

```
. lnskew0 lnstudytimek = studytime, level(95)
      Transform |         k     [95% Conf. Interval]      Skewness

 ln(studytim-k) |  -11.01181   -infinity  -.9477328     -.0000173
. swilk lnstudytimek, lnnormal
            Shapiro-Wilk W test for 3-parameter lognormal data
     Variable |    Obs       W         V         z       Prob>z

 lnstudytimek |     48    0.97064    1.337    1.261     0.10363
```

We cannot reject the hypothesis that $\ln(\text{studytime} + 11.01181)$ is distributed normally. We do specify the lnnormal option when using an estimated value of k.

◁

Saved results

swilk and sfrancia save the following in r():

Scalars

r(N)	number of observations	r(W)	W or W'
r(p)	significance	r(V)	V or V'
r(z)	z statistic		

Methods and formulas

swilk and sfrancia are implemented as ado-files.

The Shapiro–Wilk test is based on Shapiro and Wilk (1965) with a new approximation accurate for $4 \le n \le 2000$ (Royston 1992). The calculations made by swilk are based on Royston (1982, 1992, 1993b).

The Shapiro–Francia test (Shapiro and Francia 1972; Royston 1983; Royston 1993a) is an approximate test that is similar to the Shapiro–Wilk test for very large samples.

Samuel Sanford Shapiro (1930–) earned degrees in statistics and engineering from City College of New York, Columbia, and Rutgers. After employment in the U.S. Army and industry, he joined the faculty at Florida International University in 1972. Shapiro has coauthored various texts in statistics and published several papers on distributional testing and other statistical topics.

Acknowledgment

swilk and sfrancia were written by Patrick Royston of the MRC Clinical Trials Unit, London.

References

Genest, C., and G. Brackstone. 2010. A conversation with Martin Bradbury Wilk. *Statistical Science* 25: 258–273.

Gould, W. W. 1992. sg3.7: Final summary of tests of normality. *Stata Technical Bulletin* 5: 10–11. Reprinted in *Stata Technical Bulletin Reprints*, vol. 1, pp. 114–115. College Station, TX: Stata Press.

Royston, P. 1982. An extension of Shapiro and Wilks's *W* test for normality to large samples. *Applied Statistics* 31: 115–124.

——. 1983. A simple method for evaluating the Shapiro–Francia *W'* test of non-normality. *Statistician* 32: 297–300.

——. 1991a. sg3.2: Shapiro–Wilk and Shapiro–Francia tests. *Stata Technical Bulletin* 3: 19. Reprinted in *Stata Technical Bulletin Reprints*, vol. 1, p. 105. College Station, TX: Stata Press.

——. 1991b. Estimating departure from normality. *Statistics in Medicine* 10: 1283–1293.

——. 1992. Approximating the Shapiro–Wilk *W*-test for non-normality. *Statistics and Computing* 2: 117–119.

——. 1993a. A pocket-calculator algorithm for the Shapiro–Francia test for non-normality: An application to medicine. *Statistics in Medicine* 12: 181–184.

——. 1993b. A toolkit for testing for non-normality in complete and censored samples. *Statistician* 42: 37–43.

Shapiro, S. S., and R. S. Francia. 1972. An approximate analysis of variance test for normality. *Journal of the American Statistical Association* 67: 215–216.

Shapiro, S. S., and M. B. Wilk. 1965. An analysis of variance test for normality (complete samples). *Biometrika* 52: 591–611.

Also see

[R] **lnskew0** — Find zero-skewness log or Box–Cox transform

[R] **lv** — Letter-value displays

[R] **sktest** — Skewness and kurtosis test for normality

[MV] **mvtest normality** — Multivariate normality tests

Title

symmetry — Symmetry and marginal homogeneity tests

Syntax

Symmetry and marginal homogeneity tests

> symmetry *casevar controlvar* $[\textit{if}]$ $[\textit{in}]$ $[\textit{weight}]$ $[\,$ *options* $]$

Immediate form of symmetry and marginal homogeneity tests

> symmi $\#_{11}$ $\#_{12}$ $[...]$ \ $\#_{21}$ $\#_{22}$ $[...]$ $[\backslash...]$ $[\textit{if}]$ $[\textit{in}]$ $[\,$ *options* $]$

options	Description
Main	
<u>not</u>able	suppress output of contingency table
<u>con</u>trib	report contribution of each off-diagonal cell pair
<u>ex</u>act	perform exact test of table symmetry
mh	perform two marginal homogeneity tests
<u>tr</u>end	perform a test for linear trend in the (log) relative risk (RR)
cc	use continuity correction when calculating test for linear trend

fweights are allowed; see [U] **11.1.6 weight**.

Menu

symmetry

Statistics > Epidemiology and related > Other > Symmetry and marginal homogeneity test

symmi

Statistics > Epidemiology and related > Other > Symmetry and marginal homogeneity test calculator

Description

symmetry performs asymptotic symmetry and marginal homogeneity tests, as well as an exact symmetry test on $K \times K$ tables where there is a 1-to-1 matching of cases and controls (nonindependence). This testing is used to analyze matched-pair case–control data with multiple discrete levels of the exposure (outcome) variable. In genetics, the test is known as the transmission/disequilibrium test (TDT) and is used to test the association between transmitted and nontransmitted parental marker alleles to an affected child (Spieldman, McGinnis, and Ewens 1993). For 2×2 tables, the asymptotic test statistics reduce to the McNemar test statistic, and the exact symmetry test produces an exact McNemar test; see [ST] **epitab**. For many exposure variables, symmetry can optionally perform a test for linear trend in the log relative risk.

symmetry expects the data to be in the wide format; that is, each observation contains the matched case and control values in variables *casevar* and *controlvar*. Variables can be numeric or string.

symmi is the immediate form of symmetry. The symmi command uses the values specified on the command line; rows are separated by '\', and options are the same as for symmetry. See [U] **19 Immediate commands** for a general introduction to immediate commands.

2115

Options

notable suppresses the output of the contingency table. By default, symmetry displays the $n \times n$ contingency table at the top of the output.

contrib reports the contribution of each off-diagonal cell pair to the overall symmetry χ^2.

exact performs an exact test of table symmetry. This option is recommended for sparse tables. CAUTION: The exact test requires substantial amounts of time and memory for large tables.

mh performs two marginal homogeneity tests that do not require the inversion of the variance–covariance matrix.

By default, symmetry produces the Stuart–Maxwell test statistic, which requires the inversion of the nondiagonal variance–covariance matrix, $\mathbf{V}$. When the table is sparse, the matrix may not be of full rank, and then the command substitutes a generalized inverse $\mathbf{V}^*$ for $\mathbf{V}^{-1}$. mh calculates optional marginal homogeneity statistics that do not require the inversion of the variance–covariance matrix. These tests may be preferred in certain situations. See *Methods and formulas* and Bickeböller and Clerget-Darpoux (1995) for details on these test statistics.

trend performs a test for linear trend in the (log) relative risk (RR). This option is allowed only for numeric exposure (outcome) variables, and its use should be restricted to measurements on the ordinal or the interval scales.

cc specifies that the continuity correction be used when calculating the test for linear trend. This correction should be specified only when the levels of the exposure variable are equally spaced.

Remarks

symmetry and symmi may be used to analyze 1-to-1 matched case–control data with multiple discrete levels of the exposure (outcome) variable.

▷ Example 1

Consider a survey of 344 individuals (BMDP 1990, 267–270) who were asked in October 1986 whether they agreed with President Reagan's handling of foreign affairs. In January 1987, after the Iran-Contra affair became public, these same individuals were surveyed again and asked the same question. We would like to know if public opinion changed over this period.

We first describe the dataset and list a few observations.

```
. use http://www.stata-press.com/data/r12/iran

. describe
Contains data from http://www.stata-press.com/data/r12/iran.dta
  obs:           344
 vars:             2                          29 Jan 2011 02:37
 size:           688

              storage   display    value
variable name   type    format     label      variable label

before          byte    %8.0g      vlab       Public Opinion before IC
after           byte    %8.0g      vlab       Public Opinion after IC

Sorted by:
```

```
. list in 1/5
```

	before	after
1.	agree	agree
2.	agree	disagree
3.	agree	unsure
4.	disagree	agree
5.	disagree	disagree

Each observation corresponds to one of the 344 individuals. The data are in wide form so that each observation has a before and an after measurement. We now perform the test without options.

```
. symmetry before after
```

Public Opinion before IC	Public Opinion after IC			
	agree	disagree	unsure	Total
agree	47	56	38	141
disagree	28	61	31	120
unsure	26	47	10	83
Total	101	164	79	344

	chi2	df	Prob>chi2
Symmetry (asymptotic)	14.87	3	0.0019
Marginal homogeneity (Stuart-Maxwell)	14.78	2	0.0006

The test first tabulates the data in a $K \times K$ table and then performs Bowker's (1948) test for table symmetry and the Stuart–Maxwell (Stuart 1955; Maxwell 1970) test for marginal homogeneity.

Both the symmetry test and the marginal homogeneity test are highly significant, thus indicating a shift in public opinion.

An exact test of symmetry is provided for use on sparse tables. This test is computationally intensive, so it should not be used on large tables. Because we are working on a fast computer, we will run the symmetry test again and this time include the exact option. We will suppress the output of the contingency table by specifying notable and include the contrib option so that we may further examine the cells responsible for the significant result.

```
. symmetry before after, contrib exact mh notable
```

Cells	Contribution to symmetry chi-squared
n1_2 & n2_1	9.3333
n1_3 & n3_1	2.2500
n2_3 & n3_2	3.2821

	chi2	df	Prob>chi2
Symmetry (asymptotic)	14.87	3	0.0019
Marginal homogeneity (Stuart-Maxwell)	14.78	2	0.0006
Marginal homogeneity (Bickenboller)	13.53	2	0.0012
Marginal homogeneity (no diagonals)	15.25	2	0.0005

Symmetry (exact significance probability)			0.0018

The largest contribution to the symmetry χ^2 is due to cells n_{12} and n_{21}. These correspond to changes between the agree and disagree categories. Of the 344 individuals, 56 (16.3%) changed from the agree to the disagree response, whereas only 28 (8.1%) changed in the opposite direction.

For these data, the results from the exact test are similar to those from the asymptotic test.

◁

▷ Example 2

Breslow and Day (1980, 163) reprinted data from Mack et al. (1976) from a case–control study of the effect of exogenous estrogen on the risk of endometrial cancer. The data consist of 59 elderly women diagnosed with endometrial cancer and 59 disease-free control subjects living in the same community as the cases. Cases and controls were matched on age, marital status, and time living in the community. The data collected included information on the daily dose of conjugated estrogen therapy. Breslow and Day analyzed these data by creating four levels of the dose variable. Here are the data as entered into a Stata dataset:

```
. use http://www.stata-press.com/data/r12/bd163
. list, noobs divider
```

case	control	count
0	0	6
0	0.1-0.299	2
0	0.3-0.625	3
0	0.626+	1
0.1-0.299	0	9
0.1-0.299	0.1-0.299	4
0.1-0.299	0.3-0.625	2
0.1-0.299	0.626+	1
0.3-0.625	0	9
0.3-0.625	0.1-0.299	2
0.3-0.625	0.3-0.625	3
0.3-0.625	0.626+	1
0.626+	0	12
0.626+	0.1-0.299	1
0.626+	0.3-0.625	2
0.626+	0.626+	1

This dataset is in a different format from that of the previous example. Instead of each observation representing one matched pair, each observation represents possibly multiple pairs indicated by the count variable. For instance, the first observation corresponds to six matched pairs where neither the case nor the control was on estrogen, the second observation corresponds to two matched pairs where the case was not on estrogen and the control was on 0.1 to 0.299 mg/day, etc.

To use symmetry to analyze this dataset, we must specify fweight to indicate that in our data there are observations corresponding to more than one matched pair.

```
. symmetry case control [fweight=count]
```

			control		
case	0	0.1-0.299	0.3-0.625	0.626+	Total
0	6	2	3	1	12
0.1-0.299	9	4	2	1	16
0.3-0.625	9	2	3	1	15
0.626+	12	1	2	1	16
Total	36	9	10	4	59

	chi2	df	Prob>chi2
Symmetry (asymptotic)	17.10	6	0.0089
Marginal homogeneity (Stuart-Maxwell)	16.96	3	0.0007

Both the test of symmetry and the test of marginal homogeneity are highly significant, thus leading us to reject the null hypothesis that there is no effect of exposure to estrogen on the risk of endometrial cancer.

Breslow and Day perform a test for trend assuming that the estrogen exposure levels were equally spaced by recoding the exposure levels as 1, 2, 3, and 4.

We can easily reproduce their results by recoding our data in this way and by specifying the trend option. Two new numeric variables were created, ca and co, corresponding to the variables case and control, respectively. Below we list some of the data and our results from symmetry:

```
. encode case, gen(ca)
. encode control, gen(co)
. label values ca
. label values co
. list in 1/4
```

	case	control	count	ca	co
1.	0	0	6	1	1
2.	0	0.1-0.299	2	1	2
3.	0	0.3-0.625	3	1	3
4.	0	0.626+	1	1	4

```
. symmetry ca co [fw=count], notable trend cc
```

	chi2	df	Prob>chi2
Symmetry (asymptotic)	17.10	6	0.0089
Marginal homogeneity (Stuart-Maxwell)	16.96	3	0.0007
Linear trend in the (log) RR	14.43	1	0.0001

We requested the continuity correction by specifying cc. Doing so is appropriate because our coded exposure levels are equally spaced.

The test for trend was highly significant, indicating an increased risk of endometrial cancer with increased dosage of conjugated estrogen.

You must be cautious: the way in which you code the exposure variable affects the linear trend statistic. If instead of coding the levels as 1, 2, 3, and 4, we had instead used 0, 0.2, 0.46, and 0.7

(roughly the midpoint in the range of each level), we would have obtained a χ^2 statistic of 11.19 for these data.

◁

Saved results

symmetry saves the following in r():

Scalars

r(N_pair)	number of matched pairs
r(chi2)	asymptotic symmetry χ^2
r(df)	asymptotic symmetry degrees of freedom
r(p)	asymptotic symmetry p-value
r(chi2_sm)	MH (Stuart–Maxwell) χ^2
r(df_sm)	MH (Stuart–Maxwell) degrees of freedom
r(p_sm)	MH (Stuart–Maxwell) p-value
r(chi2_b)	MH (Bickenböller) χ^2
r(df_b)	MH (Bickenböller) degrees of freedom
r(p_b)	MH (Bickenböller) p-value
r(chi2_nd)	MH (no diagonals) χ^2
r(df_nd)	MH (no diagonals) degrees of freedom
r(p_nd)	MH (no diagonals) p-value
r(chi2_t)	χ^2 for linear trend
r(p_trend)	p-value for linear trend
r(p_exact)	exact symmetry p-value

Methods and formulas

symmetry and symmi are implemented as ado-files.

Methods and formulas are presented under the following headings:

Asymptotic tests
Exact symmetry test

Asymptotic tests

Consider a square table with K exposure categories, that is, K rows and K columns. Let n_{ij} be the count corresponding to row i and column j of the table, $N_{ij} = n_{ij} + n_{ji}$, for $i, j = 1, 2, \ldots, K$, and $n_{i.}$, and let $n_{.j}$ be the marginal totals for row i and column j, respectively. Asymptotic tests for symmetry and marginal homogeneity for this $K \times K$ table are calculated as follows:

The null hypothesis of complete symmetry $p_{ij} = p_{ji}$, $i \neq j$, is tested by calculating the test statistic (Bowker 1948)

$$T_{cs} = \sum_{i<j} \frac{(n_{ij} - n_{ji})^2}{n_{ij} + n_{ji}}$$

which is asymptotically distributed as χ^2 with $K(K-1)/2 - R$ degrees of freedom, where R is the number of off-diagonal cells with $N_{ij} = 0$.

The null hypothesis of marginal homogeneity, $p_{i.} = p_{.i}$, is tested by calculating the Stuart–Maxwell test statistic (Stuart 1955; Maxwell 1970),

$$T_{sm} = \mathbf{d}' \mathbf{V}^{-1} \mathbf{d}$$

where $\mathbf{d}$ is a column vector with elements equal to the differences $d_i = n_{i.} - n_{.i}$ for $i = 1, 2, \ldots, K$, and $\mathbf{V}$ is the variance–covariance matrix with elements

$$v_{ii} = n_{i.} + n_{.i} - 2n_{ii}$$
$$v_{ij} = -(n_{ij} + n_{ji}), \quad i \neq j$$

T_{sm} is asymptotically χ^2 with $K - 1$ degrees of freedom.

This test statistic properly accounts for the dependence between the table's rows and columns. When the matrix $\mathbf{V}$ is not of full rank, a generalized inverse $\mathbf{V}^*$ is substituted for $\mathbf{V}^{-1}$.

The Bickeböller and Clerget-Darpoux (1995) marginal homogeneity test statistic is calculated by

$$T_{\text{mh}} = \sum_i \frac{(n_{i.} - n_{.i})^2}{n_{i.} + n_{.i}}$$

This statistic is asymptotically distributed, under the assumption of marginal independence, as χ^2 with $K - 1$ degrees of freedom.

The marginal homogeneity (no diagonals) test statistic T_{mh}^0 is calculated in the same way as T_{mh}, except that the diagonal elements do not enter into the calculation of the marginal totals. Unlike the previous test statistic, T_{mh}^0 reduces to a McNemar test statistic for 2×2 tables. The test statistic $\{(K - 1)/2\}T_{\text{mh}}^0$ is asymptotically distributed as χ^2 with $K - 1$ degrees of freedom (Cleves, Olson, and Jacobs 1997; Spieldman and Ewens 1996).

Breslow and Day's test statistic for linear trend in the (log) of RR is

$$\frac{\left\{ \sum_{i<j} (n_{ij} - n_{ji})(X_j - X_i) - cc \right\}^2}{\sum_{i<j} (n_{ij} + n_{ji})(X_j - X_i)^2}$$

where the X_j are the doses associated with the various levels of exposure and cc is the continuity correction; it is asymptotically distributed as χ^2 with 1 degree of freedom.

The continuity correction option is applicable only when the levels of the exposure variable are equally spaced.

Exact symmetry test

The exact test is based on a permutation algorithm applied to the null distribution. The distribution of the off-diagonal elements n_{ij}, $i \neq j$, conditional on the sum of the complementary off-diagonal cells, $N_{ij} = n_{ij} + n_{ji}$, can be written as the product of $K(K - 1)/2$ binomial random variables,

$$P(\mathbf{n}) = \prod_{i<j} \binom{N_{ij}}{n_{ij}} \pi_{ij}^{n_{ij}} (1 - \pi_{ij})^{n_{ij}}$$

where $\mathbf{n}$ is a vector with elements n_{ij} and $\pi_{ij} = E(n_{ij}/N_{ij}|N_{ij})$. Under the null hypothesis of complete symmetry, $\pi_{ij} = \pi_{ji} = 1/2$, and thus the permutation distribution is given by

$$P_0(\mathbf{n}) = \prod_{i<j} \binom{N_{ij}}{n_{ij}} \left(\frac{1}{2}\right)^{N_{ij}}$$

The exact significance test is performed by evaluating

$$P_{\text{cs}} = \sum_{n \in p} P_0(\mathbf{n})$$

where $p = \{n : P_0(\mathbf{n}) < P_0(\mathbf{n}^*)\}$ and $\mathbf{n}^*$ is the observed contingency table data vector. The algorithm evaluates p_{cs} exactly. For information about permutation tests, see Good (2005, 2006).

References

Bickeböller, H., and F. Clerget-Darpoux. 1995. Statistical properties of the allelic and genotypic transmission/disequilibrium test for multiallelic markers. *Genetic Epidemiology* 12: 865–870.

BMDP. 1990. *BMDP Statistical Software Manual.* Oakland, CA: University of California Press.

Bowker, A. H. 1948. A test for symmetry in contingency tables. *Journal of the American Statistical Association* 43: 572–574.

Breslow, N. E., and N. E. Day. 1980. *Statistical Methods in Cancer Research: Vol. 1—The Analysis of Case–Control Studies.* Lyon: IARC.

Cleves, M. A. 1997. sg74: Symmetry and marginal homogeneity test/Transmission-Disequilibrium Test (TDT). *Stata Technical Bulletin* 40: 23–27. Reprinted in *Stata Technical Bulletin Reprints*, vol. 7, pp. 193–197. College Station, TX: Stata Press.

———. 1999. sg110: Hardy–Weinberg equilibrium test and allele frequency estimation. *Stata Technical Bulletin* 48: 34–37. Reprinted in *Stata Technical Bulletin Reprints*, vol. 8, pp. 280–284. College Station, TX: Stata Press.

Cleves, M. A., J. M. Olson, and K. B. Jacobs. 1997. Exact transmission-disequilibrium tests with multiallelic markers. *Genetic Epidemiology* 14: 337–347.

Cui, J. 2000. sg150: Hardy–Weinberg equilibrium test in case–control studies. *Stata Technical Bulletin* 57: 17–19. Reprinted in *Stata Technical Bulletin Reprints*, vol. 10, pp. 218–220. College Station, TX: Stata Press.

Good, P. I. 2005. *Permutation, Parametric, and Bootstrap Tests of Hypotheses: A Practical Guide to Resampling Methods for Testing Hypotheses.* 3rd ed. New York: Springer.

———. 2006. *Resampling Methods: A Practical Guide to Data Analysis.* 3rd ed. Boston: Birkhäuser.

Mack, T. M., M. C. Pike, B. E. Henderson, R. I. Pfeffer, V. R. Gerkins, M. Arthur, and S. E. Brown. 1976. Estrogens and endometrial cancer in a retirement community. *New England Journal of Medicine* 294: 1262–1267.

Mander, A. 2000. sbe38: Haplotype frequency estimation using an EM algorithm and log-linear modeling. *Stata Technical Bulletin* 57: 5–7. Reprinted in *Stata Technical Bulletin Reprints*, vol. 10, pp. 104–107. College Station, TX: Stata Press.

Maxwell, A. E. 1970. Comparing the classification of subjects by two independent judges. *British Journal of Psychiatry* 116: 651–655.

Spieldman, R. S., and W. J. Ewens. 1996. The TDT and other family-based tests for linkage disequilibrium and association. *American Journal of Human Genetics* 59: 983–989.

Spieldman, R. S., R. E. McGinnis, and W. J. Ewens. 1993. Transmission test for linkage disequilibrium: The insulin gene region and insulin-dependence diabetes mellitus (IDDM). *American Journal of Human Genetics* 52: 506–516.

Stuart, A. 1955. A test for homogeneity of the marginal distributions in a two-way classification. *Biometrika* 42: 412–416.

Also see

[ST] **epitab** — Tables for epidemiologists

Title

> **table** — Tables of summary statistics

Syntax

table *rowvar* [*colvar* [*supercolvar*]] [*if*] [*in*] [*weight*] [, *options*]

options	Description
Main	
<u>c</u>ontents(*clist*)	contents of table cells; select up to five statistics; default is contents(freq)
by(*superrowvarlist*)	superrow variables
Options	
<u>cell</u>width(*#*)	cell width
<u>c</u>sepwidth(*#*)	column-separation width
<u>stub</u>width(*#*)	stub width
<u>scsep</u>width(*#*)	supercolumn-separation width
<u>center</u>	center-align table cells; default is right-align
<u>left</u>	left-align table cells; default is right-align
cw	perform casewise deletion
row	add row totals
col	add column totals
<u>s</u>col	add supercolumn totals
<u>conc</u>ise	suppress rows with all missing entries
<u>m</u>issing	show missing statistics with period
replace	replace current data with table statistics
<u>name</u>(*string*)	name new variables with prefix *string*
<u>f</u>ormat(%*fmt*)	display format for numbers in cells; default is format(%9.0g)

by is allowed; see [D] **by**.

fweights, aweights, iweights, and pweights are allowed; see [U] **11.1.6 weight**.

pweights may not be used with sd, semean, sebinomial, or sepoisson. iweights may not be used with semean, sebinomial, or sepoisson. aweights may not be used with sebinomial or sepoisson.

where the elements of *clist* may be

freq	frequency	n *varname*	same as count
<u>mean</u> *varname*	mean of *varname*	max *varname*	maximum
sd *varname*	standard deviation	min *varname*	minimum
<u>seme</u>an *varname*	standard error of the mean (sd/sqrt(n))	<u>med</u>ian *varname*	median
		p1 *varname*	1st percentile
<u>sebin</u>omial *varname*	standard error of the mean, binomial distribution (sqrt(p(1-p)/n))	p2 *varname*	2nd percentile
		. . .	3rd–49th percentiles
<u>sepoi</u>sson *varname*	standard error of the mean, Poisson distribution (sqrt(mean))	p50 *varname*	50th percentile (median)
		. . .	51st–97th percentiles
sum *varname*	sum	p98 *varname*	98th percentile
<u>raw</u>sum *varname*	sums ignoring optionally specified weight	p99 *varname*	99th percentile
count *varname*	count of nonmissing observations	iqr *varname*	interquartile range

Rows, columns, supercolumns, and superrows are thus defined as

row 1	.
row 2	.

	supercol 1		supercol 2	
	col 1	col 2	col 1	col 2
row 1	.	.	.	.
row 2	.	.	.	.

	col 1	col 2
row 1	.	.
row 2	.	.

	supercol 1		supercol 2	
	col 1	col 2	col 1	col 2
superrow 1:				
row 1	.	.	.	.
row 2	.	.	.	.
superrow 2:				
row 1	.	.	.	.
row 2	.	.	.	.

Menu

Statistics > Summaries, tables, and tests > Tables > Table of summary statistics (table)

Description

table calculates and displays tables of statistics.

Options

⌐ Main ⌐

contents(*clist*) specifies the contents of the table's cells; if not specified, contents(freq) is used by default. contents(freq) produces a table of frequencies. contents(mean mpg) produces a table of the means of variable mpg. contents(freq mean mpg sd mpg) produces a table of frequencies together with the mean and standard deviation of variable mpg. Up to five statistics may be specified.

by(*superrowvarlist*) specifies that numeric or string variables be treated as superrows. Up to four variables may be specified in *superrowvarlist*. The by() option may be specified with the by prefix.

⌐ Options ⌐

cellwidth(*#*) specifies the width of the cell in units of digit widths; 10 means the space occupied by 10 digits, which is 0123456789. The default cellwidth() is not a fixed number, but a number chosen by table to spread the table out while presenting a reasonable number of columns across the page.

csepwidth(#) specifies the separation between columns in units of digit widths. The default is not a fixed number, but a number chosen by table according to what it thinks looks best.

stubwidth(#) specifies the width, in units of digit widths, to be allocated to the left stub of the table. The default is not a fixed number, but a number chosen by table according to what it thinks looks best.

scsepwidth(#) specifies the separation between supercolumns in units of digit widths. The default is not a fixed number, but a number chosen by table to present the results best.

center specifies that results be centered in the table's cells. The default is to right-align results. For centering to work well, you typically need to specify a display format as well. center format(%9.2f) is popular.

left specifies that column labels be left-aligned. The default is to right-align column labels to distinguish them from supercolumn labels, which are left-aligned.

cw specifies casewise deletion. If cw is not specified, all observations possible are used to calculate each of the specified statistics. cw is relevant only when you request a table containing statistics on multiple variables. For instance, contents(mean mpg mean weight) would produce a table reporting the means of variables mpg and weight. Consider an observation in which mpg is known but weight is missing. By default, that observation will be used in the calculation of the mean of mpg. If you specify cw, the observation will be excluded in the calculation of the means of both mpg and weight.

row specifies that a row be added to the table reflecting the total across the rows.

col specifies that a column be added to the table reflecting the total across columns.

scol specifies that a supercolumn be added to the table reflecting the total across supercolumns.

concise specifies that rows with all missing entries not be displayed.

missing specifies that missing statistics be shown in the table as periods (Stata's missing-value indicator). The default is that missing entries be left blank.

replace specifies that the data in memory be replaced with data containing 1 observation per cell (row, column, supercolumn, and superrow) and with variables containing the statistics designated in contents().

This option is rarely specified. If you do not specify this option, the data in memory remain unchanged.

If you do specify this option, the first statistic will be named table1, the second table2, and so on. For instance, if contents(mean mpg sd mpg) was specified, the means of mpg would be in variable table1 and the standard deviations in table2.

name(string) is relevant only if you specify replace. name() allows changing the default stub name that replace uses to name the new variables associated with the statistics. If you specify name(stat), the first statistic will be placed in variable stat1, the second in stat2, and so on.

format(%fmt) specifies the display format for presenting numbers in the table's cells. format(%9.0g) is the default; format(%9.2f) and format(%9.2fc) are popular alternatives. The width of the format you specify does not matter, except that %fmt must be valid. The width of the cells is chosen by table to present the results best. The cellwidth() option allows you to override table's choice.

Limits

Up to four variables may be specified in the by(), so with the three row, column, and supercolumn variables, seven-way tables may be displayed.

Up to five statistics may be displayed in each cell of the table.

The sum of the number of rows, columns, supercolumns, and superrows is called the number of margins. A table may contain up to 3,000 margins. Thus a one-way table may contain 3,000 rows. A two-way table could contain 2,998 rows and two columns, 2,997 rows and three columns, ..., 1,500 rows and 1,500 columns, ..., two rows and 2,998 columns. A three-way table is similarly limited by the sum of the number of rows, columns, and supercolumns. A $r \times c \times d$ table is feasible if $r + c + d \leq 3{,}000$. The limit is set in terms of the sum of the rows, columns, supercolumns, and superrows, and not, as you might expect, in terms of their product.

Remarks

Remarks are presented under the following headings:

One-way tables
Two-way tables
Three-way tables
Four-way and higher-dimensional tables

One-way tables

▷ Example 1

From the automobile dataset, here is a simple one-way table:

```
. use http://www.stata-press.com/data/r12/auto
(1978 Automobile Data)
. table rep78, contents(mean mpg)
```

Repair Record 1978	mean(mpg)
1	21
2	19.125
3	19.4333
4	21.6667
5	27.3636

We are not limited to including only one statistic:

```
. table rep78, c(n mpg  mean mpg  sd mpg  median mpg)
```

Repair Record 1978	N(mpg)	mean(mpg)	sd(mpg)	med(mpg)
1	2	21	4.24264	21
2	8	19.125	3.758324	18
3	30	19.4333	4.141325	19
4	18	21.6667	4.93487	22.5
5	11	27.3636	8.732385	30

We abbreviated contents() as c(). The format() option will allow us to better format the numbers in the table:

```
. table rep78, c(n mpg  mean mpg  sd mpg  median mpg) format(%9.2f)
```

Repair Record 1978	N(mpg)	mean(mpg)	sd(mpg)	med(mpg)
1	2	21.00	4.24	21.00
2	8	19.12	3.76	18.00
3	30	19.43	4.14	19.00
4	18	21.67	4.93	22.50
5	11	27.36	8.73	30.00

The center option will center the results under the headings:

```
. table rep78, c(n mpg  mean mpg  sd mpg  median mpg) format(%9.2f) center
```

Repair Record 1978	N(mpg)	mean(mpg)	sd(mpg)	med(mpg)
1	2	21.00	4.24	21.00
2	8	19.12	3.76	18.00
3	30	19.43	4.14	19.00
4	18	21.67	4.93	22.50
5	11	27.36	8.73	30.00

◁

Two-way tables

▷ Example 2

In example 1, when we typed 'table rep78, ...', we obtained a one-way table. If we were to type 'table rep78 foreign, ...', we would obtain a two-way table:

```
. table rep78 foreign, c(mean mpg)
```

Repair Record 1978	Car type Domestic	Foreign
1	21	
2	19.125	
3	19	23.3333
4	18.4444	24.8889
5	32	26.3333

Note the missing cells. Certain combinations of repair record and car type do not exist in our dataset.

As with one-way tables, we can specify a display format for the cells and center the numbers within the cells if we wish.

```
. table rep78 foreign, c(mean mpg) format(%9.2f) center
```

Repair Record 1978	Car type Domestic	Foreign
1	21.00	
2	19.12	
3	19.00	23.33
4	18.44	24.89
5	32.00	26.33

We can obtain row totals by specifying the row option and obtain column totals by specifying the col option. We specify both below:

```
. table rep78 foreign, c(mean mpg) format(%9.2f) center row col
```

Repair Record 1978	Car type Domestic	Foreign	Total
1	21.00		21.00
2	19.12		19.12
3	19.00	23.33	19.43
4	18.44	24.89	21.67
5	32.00	26.33	27.36
Total	19.54	25.29	21.29

table can display multiple statistics within cells, but once we move beyond one-way tables, the table becomes busy:

```
. table foreign rep78, c(mean mpg  n mpg) format(%9.2f) center
```

Car type	Repair Record 1978 1	2	3	4	5
Domestic	21.00 2	19.12 8	19.00 27	18.44 9	32.00 2
Foreign			23.33 3	24.89 9	26.33 9

This two-way table with two statistics per cell works well here. That was, in part, helped along by our interchanging the rows and columns. We turned the table around by typing table foreign rep78 rather than table rep78 foreign.

Another way to display two-way tables is to specify a row and superrow rather than a row and column. We do that below and display three statistics per cell:

```
. table foreign, by(rep78) c(mean mpg  sd mpg  n mpg) format(%9.2f) center
```

Repair Record 1978 and Car type	mean(mpg)	sd(mpg)	N(mpg)
1 Domestic Foreign	21.00	4.24	2
2 Domestic Foreign	19.12	3.76	8
3 Domestic Foreign	19.00 23.33	4.09 2.52	27 3
4 Domestic Foreign	18.44 24.89	4.59 2.71	9 9
5 Domestic Foreign	32.00 26.33	2.83 9.37	2 9

◁

Three-way tables

▷ Example 3

We have data on the prevalence of byssinosis, a form of pneumoconiosis to which workers exposed to cotton dust are susceptible. The dataset is on 5,419 workers in a large cotton mill. We know whether each worker smokes, his or her race, and the dustiness of the work area. The categorical variables are

smokes	Smoker or nonsmoker in the last five years.
race	White or other.
workplace	1 (most dusty), 2 (less dusty), 3 (least dusty).

Moreover, this dataset includes a frequency-weight variable pop. Here is a three-way table showing the fraction of workers with byssinosis:

```
. use http://www.stata-press.com/data/r12/byssin
(Byssinosis incidence)
. table workplace smokes race [fw=pop], c(mean prob)
```

Dustiness of workplace	Race and Smokes			
	— other —		— white —	
	no	yes	no	yes
least	.0107527	.0101523	.0081549	.0162774
less	.02	.0081633	.0136612	.0143149
most	.0820896	.1679105	.0833333	.2295082

This table would look better if we showed the fraction to four digits:

```
. table workplace smokes race [fw=pop], c(mean prob) format(%9.4f)
```

Dustiness of workplace	Race and Smokes			
	── other ──		── white ──	
	no	yes	no	yes
least	0.0108	0.0102	0.0082	0.0163
less	0.0200	0.0082	0.0137	0.0143
most	0.0821	0.1679	0.0833	0.2295

In this table, the rows are the dustiness of the workplace, the columns are whether the worker smokes, and the supercolumns are the worker's race.

Now we request that the table include the supercolumn totals by specifying the sctotal option, which we can abbreviate as sc:

```
. table workplace smokes race [fw=pop], c(mean prob) format(%9.4f) sc
```

Dustiness of workplace	Race and Smokes					
	── other ──		── white ──		── Total ──	
	no	yes	no	yes	no	yes
least	0.0108	0.0102	0.0082	0.0163	0.0090	0.0145
less	0.0200	0.0082	0.0137	0.0143	0.0159	0.0123
most	0.0821	0.1679	0.0833	0.2295	0.0826	0.1929

The supercolumn total is the total over race and is divided into its columns based on smokes. Here is the table with the column rather than the supercolumn totals:

```
. table workplace smokes race [fw=pop], c(mean prob) format(%9.4f) col
```

Dustiness of workplace	Race and Smokes					
	──── other ────			──── white ────		
	no	yes	Total	no	yes	Total
least	0.0108	0.0102	0.0104	0.0082	0.0163	0.0129
less	0.0200	0.0082	0.0135	0.0137	0.0143	0.0140
most	0.0821	0.1679	0.1393	0.0833	0.2295	0.1835

Here is the table with both column and supercolumn totals:

```
. table workplace smokes race [fw=pop], c(mean prob) format(%9.4f) sc col
```

Dustiness of workplace	Race and Smokes								
	──── other ────			──── white ────			──── Total ────		
	no	yes	Total	no	yes	Total	no	yes	Total
least	0.0108	0.0102	0.0104	0.0082	0.0163	0.0129	0.0090	0.0145	0.0122
less	0.0200	0.0082	0.0135	0.0137	0.0143	0.0140	0.0159	0.0123	0.0138
most	0.0821	0.1679	0.1393	0.0833	0.2295	0.1835	0.0826	0.1929	0.1570

table is struggling to keep this table from becoming too wide—notice how it divided the words in the title in the top-left stub. Here, if the table had more columns, or, if we demanded more digits, table would be forced to segment the table and present it in pieces, which it would do:

```
. table workplace smokes race [fw=pop], c(mean prob) format(%9.6f) sc col
```

Dustiness of workplace	Race and Smokes					
	other			white		
	no	yes	Total	no	yes	Total
least	0.010753	0.010152	0.010417	0.008155	0.016277	0.012949
less	0.020000	0.008163	0.013483	0.013661	0.014315	0.014035
most	0.082090	0.167910	0.139303	0.083333	0.229508	0.183521

Dustiness of workplace	Race and Smokes		
	Total		
	no	yes	Total
least	0.008990	0.014471	0.012174
less	0.015901	0.012262	0.013846
most	0.082569	0.192905	0.156951

Here three digits is probably enough, so here is the table including all the row, column, and supercolumn totals:

```
. table workplace smokes race [fw=pop], c(mean prob) format(%9.3f)
> sc col row
```

Dustiness of workplace	Race and Smokes								
	other			white			Total		
	no	yes	Total	no	yes	Total	no	yes	Total
least	0.011	0.010	0.010	0.008	0.016	0.013	0.009	0.014	0.012
less	0.020	0.008	0.013	0.014	0.014	0.014	0.016	0.012	0.014
most	0.082	0.168	0.139	0.083	0.230	0.184	0.083	0.193	0.157
Total	0.025	0.048	0.038	0.014	0.035	0.026	0.018	0.039	0.030

We can show multiple statistics:

```
. table workplace smokes race [fw=pop], c(mean prob  n prob) format(%9.3f)
> sc col row
```

Dustiness of workplace	Race and Smokes								
	other			white			Total		
	no	yes	Total	no	yes	Total	no	yes	Total
least	0.011	0.010	0.010	0.008	0.016	0.013	0.009	0.014	0.012
	465	591	1,056	981	1,413	2,394	1,446	2,004	3,450
less	0.020	0.008	0.013	0.014	0.014	0.014	0.016	0.012	0.014
	200	245	445	366	489	855	566	734	1,300
most	0.082	0.168	0.139	0.083	0.230	0.184	0.083	0.193	0.157
	134	268	402	84	183	267	218	451	669
Total	0.025	0.048	0.038	0.014	0.035	0.026	0.018	0.039	0.030
	799	1,104	1,903	1,431	2,085	3,516	2,230	3,189	5,419

◁

Four-way and higher-dimensional tables

▷ Example 4

Let's pretend that our byssinosis dataset also recorded each worker's sex (it does not, and we have made up this extra information). We obtain a four-way table just as we would a three-way table, but we specify the fourth variable as a superrow by including it in the by() option:

```
. use http://www.stata-press.com/data/r12/byssin1
(Byssinosis incidence)
. table workplace smokes race [fw=pop], by(sex) c(mean prob) format(%9.3f)
> sc col row
```

Sex and Dustiness of workplace	Race and Smokes								
	——— other ———			——— white ———			——— Total ———		
	no	yes	Total	no	yes	Total	no	yes	Total
Female									
least	0.006	0.009	0.008	0.009	0.021	0.016	0.009	0.018	0.014
less	0.020	0.008	0.010	0.015	0.015	0.015	0.016	0.012	0.014
most	0.057	0.154	0.141				0.057	0.154	0.141
Total	0.017	0.051	0.043	0.011	0.020	0.016	0.012	0.032	0.024
Male									
least	0.013	0.011	0.012	0.006	0.007	0.006	0.009	0.008	0.009
less	0.020	0.000	0.019	0.000	0.013	0.011	0.016	0.013	0.014
most	0.091	0.244	0.136	0.083	0.230	0.184	0.087	0.232	0.167
Total	0.029	0.041	0.033	0.020	0.056	0.043	0.025	0.052	0.039

If our dataset also included work group and we wanted a five-way table, we could include both the sex and work-group variables in the by() option. You may include up to four variables in by(), and so produce up to 7-way tables.

◁

Methods and formulas

table is implemented as an ado-file. The contents of cells are calculated by collapse and are displayed by tabdisp; see [D] **collapse** and [P] **tabdisp**.

Also see

[R] **summarize** — Summary statistics

[R] **tabstat** — Display table of summary statistics

[R] **tabulate oneway** — One-way tables of frequencies

[R] **tabulate twoway** — Two-way tables of frequencies

[D] **collapse** — Make dataset of summary statistics

[P] **tabdisp** — Display tables

Title

tabstat — Display table of summary statistics

Syntax

tabstat *varlist* $\left[\textit{if}\right]$ $\left[\textit{in}\right]$ $\left[\textit{weight}\right]$ $\left[\,,\ \textit{options}\right]$

options	Description
Main	
by(*varname*)	group statistics by variable
<u>s</u>tatistics(*statname* $[\dots]$)	report specified statistics
Options	
<u>l</u>abelwidth(*#*)	width for by() variable labels; default is labelwidth(16)
<u>v</u>arwidth(*#*)	variable width; default is varwidth(12)
<u>c</u>olumns(<u>v</u>ariables)	display variables in table columns; the default
<u>c</u>olumns(<u>s</u>tatistics)	display statistics in table columns
<u>f</u>ormat $\left[(\%\textit{fmt})\right]$	display format for statistics; default format is %9.0g
<u>case</u>wise	perform casewise deletion of observations
<u>not</u>otal	do not report overall statistics; use with by()
<u>m</u>issing	report statistics for missing values of by() variable
<u>nose</u>parator	do not use separator line between by() categories
<u>long</u>stub	make left table stub wider
save	save summary statistics in r()

by is allowed; see [D] **by**.

aweights and fweights are allowed; see [U] **11.1.6 weight**.

Menu

Statistics > Summaries, tables, and tests > Tables > Table of summary statistics (tabstat)

Description

tabstat displays summary statistics for a series of numeric variables in one table, possibly broken down on (conditioned by) another variable.

Without the by() option, tabstat is a useful alternative to summarize (see [R] **summarize**) because it allows you to specify the list of statistics to be displayed.

With the by() option, tabstat resembles tabulate used with its summarize() option in that both report statistics of *varlist* for the different values of *varname*. tabstat allows more flexibility in terms of the statistics presented and the format of the table.

tabstat is sensitive to the linesize (see set linesize in [R] **log**); it widens the table if possible and wraps if necessary.

Options

by(*varname*) specifies that the statistics be displayed separately for each unique value of *varname*; *varname* may be numeric or string. For instance, tabstat height would present the overall mean of height. tabstat height, by(sex) would present the mean height of males, and of females, and the overall mean height. Do not confuse the by() option with the by prefix (see [D] **by**); both may be specified.

statistics(*statname* [...]) specifies the statistics to be displayed; the default is equivalent to specifying statistics(mean). (stats() is a synonym for statistics().) Multiple statistics may be specified and are separated by white space, such as statistics(mean sd). Available statistics are

statname	Definition	*statname*	Definition
mean	mean	p1	1st percentile
count	count of nonmissing observations	p5	5th percentile
n	same as count	p10	10th percentile
sum	sum	p25	25th percentile
max	maximum	median	median (same as p50)
min	minimum	p50	50th percentile (same as median)
range	range = max − min	p75	75th percentile
sd	standard deviation	p90	90th percentile
variance	variance	p95	95th percentile
cv	coefficient of variation (sd/mean)	p99	99th percentile
semean	standard error of mean (sd/$\sqrt{n}$)	iqr	interquartile range = p75 − p25
skewness	skewness	q	equivalent to specifying p25 p50 p75
kurtosis	kurtosis		

labelwidth(#) specifies the maximum width to be used within the stub to display the labels of the by() variable. The default is labelwidth(16). 8 ≤ # ≤ 32.

varwidth(#) specifies the maximum width to be used within the stub to display the names of the variables. The default is varwidth(12). varwidth() is effective only with columns(statistics). Setting varwidth() implies longstub. 8 ≤ # ≤ 16.

columns(variables | statistics) specifies whether to display variables or statistics in the columns of the table. columns(variables) is the default when more than one variable is specified.

format and format(%*fmt*) specify how the statistics are to be formatted. The default is to use a %9.0g format.

format specifies that each variable's statistics be formatted with the variable's display format; see [D] **format**.

format(%*fmt*) specifies the format to be used for all statistics. The maximum width of the specified format should not exceed nine characters.

casewise specifies casewise deletion of observations. Statistics are to be computed for the sample that is not missing for any of the variables in *varlist*. The default is to use all the nonmissing values for each variable.

nototal is for use with by(); it specifies that the overall statistics not be reported.

missing specifies that missing values of the by() variable be treated just like any other value and that statistics should be displayed for them. The default is not to report the statistics for the by()== *missing* group. If the by() variable is a string variable, by()=="" is considered to mean missing.

noseparator specifies that a separator line between the by() categories not be displayed.

longstub specifies that the left stub of the table be made wider so that it can include names of the statistics or variables in addition to the categories of by(*varname*). The default is to describe the statistics or variables in a header. longstub is ignored if by(*varname*) is not specified.

save specifies that the summary statistics be returned in r(). The overall (unconditional) statistics are returned in matrix r(StatTotal) (rows are statistics, columns are variables). The conditional statistics are returned in the matrices r(Stat1), r(Stat2), ..., and the names of the corresponding variables are returned in the macros r(name1), r(name2), ...

Remarks

This command is probably most easily understood by going through a series of examples.

▷ Example 1

We have data on the price, weight, mileage rating, and repair record of 22 foreign and 52 domestic 1978 automobiles. We want to summarize these variables for the different origins of the automobiles.

```
. use http://www.stata-press.com/data/r12/auto
(1978 Automobile Data)
. tabstat price weight mpg rep78, by(foreign)
Summary statistics: mean
  by categories of: foreign (Car type)
```

foreign	price	weight	mpg	rep78
Domestic	6072.423	3317.115	19.82692	3.020833
Foreign	6384.682	2315.909	24.77273	4.285714
Total	6165.257	3019.459	21.2973	3.405797

More summary statistics can be requested via the statistics() option. The group totals can be suppressed with the nototal option.

```
. tabstat price weight mpg rep78, by(foreign) stat(mean sd min max) nototal
Summary statistics: mean, sd, min, max
  by categories of: foreign (Car type)
```

foreign	price	weight	mpg	rep78
Domestic	6072.423	3317.115	19.82692	3.020833
	3097.104	695.3637	4.743297	.837666
	3291	1800	12	1
	15906	4840	34	5
Foreign	6384.682	2315.909	24.77273	4.285714
	2621.915	433.0035	6.611187	.7171372
	3748	1760	14	3
	12990	3420	41	5

Although the header of the table describes the statistics running vertically in the "cells", the table may become hard to read, especially with many variables or statistics. The `longstub` option specifies that a column be added describing the contents of the cells. The `format` option can be issued to specify that `tabstat` display the statistics by using the display format of the variables rather than the overall default `%9.0g`.

```
. tabstat price weight mpg rep78, by(foreign) stat(mean sd min max) long format
```

foreign	stats	price	weight	mpg	rep78
Domestic	mean	6,072.4	3,317.1	19.8269	3.02083
	sd	3,097.1	695.364	4.7433	.837666
	min	3,291	1,800	12	1
	max	15,906	4,840	34	5
Foreign	mean	6,384.7	2,315.9	24.7727	4.28571
	sd	2,621.9	433.003	6.61119	.717137
	min	3,748	1,760	14	3
	max	12,990	3,420	41	5
Total	mean	6,165.3	3,019.5	21.2973	3.4058
	sd	2,949.5	777.194	5.7855	.989932
	min	3,291	1,760	12	1
	max	15,906	4,840	41	5

We can specify a layout of the table in which the statistics run horizontally and the variables run vertically by specifying the `col(statistics)` option.

```
. tabstat price weight mpg rep78, by(foreign) stat(min mean max) col(stat) long
```

foreign	variable	min	mean	max
Domestic	price	3291	6072.423	15906
	weight	1800	3317.115	4840
	mpg	12	19.82692	34
	rep78	1	3.020833	5
Foreign	price	3748	6384.682	12990
	weight	1760	2315.909	3420
	mpg	14	24.77273	41
	rep78	3	4.285714	5
Total	price	3291	6165.257	15906
	weight	1760	3019.459	4840
	mpg	12	21.2973	41
	rep78	1	3.405797	5

Finally, `tabstat` can also be used to enhance `summarize` so we can specify the statistics to be displayed. For instance, we can display the number of observations, the mean, the coefficient of variation, and the 25%, 50%, and 75% quantiles for a list of variables.

```
. tabstat price weight mpg rep78, stat(n mean cv q) col(stat)
```

variable	N	mean	cv	p25	p50	p75
price	74	6165.257	.478406	4195	5006.5	6342
weight	74	3019.459	.2573949	2240	3190	3600
mpg	74	21.2973	.2716543	18	20	25
rep78	69	3.405797	.290661	3	3	4

Because we did not specify the by() option, these statistics were not displayed for the subgroups of the data formed by the categories of the by() variable.   ◁

Methods and formulas

tabstat is implemented as an ado-file.

Acknowledgments

The tabstat command was written by Jeroen Weesie and Vincent Buskens of Utrecht University in The Netherlands.

Also see

[R] **summarize** — Summary statistics

[R] **table** — Tables of summary statistics

[R] **tabulate, summarize()** — One- and two-way tables of summary statistics

[D] **collapse** — Make dataset of summary statistics

Title

> **tabulate oneway** — One-way tables of frequencies

Syntax

One-way tables

> ta̲bulate *varname* $\begin{bmatrix} if \end{bmatrix}$ $\begin{bmatrix} in \end{bmatrix}$ $\begin{bmatrix} weight \end{bmatrix}$ $\begin{bmatrix} , & tabulate1\_options \end{bmatrix}$

One-way table for each variable—a convenience tool

> tab1 *varlist* $\begin{bmatrix} if \end{bmatrix}$ $\begin{bmatrix} in \end{bmatrix}$ $\begin{bmatrix} weight \end{bmatrix}$ $\begin{bmatrix} , & tab1\_options \end{bmatrix}$

tabulate1_options	Description
Main	
subpop(*varname*)	exclude observations for which *varname* = 0
mi̲ssing	treat missing values like other values
no̲freq	do not display frequencies
no̲label	display numeric codes rather than value labels
p̲lot	produce a bar chart of the relative frequencies
sort	display the table in descending order of frequency
Advanced	
generate(*stubname*)	create indicator variables for *stubname*
matcell(*matname*)	save frequencies in *matname*; programmer's option
matrow(*matname*)	save unique values of *varname* in *matname*; programmer's option

tab1_options	Description
Main	
subpop(*varname*)	exclude observations for which *varname* = 0
mi̲ssing	treat missing values like other values
no̲freq	do not display frequencies
no̲label	display numeric codes rather than value labels
p̲lot	produce a bar chart of the relative frequencies
sort	display the table in descending order of frequency

by is allowed with tabulate and tab1; see [D] **by**.

fweights, aweights, and iweights are allowed by tabulate. fweights are allowed by tab1. See
[U] **11.1.6 weight**.

Menu

tabulate oneway

Statistics > Summaries, tables, and tests > Tables > One-way tables

tabulate ..., generate()

Data > Create or change data > Other variable-creation commands > Create indicator variables

tab1

Statistics > Summaries, tables, and tests > Tables > Multiple one-way tables

Description

tabulate produces one-way tables of frequency counts.

For information about two-way tables of frequency counts along with various measures of association, including the common Pearson χ^2, the likelihood-ratio χ^2, Cramér's V, Fisher's exact test, Goodman and Kruskal's gamma, and Kendall's τ_b, see [R] **tabulate twoway**.

tab1 produces a one-way tabulation for each variable specified in *varlist*.

Also see [R] **table** and [R] **tabstat** if you want one-, two-, or n-way tables of frequencies and a wide variety of summary statistics. See [R] **tabulate, summarize()** for a description of tabulate with the summarize() option; it produces tables (breakdowns) of means and standard deviations. table is better than tabulate, summarize(), but tabulate, summarize() is faster. See [ST] **epitab** for 2×2 tables with statistics of interest to epidemiologists.

Options

    [Main]

subpop(*varname*) excludes observations for which *varname* $= 0$ in tabulating frequencies. The mathematical results of tabulate ..., subpop(myvar) are the same as tabulate ... if myvar !=0, but the table may be presented differently. The identities of the rows and columns will be determined from all the data, including the myvar $= 0$ group, so there may be entries in the table with frequency 0.

Consider tabulating answer, a variable that takes on values 1, 2, and 3, but consider tabulating it just for the male==1 subpopulation. Assume that answer is never 2 in this group. tabulate answer if male==1 produces a table with two rows: one for answer 1 and one for answer 3. There will be no row for answer 2 because answer 2 was never observed. tabulate answer, subpop(male) produces a table with three rows. The row for answer 2 will be shown as having 0 frequency.

missing requests that missing values be treated like other values in calculations of counts, percentages, and other statistics.

nofreq suppresses the printing of the frequencies.

nolabel causes the numeric codes to be displayed rather than the value labels.

plot produces a bar chart of the relative frequencies in a one-way table. (Also see [R] **histogram**.)

sort puts the table in descending order of frequency (and ascending order of the variable within equal values of frequency).

┌─ Advanced ┐

generate(*stubname*) creates a set of indicator variables (*stubname*1, *stubname*2, ...) reflecting the observed values of the tabulated variable. The generate() option may not be used with the by prefix.

matcell(*matname*) saves the reported frequencies in *matname*. This option is for use by programmers.

matrow(*matname*) saves the numeric values of the $r \times 1$ row stub in *matname*. This option is for use by programmers. matrow() may not be specified if the row variable is a string.

Limits

One-way tables may have a maximum of 12,000 rows (Stata/MP and Stata/SE), 3,000 rows (Stata/IC), or 500 rows (Small Stata).

Remarks

Remarks are presented under the following headings:

> *tabulate*
> *tab1*

For each value of a specified variable, tabulate reports the number of observations with that value. The number of times a value occurs is called its *frequency*.

tabulate

▷ Example 1

We have data summarizing the speed limit and the accident rate per million vehicle miles along various Minnesota highways in 1973. The variable containing the speed limit is called spdlimit. If we summarize the variable, we obtain its mean and standard deviation:

```
. use http://www.stata-press.com/data/r12/hiway
(Minnesota Highway Data, 1973)

. summarize spdlimit
```

Variable	Obs	Mean	Std. Dev.	Min	Max
spdlimit	39	55	5.848977	40	70

The average speed limit is 55 miles per hour. We can learn more about this variable by tabulating it:

```
. tabulate spdlimit
```

Speed limit	Freq.	Percent	Cum.
40	1	2.56	2.56
45	3	7.69	10.26
50	7	17.95	28.21
55	15	38.46	66.67
60	11	28.21	94.87
65	1	2.56	97.44
70	1	2.56	100.00
Total	39	100.00	

We see that one highway has a speed limit of 40 miles per hour, three have speed limits of 45, 7 of 50, and so on. The column labeled Percent shows the percentage of highways in the dataset that have the indicated speed limit. For instance, 38.46% of highways in our dataset have a speed limit of 55 miles per hour. The final column shows the cumulative percentage. We see that 66.67% of highways in our dataset have a speed limit of 55 miles per hour or less.

◁

▷ Example 2

The plot option places a sideways histogram alongside the table:

```
. tabulate spdlimit, plot
Speed limit |      Freq.
         40 |          1  |*
         45 |          3  |***
         50 |          7  |*******
         55 |         15  |***************
         60 |         11  |***********
         65 |          1  |*
         70 |          1  |*
      Total |         39
```

Of course, graph can produce better-looking histograms; see [R] **histogram**.

◁

▷ Example 3

tabulate labels tables using *variable* and *value labels* if they exist. To demonstrate how this works, let's add a new variable to our dataset that categorizes spdlimit into three categories. We will call this new variable spdcat:

```
. generate spdcat=recode(spdlimit,50,60,70)
```

The recode() function divides spdlimit into 50 miles per hour or below, 51–60, and above 60; see [D] **functions**. We specified the breakpoints in the arguments (spdlimit,50,60,70). The first argument is the variable to be recoded. The second argument is the first breakpoint, the third argument is the second breakpoint, and so on. We can specify as many breakpoints as we wish.

recode() used our arguments not only as the breakpoints but also to label the results. If spdlimit is less than or equal to 50, spdcat is set to 50; if spdlimit is between 51 and 60, spdcat is set to 60; otherwise, spdcat is arbitrarily set to 70. (See [U] **25 Working with categorical data and factor variables**.)

Because we just created the variable spdcat, it is not yet labeled. When we make a table using this variable, tabulate uses the variable's name to label it:

```
. tabulate spdcat
     spdcat |      Freq.     Percent        Cum.
         50 |         11       28.21       28.21
         60 |         26       66.67       94.87
         70 |          2        5.13      100.00
      Total |         39      100.00
```

Even through the table is not well labeled, recode()'s coding scheme provides us with clues as to the table's meaning. The first line of the table corresponds to 50 miles per hour and below, the next to 51 through 60 miles per hour, and the last to above 60 miles per hour.

We can improve this table by labeling the values and variables:

```
. label define scat 50 "40 to 50" 60 "55 to 60" 70 "Above 60"
. label values spdcat scat
. label variable spdcat "Speed Limit Category"
```

We define a *value label* called scat that attaches labels to the numbers 50, 60, and 70 using the label define command; see [U] **12.6.3 Value labels**. We label the value 50 as '40 to 50', because we looked back at our original tabulation in the first example and saw that the speed limit was never less than 40. Similarly, we could have labeled the last category '65 to 70' because the speed limit is never greater than 70 miles per hour.

Next we requested that Stata label the values of the new variable spdcat using the value label scat. Finally, we labeled our variable Speed Limit Category. We are now ready to tabulate the result:

```
. tabulate spdcat

Speed Limit |
   Category |      Freq.     Percent        Cum.
------------+-----------------------------------
   40 to 50 |         11       28.21       28.21
   55 to 60 |         26       66.67       94.87
   Above 60 |          2        5.13      100.00
------------+-----------------------------------
      Total |         39      100.00
```

◁

▷ Example 4

If we have missing values in our dataset, tabulate ignores them unless we explicitly indicate otherwise. We have no missing data in our example, so let's add some:

```
. replace spdcat=. in 39
(1 real change made, 1 to missing)
```

We changed the first observation on spdcat to *missing*. Let's now tabulate the result:

```
. tabulate spdcat

Speed Limit |
   Category |      Freq.     Percent        Cum.
------------+-----------------------------------
   40 to 50 |         11       28.95       28.95
   55 to 60 |         26       68.42       97.37
   Above 60 |          1        2.63      100.00
------------+-----------------------------------
      Total |         38      100.00
```

Comparing this output with that in the previous example, we see that the total frequency count is now one less than it was—38 rather than 39. Also, the 'Above 60' category now has only one observation where it used to have two, so we evidently changed a road with a high speed limit.

We want tabulate to treat missing values just as it treats numbers, so we specify the missing option:

```
. tabulate spdcat, missing
```

Speed Limit Category	Freq.	Percent	Cum.
40 to 50	11	28.21	28.21
55 to 60	26	66.67	94.87
Above 60	1	2.56	97.44
.	1	2.56	100.00
Total	39	100.00	

We now see our missing value—the last category, labeled '.', shows a frequency count of 1. The table sum is once again 39.

Let's put our dataset back as it was originally:

```
. replace spdcat=70 in 39
(1 real change made)
```

◁

❏ Technical note

tabulate also can automatically create indicator variables from categorical variables. We will briefly review that capability here, but see [U] **25 Working with categorical data and factor variables** for a complete description. Let's begin by describing our highway dataset:

```
. describe
Contains data from http://www.stata-press.com/data/r12/hiway.dta
  obs:          39                          Minnesota Highway Data, 1973
 vars:           2                          16 Nov 2010 12:39
 size:         351
```

variable name	storage type	display format	value label	variable label
spdlimit	byte	%8.0g		Speed limit
rate	float	%9.0g	rcat	Accident rate per million vehicle miles
spdcat	float	%9.0g	scat	Speed Limit Category

```
Sorted by:
     Note:  dataset has changed since last saved
```

Our dataset contains three variables. We will type tabulate spdcat, generate(spd), describe our data, and then explain what happened.

```
. tabulate spdcat, generate(spd)
```

Speed Limit Category	Freq.	Percent	Cum.
40 to 50	11	28.21	28.21
55 to 60	26	66.67	94.87
Above 60	2	5.13	100.00
Total	39	100.00	

```
. describe
Contains data from http://www.stata-press.com/data/r12/hiway.dta
  obs:            39                          Minnesota Highway Data, 1973
  vars:            6                          16 Nov 2010 12:39
  size:          468
```

variable name	storage type	display format	value label	variable label
spdlimit	byte	%8.0g		Speed limit
rate	float	%9.0g	rcat	Accident rate per million vehicle miles
spdcat	float	%9.0g	scat	Speed Limit Category
spd1	byte	%8.0g		spdcat==40 to 50
spd2	byte	%8.0g		spdcat==55 to 60
spd3	byte	%8.0g		spdcat==Above 60

```
Sorted by:
    Note:  dataset has changed since last saved
```

When we typed `tabulate` with the `generate()` option, Stata responded by producing a one-way frequency table, so it appeared that the option did nothing. Yet when we `describe` our dataset, we find that we now have *six* variables instead of the original three. The new variables are named spd1, spd2, and spd3.

When we specify the `generate()` option, we are telling Stata to not only produce the table but also create a set of indicator variables that correspond to that table. Stata adds a numeric suffix to the name we specify in the parentheses. spd1 refers to the first line of the table, spd2 to the second line, and so on. Also Stata labels the variables so that we know what they mean. spd1 is an indicator variable that is *true* (takes on the value 1) when spdcat is between 40 and 50; otherwise, it is zero. (There is an exception: if spdcat is missing, so are the spd1, spd2, and spd3 variables. This did not happen in our dataset.)

We want to prove our claim. Because we have not yet introduced two-way tabulations, we will use the `summarize` statement:

```
. summarize spdlimit if spd1==1
```

Variable	Obs	Mean	Std. Dev.	Min	Max
spdlimit	11	47.72727	3.437758	40	50

```
. summarize spdlimit if spd2==1
```

Variable	Obs	Mean	Std. Dev.	Min	Max
spdlimit	26	57.11538	2.519157	55	60

```
. summarize spdlimit if spd3==1
```

Variable	Obs	Mean	Std. Dev.	Min	Max
spdlimit	2	67.5	3.535534	65	70

Notice the indicated minimum and maximum in each of the tables above. When we restrict the sample to spd1, spdlimit is between 40 and 50; when we restrict the sample to spd2, spdlimit is between 55 and 60; when we restrict the sample to spd3, spdlimit is between 65 and 70.

Thus `tabulate` provides an easy way to create indicator (sometimes called dummy) variables. For an overview of indicator and categorical variables, see [U] **25 Working with categorical data and factor variables**.

❏

tab1

tab1 is a convenience tool. Typing

    . tab1 myvar thisvar thatvar, plot

is equivalent to typing

    . tabulate myvar, plot
    . tabulate thisvar, plot
    . tabulate thatvar, plot

Saved results

tabulate and tab1 save the following in r():

Scalars
r(N)	number of observations	r(r)	number of rows

Methods and formulas

tab1 is implemented as an ado-file.

References

Cox, N. J. 2009. Speaking Stata: I. J. Good and quasi-Bayes smoothing of categorical frequencies. *Stata Journal* 9: 306–314.

Harrison, D. A. 2006. Stata tip 34: Tabulation by listing. *Stata Journal* 6: 425–427.

Also see

[R] **table** — Tables of summary statistics

[R] **tabstat** — Display table of summary statistics

[R] **tabulate twoway** — Two-way tables of frequencies

[R] **tabulate, summarize()** — One- and two-way tables of summary statistics

[D] **collapse** — Make dataset of summary statistics

[ST] **epitab** — Tables for epidemiologists

[SVY] **svy: tabulate oneway** — One-way tables for survey data

[SVY] **svy: tabulate twoway** — Two-way tables for survey data

[XT] **xttab** — Tabulate xt data

[U] **12.6.3 Value labels**

[U] **25 Working with categorical data and factor variables**

Title

tabulate twoway — Two-way tables of frequencies

Syntax

Two-way tables

<u>ta</u>bulate *varname₁* *varname₂* $\begin{bmatrix} if \end{bmatrix}$ $\begin{bmatrix} in \end{bmatrix}$ $\begin{bmatrix} weight \end{bmatrix}$ $\begin{bmatrix} , options \end{bmatrix}$

Two-way tables for all possible combinations—a convenience tool

tab2 *varlist* $\begin{bmatrix} if \end{bmatrix}$ $\begin{bmatrix} in \end{bmatrix}$ $\begin{bmatrix} weight \end{bmatrix}$ $\begin{bmatrix} , options \end{bmatrix}$

Immediate form of two-way tabulations

tabi $\#_{11}$ $\#_{12}$ $\begin{bmatrix} \dots \end{bmatrix}$ \ $\#_{21}$ $\#_{22}$ $\begin{bmatrix} \dots \end{bmatrix}$ $\begin{bmatrix} \backslash \dots \end{bmatrix}$ $\begin{bmatrix} , options \end{bmatrix}$

options	Description
Main	
<u>chi2</u>	report Pearson's χ^2
<u>e</u>xact$\begin{bmatrix} (\#) \end{bmatrix}$	report Fisher's exact test
gamma	report Goodman and Kruskal's gamma
lrchi2	report likelihood-ratio χ^2
<u>taub</u>	report Kendall's τ_b
V	report Cramér's V
<u>cchi2</u>	report Pearson's χ^2 in each cell
<u>col</u>umn	report relative frequency within its column of each cell
<u>row</u>	report relative frequency within its row of each cell
clrchi2	report likelihood-ratio χ^2 in each cell
<u>cell</u>	report the relative frequency of each cell
expected	report expected frequency in each cell
<u>nof</u>req	do not display frequencies
<u>m</u>issing	treat missing values like other values
<u>wrap</u>	do not wrap wide tables
$\begin{bmatrix} no \end{bmatrix}$key	report/suppress cell contents key
<u>nol</u>abel	display numeric codes rather than value labels
<u>nolog</u>	do not display enumeration log for Fisher's exact test
*<u>first</u>only	show only tables that include the first variable in *varlist*
Advanced	
matcell(*matname*)	save frequencies in *matname*; programmer's option
matrow(*matname*)	save unique values of *varname₁* in *matname*; programmer's option
matcol(*matname*)	save unique values of *varname₂* in *matname*; programmer's option
‡replace	replace current data with given cell frequencies
<u>all</u>	equivalent to specifying chi2 lrchi2 V gamma taub

[*]firstonly is available only for tab2.

[‡]replace is available only for tabi.

by is allowed with tabulate and tab2; see [D] **by**.

fweights, aweights, and iweights are allowed by tabulate. fweights are allowed by tab2.
   See [U] **11.1.6 weight**.

all does not appear in the dialog box.

Menu

tabulate

Statistics > Summaries, tables, and tests > Tables > Two-way tables with measures of association

tab2

Statistics > Summaries, tables, and tests > Tables > All possible two-way tabulations

tabi

Statistics > Summaries, tables, and tests > Tables > Table calculator

Description

tabulate produces two-way tables of frequency counts, along with various measures of association, including the common Pearson's χ^2, the likelihood-ratio χ^2, Cramér's V, Fisher's exact test, Goodman and Kruskal's gamma, and Kendall's τ_b.

Line size is respected. That is, if you resize the Results window before running tabulate, the resulting two-way tabulation will take advantage of the available horizontal space. Stata for Unix(console) users can instead use the set linesize command to take advantage of this feature.

tab2 produces all possible two-way tabulations of the variables specified in *varlist*.

tabi displays the $r \times c$ table, using the values specified; rows are separated by '\'. If no options are specified, it is as if exact were specified for 2×2 tables and chi2 were specified otherwise. See [U] **19 Immediate commands** for a general description of immediate commands. See *Tables with immediate data* below for examples using tabi.

See [R] **tabulate oneway** if you want one-way tables of frequencies. See [R] **table** and [R] **tabstat** if you want one-, two-, or n-way tables of frequencies and a wide variety of summary statistics. See [R] **tabulate, summarize()** for a description of tabulate with the summarize() option; it produces tables (breakdowns) of means and standard deviations. table is better than tabulate, summarize(), but tabulate, summarize() is faster. See [ST] **epitab** for 2×2 tables with statistics of interest to epidemiologists.

Options

> Main

chi2 calculates and displays Pearson's χ^2 for the hypothesis that the rows and columns in a two-way table are independent. chi2 may not be specified if aweights or iweights are specified.

exact$\big[(\#)\big]$ displays the significance calculated by Fisher's exact test and may be applied to $r \times c$ as well as to 2×2 tables. For 2×2 tables, both one- and two-sided probabilities are displayed. For $r \times c$ tables, one-sided probabilities are displayed. The optional positive integer # is a multiplier on the amount of memory that the command is permitted to consume. The default is 1. This option should not be necessary for reasonable $r \times c$ tables. If the command terminates with error 910, try exact(2). The maximum row or column dimension allowed when computing Fisher's exact test is the maximum row or column dimension for tabulate (see help limits).

gamma displays Goodman and Kruskal's gamma along with its asymptotic standard error. gamma is appropriate only when both variables are ordinal. gamma may not be specified if aweights or iweights are specified.

lrchi2 displays the likelihood-ratio χ^2 statistic. lrchi2 may not be specified if aweights or iweights are specified.

taub displays Kendall's τ_b along with its asymptotic standard error. taub is appropriate only when both variables are ordinal. taub may not be specified if aweights or iweights are specified.

V (note capitalization) displays Cramér's V. V may not be specified if aweights or iweights are specified.

cchi2 displays each cell's contribution to Pearson's chi-squared in a two-way table.

column displays the relative frequency of each cell within its column in a two-way table.

row displays the relative frequency of each cell within its row in a two-way table.

clrchi2 displays each cell's contribution to the likelihood-ratio chi-squared in a two-way table.

cell displays the relative frequency of each cell in a two-way table.

expected displays the expected frequency of each cell in a two-way table.

nofreq suppresses the printing of the frequencies.

missing requests that missing values be treated like other values in calculations of counts, percentages, and other statistics.

wrap requests that Stata take no action on wide, two-way tables to make them readable. Unless wrap is specified, wide tables are broken into pieces to enhance readability.

$\big[$no$\big]$key suppresses or forces the display of a key above two-way tables. The default is to display the key if more than one cell statistic is requested, and otherwise to omit it. key forces the display of the key. nokey suppresses its display.

nolabel causes the numeric codes to be displayed rather than the value labels.

nolog suppresses the display of the log for Fisher's exact test. Using Fisher's exact test requires counting all tables that have a probability exceeding that of the observed table given the observed row and column totals. The log counts down each stage of the network computations, starting from the number of columns and counting down to 1, displaying the number of nodes in the network at each stage. A log is not displayed for 2×2 tables.

firstonly, available only with tab2, restricts the output to only those tables that include the first variable in *varlist*. Use this option to interact one variable with a set of others.

     Advanced

matcell(*matname*) saves the reported frequencies in *matname*. This option is for use by programmers.

matrow(*matname*) saves the numeric values of the $r \times 1$ row stub in *matname*. This option is for use by programmers. matrow() may not be specified if the row variable is a string.

matcol(*matname*) saves the numeric values of the $1 \times c$ column stub in *matname*. This option is for use by programmers. matcol() may not be specified if the column variable is a string.

replace indicates that the immediate data specified as arguments to the tabi command be left as the current data in place of whatever data were there.

The following option is available with tabulate but is not shown in the dialog box:

all is equivalent to specifying chi2 lrchi2 V gamma taub. Note the omission of exact. When all is specified, no may be placed in front of the other options. all noV requests all association measures except Cramér's V (and Fisher's exact). all exact requests all association measures, including Fisher's exact test. all may not be specified if aweights or iweights are specified.

Limits

Two-way tables may have a maximum of 1,200 rows and 80 columns (Stata/MP and Stata/SE), 300 rows and 20 columns (Stata/IC), or 160 rows and 20 columns (Small Stata). If larger tables are needed, see [R] **table**.

Remarks

Remarks are presented under the following headings:

> *tabulate*
> *Measures of association*
> *N-way tables*
> *Weighted data*
> *Tables with immediate data*
> *tab2*

For each value of a specified variable (or a set of values for a pair of variables), tabulate reports the number of observations with that value. The number of times a value occurs is called its *frequency*.

tabulate

▷ Example 1

tabulate will make two-way tables if we specify two variables following the word tabulate. In our highway dataset, we have a variable called rate that divides the accident rate into three categories: below 4, 4–7, and above 7 per million vehicle miles. Let's make a table of the speed limit category and the accident-rate category:

```
. use http://www.stata-press.com/data/r12/hiway2
(Minnesota Highway Data, 1973)

. tabulate spdcat rate
```

Speed Limit Category	Accident rate per million vehicle miles			Total
	Below 4	4-7	Above 7	
40 to 50	3	5	3	11
55 to 60	19	6	1	26
Above 60	2	0	0	2
Total	24	11	4	39

The table indicates that three stretches of highway have an accident rate below 4 and a speed limit of 40 to 50 miles per hour. The table also shows the row and column sums (called the *marginals*). The number of highways with a speed limit of 40 to 50 miles per hour is 11, which is the same result we obtained in our previous one-way tabulations.

Stata can present this basic table in several ways—16, to be precise—and we will show just a few below. It might be easier to read the table if we included the row percentages. For instance, of 11 highways in the lowest speed limit category, three are also in the lowest accident-rate category. Three-elevenths amounts to some 27.3%. We can ask Stata to fill in this information for us by using the row option:

```
. tabulate spdcat rate, row
```

Key
frequency
row percentage

Speed Limit Category	Accident rate per million vehicle miles			
	Below 4	4-7	Above 7	Total
40 to 50	3	5	3	11
	27.27	45.45	27.27	100.00
55 to 60	19	6	1	26
	73.08	23.08	3.85	100.00
Above 60	2	0	0	2
	100.00	0.00	0.00	100.00
Total	24	11	4	39
	61.54	28.21	10.26	100.00

The number listed below each frequency is the percentage of cases that each cell represents out of its row. That is easy to remember because we see 100% listed in the "Total" column. The bottom row is also informative. We see that 61.54% of all the highways in our dataset fall into the lowest accident-rate category, that 28.21% are in the middle category, and that 10.26% are in the highest.

tabulate can calculate column percentages and cell percentages, as well. It does so when we specify the column or cell options, respectively. We can even specify them together. Below is a table that includes everything:

```
. tabulate spdcat rate, row column cell
```

```
┌─────────────────┐
│ Key             │
├─────────────────┤
│     frequency   │
│   row percentage│
│ column percentage│
│  cell percentage│
└─────────────────┘
```

Speed Limit Category	Accident rate per million vehicle miles			Total
	Below 4	4-7	Above 7	
40 to 50	3	5	3	11
	27.27	45.45	27.27	100.00
	12.50	45.45	75.00	28.21
	7.69	12.82	7.69	28.21
55 to 60	19	6	1	26
	73.08	23.08	3.85	100.00
	79.17	54.55	25.00	66.67
	48.72	15.38	2.56	66.67
Above 60	2	0	0	2
	100.00	0.00	0.00	100.00
	8.33	0.00	0.00	5.13
	5.13	0.00	0.00	5.13
Total	24	11	4	39
	61.54	28.21	10.26	100.00
	100.00	100.00	100.00	100.00
	61.54	28.21	10.26	100.00

The number at the top of each cell is the frequency count. The second number is the row percentage—they sum to 100% going across the table. The third number is the column percentage—they sum to 100% going down the table. The bottom number is the cell percentage—they sum to 100% going down all the columns and across all the rows. For instance, highways with a speed limit above 60 miles per hour and in the lowest accident rate category account for 100% of highways with a speed limit above 60 miles per hour; 8.33% of highways in the lowest accident-rate category; and 5.13% of all our data.

A fourth option, nofreq, tells Stata not to print the frequency counts. To construct a table consisting of only row percentages, we type

```
. tabulate spdcat rate, row nofreq
```

Speed Limit Category	Accident rate per million vehicle miles			Total
	Below 4	4-7	Above 7	
40 to 50	27.27	45.45	27.27	100.00
55 to 60	73.08	23.08	3.85	100.00
Above 60	100.00	0.00	0.00	100.00
Total	61.54	28.21	10.26	100.00

Measures of association

▷ Example 2

tabulate will calculate the Pearson χ^2 test for the independence of the rows and columns if we specify the chi2 option. Suppose that we have 1980 census data on 956 cities in the United States and wish to compare the age distribution across regions of the country. Assume that agecat is the median age in each city and that region denotes the region of the country in which the city is located.

```
. use http://www.stata-press.com/data/r12/citytemp2
(City Temperature Data)
. tabulate region agecat, chi2
```

Census Region	agecat 19-29	30-34	35+	Total
NE	46	83	37	166
N Cntrl	162	92	30	284
South	139	68	43	250
West	160	73	23	256
Total	507	316	133	956

```
          Pearson chi2(6) =  61.2877   Pr = 0.000
```

We obtain the standard two-way table and, at the bottom, a summary of the χ^2 test. Stata informs us that the χ^2 associated with this table has 6 degrees of freedom and is 61.29. The observed differences are significant.

The table is, perhaps, easier to understand if we suppress the frequencies and print just the row percentages:

```
. tabulate region agecat, row nofreq chi2
```

Census Region	agecat 19-29	30-34	35+	Total
NE	27.71	50.00	22.29	100.00
N Cntrl	57.04	32.39	10.56	100.00
South	55.60	27.20	17.20	100.00
West	62.50	28.52	8.98	100.00
Total	53.03	33.05	13.91	100.00

```
          Pearson chi2(6) =  61.2877   Pr = 0.000
```

◁

▷ Example 3

We have data on dose level and outcome for a set of patients and wish to evaluate the association between the two variables. We can obtain all the association measures by specifying the all and exact options:

```
. use http://www.stata-press.com/data/r12/dose

. tabulate dose function, all exact
Enumerating sample-space combinations:
stage 3:  enumerations = 1
stage 2:  enumerations = 9
stage 1:  enumerations = 0
```

		Function		
Dosage	< 1 hr	1 to 4	4+	Total
1/day	20	10	2	32
2/day	16	12	4	32
3/day	10	16	6	32
Total	46	38	12	96

```
            Pearson chi2(4) =   6.7780    Pr = 0.148
    likelihood-ratio chi2(4) =   6.9844    Pr = 0.137
                Cramer's V =   0.1879
                     gamma =   0.3689   ASE = 0.129
            Kendall's tau-b =   0.2378   ASE = 0.086
             Fisher's exact =             0.145
```

We find evidence of association but not enough to be truly convincing.

If we had not also specified the exact option, we would not have obtained Fisher's exact test. Stata can calculate this statistic both for 2×2 tables and for $r \times c$. For 2×2 tables, the calculation is almost instant. On more general tables, however, the calculation can take longer.

We carefully constructed our example so that all would be meaningful. Kendall's τ_b and Goodman and Kruskal's gamma are relevant only when both dimensions of the table can be ordered, say, from low to high or from worst to best. The other statistics, however, are always applicable.

◁

❏ Technical note

Be careful when attempting to compute the p-value for Fisher's exact test because the number of tables that contribute to the p-value can be extremely large and a solution may not be feasible. The errors that are indicative of this situation are errors 910, exceeded memory limitations, and 1401, integer overflow due to large row-margin frequencies. If execution terminates because of memory limitations, use exact(2) to permit the algorithm to consume twice the memory, exact(3) for three times the memory, etc. The default memory usage should be sufficient for reasonable tables.

❏

N-way tables

If you need more than two-way tables, your best alternative to is use table, not tabulate; see [R] **table**.

The technical note below shows you how to use tabulate to create a sequence of two-way tables that together form, in effect, a three-way table, but using table is easy and produces prettier results:

```
. use http://www.stata-press.com/data/r12/birthcat
(City data)
. table birthcat region agecat, c(freq)
```

birthcat	agecat and Census Region							
	19-29				30-34			
	NE N Cntrl	South	West		NE N Cntrl	South	West	
29-136	11	23	11	11	34	27	10	8
137-195	31	97	65	46	48	58	45	42
196-529	4	38	59	91	1	3	12	21

birthcat	agecat and Census Region			
	35+			
	NE N Cntrl	South	West	
29-136	34	26	27	18
137-195	3	4	7	4
196-529			4	

❏ Technical note

We can make n-way tables by combining the by *varlist*: prefix with tabulate. Continuing with the dataset of 956 cities, say that we want to make a table of age category by birth-rate category by region of the country. The birth-rate category variable is named birthcat in our dataset. To make separate tables for each age category, we would type

```
. by agecat, sort: tabulate birthcat region
```

```
-> agecat = 19-29
```

birthcat	Census Region				Total
	NE	N Cntrl	South	West	
29-136	11	23	11	11	56
137-195	31	97	65	46	239
196-529	4	38	59	91	192
Total	46	158	135	148	487

```
-> agecat = 30-34
```

birthcat	Census Region				Total
	NE	N Cntrl	South	West	
29-136	34	27	10	8	79
137-195	48	58	45	42	193
196-529	1	3	12	21	37
Total	83	88	67	71	309

```
-> agecat = 35+
```

		Census Region			
birthcat	NE	N Cntrl	South	West	Total
29-136	34	26	27	18	105
137-195	3	4	7	4	18
196-529	0	0	4	0	4
Total	37	30	38	22	127

❏

Weighted data

▷ Example 4

tabulate can process weighted as well as unweighted data. As with all Stata commands, we indicate the weight by specifying the [*weight*] modifier; see [U] **11.1.6 weight**.

Continuing with our dataset of 956 cities, we also have a variable called pop, the population of each city. We can make a table of region by age category, weighted by population, by typing

```
. tabulate region agecat [freq=pop]
```

Census		agecat		
Region	19-29	30-34	35+	Total
NE	4,721,387	10,421,387	5,323,610	20,466,384
N Cntrl	16,901,550	8,964,756	4,015,593	29,881,899
South	13,894,254	7,686,531	4,141,863	25,722,648
West	16,698,276	7,755,255	2,375,118	26,828,649
Total	52,215,467	34,827,929	15,856,184	102899580

If we specify the cell, column, or row options, they will also be appropriately weighted. Below we repeat the table, suppressing the counts and substituting row percentages:

```
. tabulate region agecat [freq=pop], nofreq row
```

Census		agecat		
Region	19-29	30-34	35+	Total
NE	23.07	50.92	26.01	100.00
N Cntrl	56.56	30.00	13.44	100.00
South	54.02	29.88	16.10	100.00
West	62.24	28.91	8.85	100.00
Total	50.74	33.85	15.41	100.00

◁

Tables with immediate data

▷ Example 5

tabi ignores the dataset in memory and uses as the table the values that we specify on the command line:

```
. tabi 30 18 \ 38 14
```

	col		
row	1	2	Total
1	30	18	48
2	38	14	52
Total	68	32	100

```
              Fisher's exact =                0.289
      1-sided Fisher's exact =                0.179
```

We may specify any of the options of tabulate and are not limited to 2×2 tables:

```
. tabi 30 18 38 \ 13 7 22, chi2 exact
Enumerating sample-space combinations:
stage 3:   enumerations = 1
stage 2:   enumerations = 3
stage 1:   enumerations = 0
```

	col			
row	1	2	3	Total
1	30	18	38	86
2	13	7	22	42
Total	43	25	60	128

```
          Pearson chi2(2) =   0.7967   Pr = 0.671
            Fisher's exact =            0.707
. tabi 30 13 \ 18 7 \ 38 22, all exact col
```

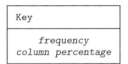

```
Key

      frequency
  column percentage
```

```
Enumerating sample-space combinations:
stage 3:   enumerations = 1
stage 2:   enumerations = 3
stage 1:   enumerations = 0
```

	col		
row	1	2	Total
1	30	13	43
	34.88	30.95	33.59
2	18	7	25
	20.93	16.67	19.53
3	38	22	60
	44.19	52.38	46.88
Total	86	42	128
	100.00	100.00	100.00

```
            Pearson chi2(2) =   0.7967   Pr = 0.671
    likelihood-ratio chi2(2) =   0.7985   Pr = 0.671
                 Cramer's V =   0.0789
                      gamma =   0.1204   ASE = 0.160
            Kendall's tau-b =   0.0630   ASE = 0.084
            Fisher's exact =                0.707
```

For 2×2 tables, both one- and two-sided Fisher's exact probabilities are displayed; this is true of both `tabulate` and `tabi`. See *Cumulative incidence data* and *Case–control data* in [ST] **epitab** for more discussion on the relationship between one- and two-sided probabilities.

◁

❏ Technical note

tabi, as with all immediate commands, leaves any data in memory undisturbed. With the `replace` option, however, the data in memory are replaced by the data from the table:

```
. tabi 30 18 \ 38 14, replace
           |        col
      row  |      1         2  |    Total
-----------+--------------------+----------
        1  |     30        18  |       48
        2  |     38        14  |       52
-----------+--------------------+----------
    Total  |     68        32  |      100
              Fisher's exact =              0.289
      1-sided Fisher's exact =              0.179
. list
```

```
     |  row   col   pop  |
     |--------------------|
  1. |   1     1    30   |
  2. |   1     2    18   |
  3. |   2     1    38   |
  4. |   2     2    14   |
```

With this dataset, you could re-create the above table by typing

```
. tabulate row col [freq=pop], exact
           |        col
      row  |      1         2  |    Total
-----------+--------------------+----------
        1  |     30        18  |       48
        2  |     38        14  |       52
-----------+--------------------+----------
    Total  |     68        32  |      100
              Fisher's exact =              0.289
      1-sided Fisher's exact =              0.179
```

❏

tab2

tab2 is a convenience tool. Typing

```
. tab2 myvar thisvar thatvar, chi2
```

is equivalent to typing

```
. tabulate myvar thisvar, chi2
. tabulate myvar thatvar, chi2
. tabulate thisvar thatvar, chi2
```

Saved results

tabulate, tab2, and tabi save the following in r():

Scalars

r(N)	number of observations	r(p_exact)	Fisher's exact p
r(r)	number of rows	r(chi2_lr)	likelihood-ratio χ^2
r(c)	number of columns	r(p_lr)	significance of likelihood-ratio χ^2
r(chi2)	Pearson's χ^2	r(CramersV)	Cramér's V
r(p)	significance of Pearson's χ^2	r(ase_gam)	ASE of gamma
r(gamma)	gamma	r(ase_taub)	ASE of τ_b
r(p1_exact)	one-sided Fisher's exact p	r(taub)	τ_b

r(p1_exact) is defined only for 2×2 tables. Also, the matrow(), matcol(), and matcell() options allow you to obtain the row values, column values, and frequencies, respectively.

Methods and formulas

tab2 and tabi are implemented as ado-files.

Let n_{ij}, $i = 1, \ldots, I$ and $j = 1, \ldots, J$, be the number of observations in the ith row and jth column. If the data are not weighted, n_{ij} is just a count. If the data are weighted, n_{ij} is the sum of the weights of all data corresponding to the (i, j) cell.

Define the row and column marginals as

$$n_{i.} = \sum_{j=1}^{J} n_{ij} \qquad n_{.j} = \sum_{i=1}^{I} n_{ij}$$

and let $n = \sum_i \sum_j n_{ij}$ be the overall sum. Also, define the concordance and discordance as

$$A_{ij} = \sum_{k>i}\sum_{l>j} n_{kl} + \sum_{k<i}\sum_{l<j} n_{kl} \qquad D_{ij} = \sum_{k>i}\sum_{l<j} n_{kl} + \sum_{k<i}\sum_{l>j} n_{kl}$$

along with twice the number of concordances $P = \sum_i \sum_j n_{ij} A_{ij}$ and twice the number of discordances $Q = \sum_i \sum_j n_{ij} D_{ij}$.

The Pearson χ^2 statistic with $(I - 1)(J - 1)$ degrees of freedom (so called because it is based on Pearson (1900); see Conover [1999, 240] and Fienberg [1980, 9]) is defined as

$$X^2 = \sum_i \sum_j \frac{(n_{ij} - m_{ij})^2}{m_{ij}}$$

where $m_{ij} = n_{i.} n_{.j} / n$.

The likelihood-ratio χ^2 statistic with $(I-1)(J-1)$ degrees of freedom (Fienberg 1980, 40) is defined as

$$G^2 = 2 \sum_i \sum_j n_{ij} \ln(n_{ij}/m_{ij})$$

Cramér's V (Cramér 1946) is a measure of association designed so that the attainable upper bound is 1. For 2×2 tables, $-1 \leq V \leq 1$, and otherwise, $0 \leq V \leq 1$.

$$V = \begin{cases} (n_{11}n_{22} - n_{12}n_{21})/(n_{1.}n_{2.}n_{.1}n_{.2})^{1/2} & \text{for } 2 \times 2 \\ \{(X^2/n)/\min(I-1, J-1)\}^{1/2} & \text{otherwise} \end{cases}$$

Gamma (Goodman and Kruskal 1954, 1959, 1963, 1972; also see Agresti [2010,186–188]) ignores tied pairs and is based only on the number of concordant and discordant pairs of observations, $-1 \leq \gamma \leq 1$,

$$\gamma = (P - Q)/(P + Q)$$

with asymptotic variance

$$16 \sum_i \sum_j n_{ij}(QA_{ij} - PD_{ij})^2/(P + Q)^4$$

Kendall's τ_b (Kendall 1945; also see Agresti 2010, 188–189), $-1 \leq \tau_b \leq 1$, is similar to gamma, except that it uses a correction for ties,

$$\tau_b = (P - Q)/(w_r w_c)^{1/2}$$

with asymptotic variance

$$\frac{\sum_i \sum_j n_{ij}(2w_r w_c d_{ij} + \tau_b v_{ij})^2 - n^3 \tau_b^2 (w_r + w_c)^2}{(w_r w_c)^4}$$

where

$$w_r = n^2 - \sum_i n_{i.}^2$$

$$w_c = n^2 - \sum_j n_{.j}^2$$

$$d_{ij} = A_{ij} - D_{ij}$$

$$v_{ij} = n_{i.}w_c + n_{.j}w_r$$

Fisher's exact test (Fisher 1935; Finney 1948; see Zelterman and Louis [1992, 293–301] for the 2×2 case) yields the probability of observing a table that gives at least as much evidence of association as the one actually observed under the assumption of no association. Holding row and column marginals fixed, the hypergeometric probability P of every possible table A is computed, and the

$$P = \sum_{T \in A} \Pr(T)$$

where A is the set of all tables with the same marginals as the observed table, T^*, such that $\Pr(T) \leq \Pr(T^*)$. For 2×2 tables, the one-sided probability is calculated by further restricting A to tables in the same tail as T^*. The first algorithm extending this calculation to $r \times c$ tables was Pagano and Halvorsen (1981); the one implemented here is the FEXACT algorithm by Mehta and Patel (1986). This is a search-tree clipping method originally published by Mehta and Patel (1983) with further refinements by Joe (1988) and Clarkson, Fan, and Joe (1993). Fisher's exact test is a permutation test. For more information on permutation tests, see Good (2005 and 2006) and Pesarin (2001).

References

Agresti, A. 2010. *Analysis of Ordinal Categorical Data*. 2nd ed. Hoboken, NJ: Wiley.

Campbell, M. J., D. Machin, and S. J. Walters. 2007. *Medical Statistics: A Textbook for the Health Sciences*. 4th ed. Chichester, UK: Wiley.

Clarkson, D. B., Y.-A. Fan, and H. Joe. 1993. A remark on Algorithm 643: FEXACT: An algorithm for performing Fisher's exact test in $r \times c$ contingency tables. *ACM Transactions on Mathematical Software* 19: 484–488.

Conover, W. J. 1999. *Practical Nonparametric Statistics*. 3rd ed. New York: Wiley.

Cox, N. J. 1996. sg57: An immediate command for two-way tables. *Stata Technical Bulletin* 33: 7–9. Reprinted in *Stata Technical Bulletin Reprints*, vol. 6, pp. 140–143. College Station, TX: Stata Press.

———. 1999. sg113: Tabulation of modes. *Stata Technical Bulletin* 50: 26–27. Reprinted in *Stata Technical Bulletin Reprints*, vol. 9, pp. 180–181. College Station, TX: Stata Press.

———. 2003. sg113_1: Software update: Tabulation of modes. *Stata Journal* 3: 211.

———. 2009. Speaking Stata: I. J. Good and quasi-Bayes smoothing of categorical frequencies. *Stata Journal* 9: 306–314.

Cramér, H. 1946. *Mathematical Methods of Statistics*. Princeton: Princeton University Press.

Fienberg, S. E. 1980. *The Analysis of Cross-Classified Categorical Data*. 2nd ed. Cambridge, MA: MIT Press.

Finney, D. J. 1948. The Fisher–Yates test of significance in 2×2 contingency tables. *Biometrika* 35: 145–156.

Fisher, R. A. 1935. The logic of inductive inference. *Journal of the Royal Statistical Society* 98: 39–82.

Good, P. I. 2005. *Permutation, Parametric, and Bootstrap Tests of Hypotheses: A Practical Guide to Resampling Methods for Testing Hypotheses*. 3rd ed. New York: Springer.

———. 2006. *Resampling Methods: A Practical Guide to Data Analysis*. 3rd ed. Boston: Birkhäuser.

Goodman, L. A., and W. H. Kruskal. 1954. Measures of association for cross classifications. *Journal of the American Statistical Association* 49: 732–764.

———. 1959. Measures of association for cross classifications II: Further discussion and references. *Journal of the American Statistical Association* 54: 123–163.

———. 1963. Measures of association for cross classifications III: Approximate sampling theory. *Journal of the American Statistical Association* 58: 310–364.

———. 1972. Measures of association for cross classifications IV: Simplification of asymptotic variances. *Journal of the American Statistical Association* 67: 415–421.

Harrison, D. A. 2006. Stata tip 34: Tabulation by listing. *Stata Journal* 6: 425–427.

Jann, B. 2008. Multinomial goodness-of-fit: Large-sample tests with survey design correction and exact tests for small samples. *Stata Journal* 8: 147–169.

Joe, H. 1988. Extreme probabilities for contingency tables under row and column independence with application to Fisher's exact test. *Communications in Statistics, Theory and Methods* 17: 3677–3685.

Judson, D. H. 1992. sg12: Extended tabulate utilities. *Stata Technical Bulletin* 10: 22–23. Reprinted in *Stata Technical Bulletin Reprints*, vol. 2, pp. 140–141. College Station, TX: Stata Press.

Kendall, M. G. 1945. The treatment of ties in rank problems. *Biometrika* 33: 239–251.

Mehta, C. R., and N. R. Patel. 1983. A network algorithm for performing Fisher's exact test in $r \times c$ contingency tables. *Journal of the American Statistical Association* 78: 427–434.

——. 1986. Algorithm 643 FEXACT: A FORTRAN subroutine for Fisher's exact test on unordered $r \times c$ contingency tables. *ACM Transactions on Mathematical Software* 12: 154–161.

Newson, R. 2002. Parameters behind "nonparametric" statistics: Kendall's tau, Somers' D and median differences. *Stata Journal* 2: 45–64.

Pagano, M., and K. T. Halvorsen. 1981. An algorithm for finding the exact significance levels of $r \times c$ contingency tables. *Journal of the American Statistical Association* 76: 931–934.

Pearson, K. 1900. On the criterion that a given system of deviations from the probable in the case of a correlated system of variables is such that it can be reasonably supposed to have arisen from random sampling. *Philosophical Magazine*, Series 5 50: 157–175.

Pesarin, F. 2001. *Multivariate Permutation Tests: With Applications in Biostatistics*. Chichester, UK: Wiley.

Weesie, J. 2001. dm91: Patterns of missing values. *Stata Technical Bulletin* 61: 5–7. Reprinted in *Stata Technical Bulletin Reprints*, vol. 10, pp. 49–51. College Station, TX: Stata Press.

Wolfe, R. 1999. sg118: Partitions of Pearson's χ^2 for analyzing two-way tables that have ordered columns. *Stata Technical Bulletin* 51: 37–40. Reprinted in *Stata Technical Bulletin Reprints*, vol. 9, pp. 203–207. College Station, TX: Stata Press.

Zelterman, D., and T. A. Louis. 1992. Contingency tables in medical studies. In *Medical Uses of Statistics*, 2nd ed, ed. J. C. Bailar III and F. Mosteller, 293–310. Boston: Dekker.

Also see

[R] **table** — Tables of summary statistics

[R] **tabstat** — Display table of summary statistics

[R] **tabulate oneway** — One-way tables of frequencies

[R] **tabulate, summarize()** — One- and two-way tables of summary statistics

[D] **collapse** — Make dataset of summary statistics

[ST] **epitab** — Tables for epidemiologists

[SVY] **svy: tabulate oneway** — One-way tables for survey data

[SVY] **svy: tabulate twoway** — Two-way tables for survey data

[XT] **xttab** — Tabulate xt data

[U] **12.6.3 Value labels**

[U] **25 Working with categorical data and factor variables**

Title

tabulate, summarize() — One- and two-way tables of summary statistics	

Syntax

<u>ta</u>bulate *varname*$_1$ [*varname*$_2$] [*if*] [*in*] [*weight*] [, *options*]

options	Description
Main	
<u>summarize</u>(*varname*$_3$)	report summary statistics for *varname*$_3$
[<u>no</u>]<u>me</u>ans	include or suppress means
[<u>no</u>]<u>st</u>andard	include or suppress standard deviations
[<u>no</u>]<u>f</u>req	include or suppress frequencies
[<u>no</u>]<u>o</u>bs	include or suppress number of observations
<u>no</u>label	show numeric codes, not labels
<u>wrap</u>	do not break wide tables
<u>m</u>issing	treat missing values of *varname*$_1$ and *varname*$_2$ as categories

by is allowed; see [D] **by**.

aweights and fweights are allowed; see [U] **11.1.6 weight**.

Menu

Statistics > Summaries, tables, and tests > Tables > One/two-way table of summary statistics

Description

tabulate, summarize() produces one- and two-way tables (breakdowns) of means and standard deviations. See [R] **tabulate oneway** and [R] **tabulate twoway** for one- and two-way frequency tables. See [R] **table** for a more flexible command that produces one-, two-, and n-way tables of frequencies and a wide variety of summary statistics. table is better, but tabulate, summarize() is faster. Also see [R] **tabstat** for yet another alternative.

Options

   Main

summarize(*varname*$_3$) identifies the name of the variable for which summary statistics are to be reported. If you do not specify this option, a table of frequencies is produced; see [R] **tabulate oneway** and [R] **tabulate twoway**. The description here concerns tabulate when this option is specified.

2162

[no]means includes or suppresses only the means from the table.

The summarize() table normally includes the mean, standard deviation, frequency, and, if the data are weighted, number of observations. Individual elements of the table may be included or suppressed by the [no]means, [no]standard, [no]freq, and [no]obs options. For example, typing

   . tabulate category, summarize(myvar) means standard

produces a summary table by category containing only the means and standard deviations of myvar. You could also achieve the same result by typing

   . tabulate category, summarize(myvar) nofreq

[no]standard includes or suppresses only the standard deviations from the table; see [no]means option above.

[no]freq includes or suppresses only the frequencies from the table; see [no]means option above.

[no]obs includes or suppresses only the reported number of observations from the table. If the data are not weighted, the number of observations is identical to the frequency, and by default only the frequency is reported. If the data are weighted, the frequency refers to the sum of the weights. See [no]means option above.

nolabel causes the numeric codes to be displayed rather than the label values.

wrap requests that no action be taken on wide tables to make them readable. Unless wrap is specified, wide tables are broken into pieces to enhance readability.

missing requests that missing values of *varname*$_1$ and *varname*$_2$ be treated as categories rather than as observations to be omitted from the analysis.

Remarks

tabulate with the summarize() option produces one- and two-way tables of summary statistics. When combined with the by prefix, it can produce n-way tables as well.

Remarks are presented under the following headings:

   One-way tables
   Two-way tables

One-way tables

▷ Example 1

We have data on 74 automobiles. Included in our dataset are the variables foreign, which marks domestic and foreign cars, and mpg, the car's mileage rating. Typing tabulate foreign displays a breakdown of the number of observations we have by the values of the foreign variable.

```
. use http://www.stata-press.com/data/r12/auto
(1978 Automobile Data)

. tabulate foreign
```

Car type	Freq.	Percent	Cum.
Domestic	52	70.27	70.27
Foreign	22	29.73	100.00
Total	74	100.00	

We discover that we have 52 domestic cars and 22 foreign cars in our dataset. If we add the summarize(*varname*) option, however, tabulate produces a table of summary statistics for *varname*:

```
. tabulate foreign, summarize(mpg)
                 Summary of Mileage (mpg)
    Car type        Mean   Std. Dev.        Freq.

    Domestic   19.826923   4.7432972           52
     Foreign   24.772727   6.6111869           22

       Total   21.297297   5.7855032           74
```

We also discover that the average gas mileage for domestic cars is about 20 mpg and the average foreign is almost 25 mpg. Overall, the average is 21 mpg in our dataset.

◁

❏ Technical note

We might now wonder if the difference in gas mileage between foreign and domestic cars is statistically significant. We can use the oneway command to find out; see [R] **oneway**. To obtain an analysis-of-variance table of mpg on foreign, we type

```
. oneway mpg foreign
                        Analysis of Variance
    Source              SS         df       MS            F      Prob > F

Between groups      378.153515      1    378.153515     13.18     0.0005
Within groups       2065.30594     72    28.6848048

    Total           2443.45946     73    33.4720474
    Bartlett's test for equal variances:  chi2(1) =    3.4818  Prob>chi2 = 0.062
```

The F statistic is 13.18, and the difference between foreign and domestic cars' mileage ratings is significant at the 0.05% level.

There are several ways that we could have statistically compared mileage ratings—see, for instance, [R] **anova**, [R] **oneway**, [R] **regress**, and [R] **ttest**—but oneway seemed the most convenient.

❏

Two-way tables

▷ Example 2

tabulate, summarize can be used to obtain two-way as well as one-way breakdowns. For instance, we obtained summary statistics on mpg decomposed by foreign by typing tabulate foreign, summarize(mpg). We can specify up to two variables before the comma:

```
. generate wgtcat = autocode(weight,4,1760,4840)
. tabulate wgtcat foreign, summarize(mpg)
```
           Means, Standard Deviations and Frequencies of Mileage (mpg)

| | Car type | | |
wgtcat	Domestic	Foreign	Total
2530	28.285714	27.0625	27.434783
	3.0937725	5.9829619	5.2295149
	7	16	23
3300	21.75	19.6	21.238095
	2.4083189	3.4351128	2.7550819
	16	5	21
4070	17.26087	14	17.125
	1.8639497	0	1.9406969
	23	1	24
4840	14.666667	.	14.666667
	3.32666	.	3.32666
	6	0	6
Total	19.826923	24.772727	21.297297
	4.7432972	6.6111869	5.7855032
	52	22	74

In addition to the means, standard deviations, and frequencies for each weight–mileage cell, also reported are the summary statistics by weight, by mileage, and overall. For instance, the last row of the table reveals that the average mileage of domestic cars is 19.83 and that of foreign cars is 24.77 — domestic cars yield poorer mileage than foreign cars. But we now see that domestic cars yield better gas mileage within weight class — the reason domestic cars yield poorer gas mileage is because they are, on average, heavier.

◁

▷ Example 3

If we do not specify the statistics to be included in a table, tabulate reports the mean, standard deviation, and frequency. We can specify the statistics that we want to see using the means, standard, and freq options:

```
. tabulate wgtcat foreign, summarize(mpg) means
```
                        Means of Mileage (mpg)

| | Car type | | |
wgtcat	Domestic	Foreign	Total
2530	28.285714	27.0625	27.434783
3300	21.75	19.6	21.238095
4070	17.26087	14	17.125
4840	14.666667	.	14.666667
Total	19.826923	24.772727	21.297297

When we specify one or more of the means, standard, and freq options, only those statistics are displayed. Thus we could obtain a table containing just the means and standard deviations by typing means standard after the summarize(mpg) option. We can also suppress selected statistics by placing no in front of the option name. Another way of obtaining only the means and standard deviations is to add the nofreq option:

```
. tabulate wgtcat foreign, summarize(mpg) nofreq
          Means and Standard Deviations of Mileage (mpg)
```

wgtcat	Car type Domestic	Foreign	Total
2530	28.285714 3.0937725	27.0625 5.9829619	27.434783 5.2295149
3300	21.75 2.4083189	19.6 3.4351128	21.238095 2.7550819
4070	17.26087 1.8639497	14 0	17.125 1.9406969
4840	14.666667 3.32666	. .	14.666667 3.32666
Total	19.826923 4.7432972	24.772727 6.6111869	21.297297 5.7855032

◁

Also see

[R] **table** — Tables of summary statistics

[R] **tabstat** — Display table of summary statistics

[R] **tabulate oneway** — One-way tables of frequencies

[R] **tabulate twoway** — Two-way tables of frequencies

[D] **collapse** — Make dataset of summary statistics

[SVY] **svy: tabulate oneway** — One-way tables for survey data

[SVY] **svy: tabulate twoway** — Two-way tables for survey data

[U] **12.6 Dataset, variable, and value labels**

[U] **25 Working with categorical data and factor variables**

Title

> **test** — Test linear hypotheses after estimation

Syntax

Basic syntax

$$\underline{\text{test}} \ \textit{coeflist} \qquad\qquad\qquad\qquad\qquad\qquad (\textit{Syntax 1})$$

$$\underline{\text{test}} \ \textit{exp=exp}\left[=\ldots\right] \qquad\qquad\qquad\qquad\qquad (\textit{Syntax 2})$$

$$\underline{\text{test}} \ \left[\textit{eqno}\right] \ \left[: \textit{coeflist}\right] \qquad\qquad\qquad\qquad (\textit{Syntax 3})$$

$$\underline{\text{test}} \ \left[\textit{eqno=eqno}\left[=\ldots\right]\right] \ \left[: \textit{coeflist}\right] \qquad\qquad (\textit{Syntax 4})$$

$$\text{testparm} \ \textit{varlist} \ \left[, \ \underline{\text{e}}\text{qual} \ \underline{\text{equation}}(\textit{eqno})\right]$$

Full syntax

$$\underline{\text{test}} \ (\textit{spec}) \ \left[(\textit{spec}) \ \ldots\right] \ \left[, \ \textit{test\_options}\right]$$

test_options	Description
Options	
<u>m</u>test$\left[(opt)\right]$	test each condition separately
coef	report estimated constrained coefficients
<u>a</u>ccumulate	test hypothesis jointly with previously tested hypotheses
<u>not</u>est	suppress the output
common	test only variables common to all the equations
<u>cons</u>tant	include the constant in coefficients to be tested
nosvyadjust	compute unadjusted Wald tests for survey results
<u>min</u>imum	perform test with the constant, drop terms until the test becomes nonsingular, and test without the constant on the remaining terms; highly technical
matvlc(*matname*)	save the variance–covariance matrix; programmer's option

coeflist and *varlist* may contain factor variables and time-series operators; see [U] **11.4.3 Factor variables** and [U] **11.4.4 Time-series varlists**.
matvlc(*matname*) does not appear in the dialog box.

Syntax 1 tests that coefficients are 0.

Syntax 2 tests that linear expressions are equal.

Syntax 3 tests that coefficients in *eqno* are 0.

Syntax 4 tests equality of coefficients between equations.

spec is one of

> *coeflist*
> *exp*=*exp* [=*exp*]
> [*eqno*] [: *coeflist*]
> [*eqno*₁=*eqno*₂ [=...]] [: *coeflist*]

coeflist is

> *coef* [*coef* ...]
> [*eqno*]*coef* [[*eqno*]*coef*...]
> [*eqno*] _b[*coef*] [[*eqno*] _b[*coef*]...]

exp is a linear expression containing

> *coef*
> _b[*coef*]
> _b[*eqno*:*coef*]
> [*eqno*]*coef*
> [*eqno*] _b[*coef*]

eqno is

> # #
> *name*

coef identifies a coefficient in the model. *coef* is typically a variable name, a level indicator, an interaction indicator, or an interaction involving continuous variables. Level indicators identify one level of a factor variable and interaction indicators identify one combination of levels of an interaction; see [U] **11.4.3 Factor variables**. *coef* may contain time-series operators; see [U] **11.4.4 Time-series varlists**.

Distinguish between [ ], which are to be typed, and [ ], which indicate optional arguments.

Although not shown in the syntax diagram, parentheses around *spec* are required only with multiple specifications. Also, the diagram does not show that test may be called without arguments to redisplay the results from the last test.

anova and manova (see [R] **anova** and [MV] **manova**) allow the test syntax above plus more (see [R] **anova postestimation** for test after anova; see [MV] **manova postestimation** for test after manova).

Menu

test

Statistics > Postestimation > Tests > Test linear hypotheses

testparm

Statistics > Postestimation > Tests > Test parameters

Description

test performs Wald tests of simple and composite linear hypotheses about the parameters of the most recently fit model.

test supports svy estimators (see [SVY] **svy estimation**), carrying out an adjusted Wald test by default in such cases. test can be used with svy estimation results, see [SVY] **svy postestimation**.

testparm provides a useful alternative to test that permits *varlist* rather than a list of coefficients (which is often nothing more than a list of variables), allowing the use of standard Stata notation, including '–' and '*', which are given the *exp*ression interpretation by test.

test and testparm perform Wald tests. For likelihood-ratio tests, see [R] **lrtest**. For Wald-type tests of nonlinear hypotheses, see [R] **testnl**. To display estimates for one-dimensional linear or nonlinear expressions of coefficients, see [R] **lincom** and [R] **nlcom**.

See [R] **anova postestimation** for additional test syntax allowed after anova.

See [MV] **manova postestimation** for additional test syntax allowed after manova.

Options for testparm

equal tests that the variables appearing in *varlist*, which also appear in the previously fit model, are equal to each other rather than jointly equal to zero.

equation(*eqno*) is relevant only for multiple-equation models, such as mvreg, mlogit, and heckman. It specifies the equation for which the all-zero or all-equal hypothesis is tested. equation(#1) specifies that the test be conducted regarding the first equation #1. equation(price) specifies that the test concern the equation named price.

Options for test

⌐ Options ⌐

mtest [(*opt*)] specifies that tests be performed for each condition separately. *opt* specifies the method for adjusting *p*-values for multiple testing. Valid values for *opt* are

bonferroni	Bonferroni's method
holm	Holm's method
sidak	Šidák's method
noadjust	no adjustment is to be made

Specifying mtest without an argument is equivalent to mtest(noadjust).

coef specifies that the constrained coefficients be displayed.

accumulate allows a hypothesis to be tested jointly with the previously tested hypotheses.

notest suppresses the output. This option is useful when you are interested only in the joint test of several hypotheses, specified in a subsequent call of test, accumulate.

common specifies that when you use the [*eqno*₁=*eqno*₂ [=...]] form of *spec*, the variables common to the equations *eqno*₁, *eqno*₂, etc., be tested. The default action is to complain if the equations have variables not in common.

constant specifies that _cons be included in the list of coefficients to be tested when using the [*eqno*₁=*eqno*₂ [=...]] or [*eqno*] forms of *spec*. The default is not to include _cons.

nosvyadjust is for use with svy estimation commands; see [SVY] **svy estimation**. It specifies that the Wald test be carried out without the default adjustment for the design degrees of freedom. That is, the test is carried out as $W/k \sim F(k, d)$ rather than as $(d - k + 1)W/(kd) \sim F(k, d - k + 1)$, where k = the dimension of the test and d = the total number of sampled PSUs minus the total number of strata.

minimum is a highly technical option. It first performs the test with the constant added. If this test is singular, coefficients are dropped until the test becomes nonsingular. Then the test without the constant is performed with the remaining terms.

The following option is available with test but is not shown in the dialog box:

matvlc(*matname*), a programmer's option, saves the variance–covariance matrix of the linear combinations involved in the suite of tests. For the test of the linear constraints $Lb = c$, *matname* contains $L\mathbf{V}L'$, where $\mathbf{V}$ is the estimated variance–covariance matrix of b.

Remarks

Remarks are presented under the following headings:

> *Introductory examples*
> *Special syntaxes after multiple-equation estimation*
> *Constrained coefficients*
> *Multiple testing*

Introductory examples

test performs F or χ^2 tests of linear restrictions applied to the most recently fit model (for example, regress or svy: regress in the linear regression case; logit, stcox, svy: logit, ... in the single-equation maximum-likelihood case; and mlogit, mvreg, streg, ... in the multiple-equation maximum-likelihood case). test may be used after *any* estimation command, although for maximum likelihood techniques, test produces a Wald test that depends only on the estimate of the covariance matrix—you may prefer to use the more computationally expensive likelihood-ratio test; see [U] **20 Estimation and postestimation commands** and [R] **lrtest**.

There are several variations on the syntax for test. The second syntax,

$$\text{test } exp\text{=}exp\left[\text{=}\ldots\right]$$

is allowed after any form of estimation. After fitting a model of *depvar* on x1, x2, and x3, typing test x1+x2=x3 tests the restriction that the coefficients on x1 and x2 sum to the coefficient on x3. The expressions can be arbitrarily complicated; for instance, typing test x1+2*(x2+x3)=x2+3*x3 is the same as typing test x1+x2=x3.

As a convenient shorthand, test also allows you to specify equality for multiple expressions; for example, test x1+x2 = x3+x4 = x5+x6 tests that the three specified pairwise sums of coefficients are equal.

test understands that when you type x1, you are referring to the coefficient on x1. You could also more explicitly type test _b[x1]+_b[x2]=_b[x3]; or you could test _coef[x1]+_coef[x2]=_coef[x3], or test [#1]x1+[#1]x2=[#1]x3, or many other things because there is more than one way to refer to an estimated coefficient; see [U] **13.5 Accessing coefficients and standard errors**. The shorthand involves less typing. On the other hand, you must be more explicit after estimation of multiple-equation models because there may be more than one coefficient associated with an independent variable. You might type, for instance, test [#2]x1+[#2]x2=[#2]x3 to test the constraint in equation 2 or, more readably, test [ford]x1+[ford]x2=[ford]x3, meaning that Stata will test the constraint on the equation corresponding to ford, which might be equation 2. ford would be an equation name after, say, sureg, or, after mlogit, ford would be one of the outcomes. For mlogit, you could also type test [2]x1+[2]x2=[2]x3—note the lack of the #—meaning not equation 2, but the equation corresponding to the numeric outcome 2. You can even test constraints across equations: test [ford]x1+[ford]x2=[buick]x3.

The syntax

$$\text{test } \textit{coeflist}$$

is available after all estimation commands and is a convenient way to test that multiple coefficients are zero following estimation. A *coeflist* can simply be a list of variable names,

$$\text{test } \textit{varname} \left[\textit{varname} \dots \right]$$

and it is most often specified that way. After you have fit a model of depvar on x1, x2, and x3, typing test x1 x3 tests that the coefficients on x1 and x3 are jointly zero. After multiple-equation estimation, this would test that the coefficients on x1 and x3 are zero in all equations that contain them. You can also be more explicit and type, for instance, test [ford]x1 [ford]x3 to test that the coefficients on x1 and x3 are zero in the equation for ford.

In the multiple-equation case, there are more alternatives. You could also test that the coefficients on x1 and x3 are zero in the equation for ford by typing test [ford]: x1 x3. You could test that all coefficients except the coefficient on the constant are zero in the equation for ford by typing test [ford]. You could test that the coefficients on x1 and x3 in the equation for ford are equal to the corresponding coefficients in the equation corresponding to buick by typing test [ford=buick]: x1 x3. You could test that all the corresponding coefficients except the constant in three equations are equal by typing test [ford=buick=volvo].

testparm is much like the first syntax of test. Its usefulness will be demonstrated below.

The examples below use regress, but what is said applies equally after any single-equation estimation command (such as logistic). It also applies after multiple-equation estimation commands as long as references to coefficients are qualified with an equation name or number in square brackets placed before them. The convenient syntaxes for dealing with tests of many coefficients in multiple-equation models are demonstrated in *Special syntaxes after multiple-equation estimation* below.

▷ Example 1

We have 1980 census data on the 50 states recording the birth rate in each state (brate), the median age (medage), and the region of the country in which each state is located.

The region variable is 1 if the state is in the Northeast, 2 if the state is in the North Central, 3 if the state is in the South, and 4 if the state is in the West. We estimate the following regression:

```
. use http://www.stata-press.com/data/r12/census3
(1980 Census data by state)

. regress brate medage c.medage#c.medage i.region
```

Source	SS	df	MS
Model	38803.4208	5	7760.68416
Residual	3393.39921	44	77.1227094
Total	42196.82	49	861.159592

```
Number of obs =      50
F(  5,    44) =  100.63
Prob > F      =  0.0000
R-squared     =  0.9196
Adj R-squared =  0.9104
Root MSE      =   8.782
```

brate	Coef.	Std. Err.	t	P>|t|	[95% Conf. Interval]
medage	-109.0958	13.52452	-8.07	0.000	-136.3527 -81.83892
c.medage# c.medage	1.635209	.2290536	7.14	0.000	1.173582 2.096836
region					
2	15.00283	4.252067	3.53	0.001	6.433353 23.57231
3	7.366445	3.953335	1.86	0.069	-.6009775 15.33387
4	21.39679	4.650601	4.60	0.000	12.02412 30.76946
_cons	1947.611	199.8405	9.75	0.000	1544.859 2350.363

`test` can now be used to perform a variety of statistical tests. Specify the `coeflegend` option with your estimation command to see a legend of the coefficients and how to specify them; see [R] **estimation options**. We can test the hypothesis that the coefficient on `3.region` is zero by typing

```
. test 3.region=0

 ( 1)  3.region = 0

       F(  1,    44) =    3.47
            Prob > F =    0.0691
```

The F statistic with 1 numerator and 44 denominator degrees of freedom is 3.47. The significance level of the test is 6.91%—we can reject the hypothesis at the 10% level but not at the 5% level.

This result from `test` is identical to one presented in the output from `regress`, which indicates that the t statistic on the `3.region` coefficient is 1.863 and that its significance level is 0.069. The t statistic presented in the output can be used to test the hypothesis that the corresponding coefficient is zero, although it states the test in slightly different terms. The F distribution with 1 numerator degree of freedom is, however, identical to the t^2 distribution. We note that $1.863^2 \approx 3.47$ and that the significance levels in each test agree, although one extra digit is presented by the `test` command.

◁

❏ Technical note

After all estimation commands, including those that use the maximum likelihood method, the test that one variable is zero is identical to that reported by the command's output. The tests are performed in the same way—using the estimated covariance matrix—and are known as Wald tests. If the estimation command reports significance levels and confidence intervals using z rather than t statistics, `test` reports results using the χ^2 rather than the F statistic.

❏

▷ Example 2

If that were all `test` could do, it would be useless. We can use `test`, however, to perform other tests. For instance, we can `test` the hypothesis that the coefficient on `2.region` is 21 by typing

```
. test 2.region=21
( 1)   2.region = 21
       F(  1,    44) =     1.99
            Prob > F =    0.1654
```

We find that we cannot reject that hypothesis, or at least we cannot reject it at any significance level below 16.5%.

◁

▷ Example 3

The previous test is useful, but we could almost as easily perform it by hand using the results presented in the regression output if we were well read on our statistics. We could type

```
. display Ftail(1,44,((_coef[2.region]-21)/4.252068)^2)
.16544873
```

So, now let's `test` something a bit more difficult: whether the coefficient on `2.region` is the same as the coefficient on `4.region`:

```
. test 2.region=4.region
( 1)   2.region - 4.region = 0
       F(  1,    44) =     2.84
            Prob > F =    0.0989
```

We find that we cannot reject the equality hypothesis at the 5% level, but we can at the 10% level.

◁

▷ Example 4

When we tested the equality of the `2.region` and `4.region` coefficients, Stata rearranged our algebra. When Stata displayed its interpretation of the specified test, it indicated that we were testing whether `2.region` *minus* `4.region` is zero. The rearrangement is innocuous and, in fact, allows Stata to perform much more complicated algebra, for instance,

```
. test 2*(2.region-3*(3.region-4.region))=3.region+2.region+6*(4.region-3.region)
( 1)   2.region - 3.region = 0
       F(  1,    44) =     5.06
            Prob > F =    0.0295
```

Although we requested what appeared to be a lengthy hypothesis, once Stata simplified the algebra, it realized that all we wanted to do was test whether the coefficient on `2.region` is the same as the coefficient on `3.region`.

◁

❏ Technical note

Stata's ability to simplify and test complex hypotheses is limited to *linear* hypotheses. If you attempt to `test` a nonlinear hypothesis, you will be told that it is not possible:

```
. test 2.region/3.region=2.region+3.region
not possible with test
r(131);
```

To test a nonlinear hypothesis, see [R] **testnl**.

❏

▷ Example 5

The real power of test is demonstrated when we test *joint* hypotheses. Perhaps we wish to test whether the region variables, taken as a whole, are significant by testing whether the coefficients on 2.region, 3.region, and 4.region are simultaneously zero. test allows us to specify multiple conditions to be tested, each embedded within parentheses.

```
. test (2.region=0) (3.region=0) (4.region=0)
 ( 1)  2.region = 0
 ( 2)  3.region = 0
 ( 3)  4.region = 0
       F(  3,    44) =     8.85
            Prob > F =    0.0001
```

test displays the set of conditions and reports an F statistic of 8.85. test also reports the degrees of freedom of the test to be 3, the "dimension" of the hypothesis, and the residual degrees of freedom, 44. The significance level of the test is close to 0, so we can strongly reject the hypothesis of no difference between the regions.

An alternative method to specify simultaneous hypotheses uses the convenient shorthand of conditions with multiple equality operators.

```
. test 2.region=3.region=4.region=0
 ( 1)  2.region - 3.region = 0
 ( 2)  2.region - 4.region = 0
 ( 3)  2.region = 0
       F(  3,    44) =     8.85
            Prob > F =    0.0001
```

◁

❏ Technical note

Another method to test simultaneous hypotheses is to specify a test for each constraint and accumulate it with the previous constraints:

```
. test 2.region=0
 ( 1)  2.region = 0
       F(  1,    44) =    12.45
            Prob > F =     0.0010
. test 3.region=0, accumulate
 ( 1)  2.region = 0
 ( 2)  3.region = 0
       F(  2,    44) =     6.42
            Prob > F =     0.0036
. test 4.region=0, accumulate
 ( 1)  2.region = 0
 ( 2)  3.region = 0
 ( 3)  4.region = 0
       F(  3,    44) =     8.85
            Prob > F =     0.0001
```

We tested the hypothesis that the coefficient on 2.region was zero by typing test 2.region=0. We then tested whether the coefficient on 3.region was also zero by typing test 3.region=0, accumulate. The accumulate option told Stata that this was not the start of a new test but a continuation of a previous one. Stata responded by showing us the two equations and reporting an F statistic of 6.42. The significance level associated with those two coefficients being zero is 0.36%.

When we added the last constraint `test 4.region=0, accumulate`, we discovered that the three region variables are significant. If all we wanted was the overall significance and we did not want to bother seeing the interim results, we could have used the `notest` option:

```
. test 2.region=0, notest
 ( 1)   2.region = 0
. test 3.region=0, accumulate notest
 ( 1)   2.region = 0
 ( 2)   3.region = 0
. test 4.region=0, accumulate
 ( 1)   2.region = 0
 ( 2)   3.region = 0
 ( 3)   4.region = 0
       F(  3,    44) =     8.85
            Prob > F =    0.0001
```

❑

▷ Example 6

Because tests that coefficients are zero are so common in applied statistics, the `test` command has a more convenient syntax to accommodate this case:

```
. test 2.region 3.region 4.region
 ( 1)   2.region = 0
 ( 2)   3.region = 0
 ( 3)   4.region = 0
       F(  3,    44) =     8.85
            Prob > F =    0.0001
```

◁

▷ Example 7

We will now show how to use `testparm`. In its first syntax, `test` accepts a list of variable names but not a *varlist*.

```
. test i(2/4).region
i not found
r(111);
```

In the varlist, `i(2/4).region` means all the level variables from `2.region` through `4.region`, yet we received an error. `test` does not actually understand varlists, but `testparm` does. In fact, it understands only varlists.

```
. testparm i(2/4).region
 ( 1)   2.region = 0
 ( 2)   3.region = 0
 ( 3)   4.region = 0
       F(  3,    44) =     8.85
            Prob > F =    0.0001
```

Another way to test all the `region` variables is to type `testparm i.region`.

That `testparm` accepts varlists has other advantages that do not involve factor variables. Suppose that we have a dataset that has dummy variables `reg2`, `reg3`, and `reg4`, rather than the categorical variable `region`.

```
. use http://www.stata-press.com/data/r12/census4
(birth rate, median age)
. regress brate medage c.medage#c.medage reg2 reg3 reg4
  (output omitted )
. test reg2-reg4
- not found
r(111);
```

In a varlist, reg2-reg4 means variables reg2 and reg4 and all the variables between, yet we received an error. test is confused because the - has two meanings: it means subtraction in an expression and "through" in a *varlist*. Similarly, '*' means "any set of characters" in a *varlist* and multiplication in an expression. testparm avoids this confusion—it allows only a *varlist*.

```
. testparm reg2-reg4
 ( 1)   reg2 = 0
 ( 2)   reg3 = 0
 ( 3)   reg4 = 0
        F(  3,    44) =     8.85
             Prob > F =    0.0001
```

testparm has another advantage. We have five variables in our dataset that start with the characters reg: region, reg1, reg2, reg3, and reg4. reg* thus means those five variables:

```
. describe reg*
              storage  display          value
variable name    type   format          label    variable label

region           int    %8.0g           region   Census Region
reg1             byte   %9.0g                    region==NE
reg2             byte   %9.0g                    region==N Cntrl
reg3             byte   %9.0g                    region==South
reg4             byte   %9.0g                    region==West
```

We cannot type test reg* because, in an expression, '*' means multiplication, but here is what would happen if we attempted to test all the variables that begin with reg:

```
. test region reg1 reg2 reg3 reg4
region not found
r(111);
```

The variable region was not included in our model, so it was not found. However, with testparm,

```
. testparm reg*
 ( 1)   reg2 = 0
 ( 2)   reg3 = 0
 ( 3)   reg4 = 0
        F(  3,    44) =     8.85
             Prob > F =    0.0001
```

That is, testparm took reg* to mean all variables that start with reg that were in our model.

◁

❑ Technical note

Actually, reg* means what it always does—all variables in our dataset that begin with reg—in this case, region reg1 reg2 reg3 reg4. testparm just ignores any variables you specify that are not in the model.

❑

▷ Example 8

We just used test (testparm, actually, but it does not matter) to test the hypothesis that reg2, reg3, and reg4 are jointly zero. We can review the results of our last test by typing test without arguments:

```
. test
 ( 1)  reg2 = 0
 ( 2)  reg3 = 0
 ( 3)  reg4 = 0
       F(  3,    44) =     8.85
            Prob > F =    0.0001
```

◁

❑ Technical note

test does not care how we build joint hypotheses; we may freely mix different forms of syntax. (We can even start with testparm, but we cannot use it thereafter because it does not have an accumulate option.)

Say that we type test reg2 reg3 reg4 to test that the coefficients on our region dummies are jointly zero. We could then add a fourth constraint, say, that medage = 100, by typing test medage=100, accumulate. Or, if we had introduced the medage constraint first (our first test command had been test medage=100), we could then add the region dummy test by typing test reg2 reg3 reg4, accumulate or test (reg2=0) (reg3=0) (reg4=0), accumulate.

Remember that all previous tests are cleared when we do not specify the accumulate option. No matter what tests we performed in the past, if we type test medage c.medage#c.medage, omitting the accumulate option, we would test that medage and c.medage#c.medage are jointly zero.

❑

▷ Example 9

Let's return to our census3.dta dataset and test the hypothesis that all the included regions have the *same* coefficient—that the Northeast is significantly different from the rest of the nation:

```
. use http://www.stata-press.com/data/r12/census3
(1980 Census data by state)
. regress brate medage c.medage#c.medage i.region
 (output omitted )
. test 2.region=3.region=4.region
 ( 1)  2.region - 3.region = 0
 ( 2)  2.region - 4.region = 0
       F(  2,    44) =     8.23
            Prob > F =    0.0009
```

We find that they are not all the same. The syntax 2.region=3.region=4.region with multiple = operators is just a convenient shorthand for typing that the first expression equals the second expression and that the first expression equals the third expression,

```
. test (2.region=3.region) (2.region=4.region)
```

We performed the test for equality of the three regions by imposing two constraints: region 2 has the same coefficient as region 3, and region 2 has the same coefficient as region 4. Alternatively, we could have tested that the coefficients on regions 2 and 3 are the same and that the coefficients on regions 3 and 4 are the same. We would obtain the same results in either case.

To test for equality of the three regions, we might, likely by mistake, type equality constraints for *all* pairs of regions:

```
. test (2.region=3.region) (2.region=4.region) (3.region=4.region)
 ( 1)   2.region - 3.region = 0
 ( 2)   2.region - 4.region = 0
 ( 3)   3.region - 4.region = 0
        Constraint 3 dropped
        F(  2,    44) =    8.23
             Prob > F =    0.0009
```

Equality of regions 2 and 3 and of regions 2 and 4, however, implies equality of regions 3 and 4. test recognized that the last constraint is implied by the other constraints and hence dropped it.

◁

❑ Technical note

Generally, Stata uses = for assignment, as in **gen** *newvar* = *exp*, and == as the operator for testing equality in expressions. For your convenience, test allows both = and == to be used.

❑

▷ Example 10

The test for the equality of the regions is also possible with the **testparm** command. When we include the **equal** option, **testparm** tests that the coefficients of all the variables specified are equal:

```
. testparm i(2/4).region, equal
 ( 1) - 2.region + 3.region = 0
 ( 2) - 2.region + 4.region = 0
        F(  2,    44) =    8.23
             Prob > F =    0.0009
```

We can also obtain the equality test by accumulating single equality tests.

```
. test 2.region=3.region, notest
 ( 1)   2.region - 3.region = 0
. test 2.region=4.region, accum
 ( 1)   2.region - 3.region = 0
 ( 2)   2.region - 4.region = 0
        F(  2,    44) =    8.23
             Prob > F =    0.0009
```

◁

❑ Technical note

If we specify a set of inconsistent constraints, test will tell us by dropping the constraint or constraints that led to the inconsistency. For instance, let's test that the coefficients on region 2 and region 4 are the same, add the test that the coefficient on region 2 is 20, and finally add the test that the coefficient on region 4 is 21:

```
. test (2.region=4.region) (2.region=20) (4.region=21)
 ( 1)   2.region - 4.region = 0
 ( 2)   2.region = 20
 ( 3)   4.region = 21
        Constraint 2 dropped
        F(  2,    44) =    1.82
             Prob > F =    0.1737
```

test informed us that it was dropping constraint 2. All three equations cannot be simultaneously true, so test drops whatever it takes to get back to something that makes sense.

❑

Special syntaxes after multiple-equation estimation

Everything said above about tests after single-equation estimation applies to tests after multiple-equation estimation, as long as you remember to specify the equation name. To demonstrate, let's estimate a seemingly unrelated regression by using sureg; see [R] **sureg**.

```
. use http://www.stata-press.com/data/r12/auto
(1978 Automobile Data)
. sureg (price foreign mpg displ) (weight foreign length)
Seemingly unrelated regression
```

Equation	Obs	Parms	RMSE	"R-sq"	chi2	P
price	74	3	2165.321	0.4537	49.64	0.0000
weight	74	2	245.2916	0.8990	661.84	0.0000

	Coef.	Std. Err.	z	P>|z|	[95% Conf. Interval]	
price						
foreign	3058.25	685.7357	4.46	0.000	1714.233	4402.267
mpg	-104.9591	58.47209	-1.80	0.073	-219.5623	9.644042
displacement	18.18098	4.286372	4.24	0.000	9.779842	26.58211
_cons	3904.336	1966.521	1.99	0.047	50.0263	7758.645
weight						
foreign	-147.3481	75.44314	-1.95	0.051	-295.2139	.517755
length	30.94905	1.539895	20.10	0.000	27.93091	33.96718
_cons	-2753.064	303.9336	-9.06	0.000	-3348.763	-2157.365

To test the significance of foreign in the price equation, we could type

```
. test [price]foreign
 ( 1)  [price]foreign = 0
           chi2( 1) =    19.89
         Prob > chi2 =    0.0000
```

which is the same result reported by sureg: $4.460^2 \approx 19.89$. To test foreign in both equations, we could type

```
. test [price]foreign [weight]foreign
 ( 1)  [price]foreign = 0
 ( 2)  [weight]foreign = 0
           chi2( 2) =    31.61
         Prob > chi2 =    0.0000
```

or

```
. test foreign
 ( 1)  [price]foreign = 0
 ( 2)  [weight]foreign = 0
           chi2( 2) =    31.61
         Prob > chi2 =    0.0000
```

This last syntax—typing the variable name by itself—tests the coefficients in all equations in which they appear. The variable length appears in only the weight equation, so typing

```
. test length
 ( 1)  [weight]length = 0
           chi2( 1) =   403.94
         Prob > chi2 =    0.0000
```

yields the same result as typing `test [weight]length`. We may also specify a linear expression rather than a list of coefficients:

```
. test mpg=displ

 ( 1)  [price]mpg - [price]displ = 0

        chi2(  1) =     4.85
      Prob > chi2 =    0.0277
```

or

```
. test [price]mpg = [price]displ

 ( 1)  [price]mpg - [price]displ = 0

        chi2(  1) =     4.85
      Prob > chi2 =    0.0277
```

A variation on this syntax can be used to test cross-equation constraints:

```
. test [price]foreign = [weight]foreign

 ( 1)  [price]foreign - [weight]foreign = 0

        chi2(  1) =    23.07
      Prob > chi2 =    0.0000
```

Typing an equation name in square brackets by itself tests all the coefficients except the intercept in that equation:

```
. test [price]

 ( 1)  [price]foreign = 0
 ( 2)  [price]mpg = 0
 ( 3)  [price]displacement = 0

        chi2(  3) =    49.64
      Prob > chi2 =    0.0000
```

Typing an equation name in square brackets, a colon, and a list of variable names tests those variables in the specified equation:

```
. test [price]: foreign displ

 ( 1)  [price]foreign = 0
 ( 2)  [price]displacement = 0

        chi2(  2) =    25.19
      Prob > chi2 =    0.0000
```

test [*eqname*$_1$=*eqname*$_2$] tests that all the coefficients in the two equations are equal. We cannot use that syntax here because there are different variables in the model:

```
. test [price=weight]
variables differ between equations
(to test equality of coefficients in common, specify option -common-)
r(111);
```

The common option specifies a test of the equality coefficients common to the equations price and weight,

```
. test [price=weight], common

 ( 1)  [price]foreign - [weight]foreign = 0

        chi2(  1) =    23.07
      Prob > chi2 =    0.0000
```

By default, `test` does not include the constant, the coefficient of the constant variable `_cons`, in the test. The `cons` option specifies that the constant be included.

```
. test [price=weight], common cons
 ( 1)  [price]foreign - [weight]foreign = 0
 ( 2)  [price]_cons - [weight]_cons = 0
           chi2(  2) =    51.23
          Prob > chi2 =  0.0000
```

We can also use a modification of this syntax with the model if we also type a colon and the names of the variables we want to test:

```
. test [price=weight]: foreign
 ( 1)  [price]foreign - [weight]foreign = 0
           chi2(  1) =    23.07
          Prob > chi2 =     0.0000
```

We have only one variable in common between the two equations, but if there had been more, we could have listed them.

Finally, a simultaneous test of multiple constraints may be specified just as after single-equation estimation.

```
. test ([price]: foreign) ([weight]: foreign)
 ( 1)  [price]foreign = 0
 ( 2)  [weight]foreign = 0
           chi2(  2) =    31.61
          Prob > chi2 =  0.0000
```

`test` can also test for equality of coefficients across more than two equations. For instance, `test [eq1=eq2=eq3]` specifies a test that the coefficients in the three equations eq1, eq2, and eq3 are equal. This requires that the same variables be included in the three equations. If some variables are entered only in some of the equations, you can type `test [eq1=eq2=eq3], common` to test that the coefficients of the variables common to all three equations are equal. Alternatively, you can explicitly list the variables for which equality of coefficients across the equations is to be tested. For instance, `test [eq1=eq2=eq3]: time money` tests that the coefficients of the variables `time` and `money` do not differ between the equations.

❏ Technical note

`test [eq1=eq2=eq3], common` tests the equality of the coefficients common to all equations, but it does *not* test the equality of all common coefficients. Consider the case where

eq1	contains the variables var1 var2 var3
eq2	contains the variables var1 var2 var4
eq3	contains the variables var1 var3 var4

Obviously, only `var1` is common to all three equations. Thus `test [eq1=eq2=eq3], common` tests that the coefficients of `var1` do not vary across the equations, so it is equivalent to `test [eq1=eq2=eq3]: var1`. To perform a test of the coefficients of variables common to two equations, you could explicitly list the constraints to be tested,

```
. test ([eq1=eq2=eq3]:var1) ([eq1=eq2]:var2) ([eq1=eq3]:var3) ([eq2=eq3]:var4)
```

or use `test` with the `accumulate` option, and maybe also with the `notest` option, to form the appropriate joint hypothesis:

```
. test [eq1=eq2], common notest
. test [eq1=eq3], common accumulate notest
. test [eq2=eq3], common accumulate
```

❏

Constrained coefficients

If the test indicates that the data do not allow you to conclude that the constraints are not satisfied, you may want to inspect the constrained coefficients. The `coef` option specified that the constrained results, estimated by GLS, are shown.

```
. test [price=weight], common coef

 ( 1)   [price]foreign - [weight]foreign = 0

          chi2(  1) =    23.07
        Prob > chi2 =     0.0000
```

Constrained coefficients

	Coef.	Std. Err.	z	P>|z|	[95% Conf. Interval]	
price						
foreign	-216.4015	74.06083	-2.92	0.003	-361.558	-71.2449
mpg	-121.5717	58.36972	-2.08	0.037	-235.9742	-7.169116
displacement	7.632566	3.681114	2.07	0.038	.4177148	14.84742
_cons	7312.856	1834.034	3.99	0.000	3718.215	10907.5
weight						
foreign	-216.4015	74.06083	-2.92	0.003	-361.558	-71.2449
length	30.34875	1.534815	19.77	0.000	27.34057	33.35693
_cons	-2619.719	302.6632	-8.66	0.000	-3212.928	-2026.51

The constrained coefficient of `foreign` is −216.40 with standard error 74.06 in equations `price` and `weight`. The other coefficients and their standard errors are affected by imposing the equality constraint of the two coefficients of `foreign` because the unconstrained estimates of these two coefficients were correlated with the estimates of the other coefficients.

❏ Technical note

The two-step constrained coefficients b_c displayed by `test, coef` are asymptotically equivalent to the one-stage constrained estimates that are computed by specifying the constraints during estimation using the `constraint()` option of estimation commands (Gourieroux and Monfort 1995, chap. 10). Generally, one-step constrained estimates have better small-sample properties. For inspection and interpretation, however, two-step constrained estimates are a convenient alternative. Moreover, some estimation commands (for example, `stcox`, many `xt` estimators) do not have a `constraint()` option.

❏

Multiple testing

When performing the test of a joint hypothesis, you might want to inspect the underlying 1-degree-of-freedom hypotheses. Which constraint "is to blame"? test displays the univariate as well as the simultaneous test if the mtest option is specified. For example,

```
. test [price=weight], common cons mtest
 ( 1)   [price]foreign - [weight]foreign = 0
 ( 2)   [price]_cons - [weight]_cons = 0
```

	chi2	df	p
(1)	23.07	1	0.0000 #
(2)	11.17	1	0.0008 #
all	51.23	2	0.0000

```
                # unadjusted p-values
```

Both coefficients seem to contribute to the highly significant result. The 1-degree-of-freedom test shown here is identical to those if test had been invoked to test just this simple hypotheses. There is, of course, a real risk in inspecting these simple hypotheses. Especially in high-dimensional hypotheses, you may easily find one hypothesis that happens to be significant. Multiple testing procedures are designed to provide some safeguard against this risk. p-values of the univariate hypotheses are modified so that the probability of falsely rejecting one of the null hypotheses is bounded. test provides the methods based on Bonferroni, Šidák, and Holm.

```
. test [price=weight], common cons mtest(b)
 ( 1)   [price]foreign - [weight]foreign = 0
 ( 2)   [price]_cons - [weight]_cons = 0
```

	chi2	df	p
(1)	23.07	1	0.0000 #
(2)	11.17	1	0.0017 #
all	51.23	2	0.0000

```
            # Bonferroni-adjusted p-values
```

Saved results

test and testparm save the following in r():

Scalars
r(p)	two-sided p-value	r(chi2)	χ^2
r(F)	F statistic	r(ss)	sum of squares (test)
r(df)	test constraints degrees of freedom	r(rss)	residual sum of squares
r(df_r)	residual degrees of freedom	r(drop)	1 if constraints were dropped, 0
r(dropped_i)	index of ith constraint dropped		otherwise

Macros
r(mtmethod)	method of adjustment for multiple testing

Matrices
r(mtest)	multiple test results

r(ss) and r(rss) are defined only when test is used for testing effects after anova.

Methods and formulas

test and testparm are implemented as ado-files.

test and testparm perform Wald tests. Let the estimated coefficient vector be $\mathbf{b}$ and the estimated variance–covariance matrix be $\mathbf{V}$. Let $\mathbf{Rb} = \mathbf{r}$ denote the set of q linear hypotheses to be tested jointly.

The Wald test statistic is (Judge et al. 1985, 20–28)

$$W = (\mathbf{Rb} - \mathbf{r})'(\mathbf{RVR'})^{-1}(\mathbf{Rb} - \mathbf{r})$$

If the estimation command reports its significance levels using Z statistics, a chi-squared distribution with q degrees of freedom,

$$W \sim \chi_q^2$$

is used for computation of the significance level of the hypothesis test.

If the estimation command reports its significance levels using t statistics with d degrees of freedom, an F statistic,

$$F = \frac{1}{q}W$$

is computed, and an F distribution with q numerator degrees of freedom and d denominator degrees of freedom computes the significance level of the hypothesis test.

The two-step constrained estimates b_c displayed by test with the coef option are the GLS estimates of the unconstrained estimates b subject to the specified constraints $Rb = c$ (Gourieroux and Monfort 1995, chap. 10),

$$\mathbf{b_c} = \mathbf{b} - \mathbf{R'}(\mathbf{RVR'})^{-1}\mathbf{R}(\mathbf{Rb} - \mathbf{r})$$

with variance–covariance matrix

$$\mathbf{V_c} = \mathbf{V} - \mathbf{VR'}(\mathbf{RVR'})^{-1}\mathbf{RV}$$

If test displays a Wald test for joint (simultaneous) hypotheses, it can also display all 1-degree-of-freedom tests, with p-values adjusted for multiple testing. Let $p_1, p_2, \ldots, p_k$ be the unadjusted p-values of these 1-degree-of-freedom tests. The Bonferroni-adjusted p-values are defined as $p_i^b = \min(1, kp_i)$. The Šidák-adjusted p-values are $p_i^s = 1 - (1 - p_i)^k$. Holm's method for adjusting p-values is defined as $p_i^h = \min(1, k_ip_i)$, where k_i is the number of p-values at least as large as p_i. Note that $p_i^h < p_i^b$, reflecting that Holm's method is strictly less conservative than the widely used Bonferroni method.

If test is used after a svy command, it carries out an adjusted Wald test—this adjustment should not be confused with the adjustment for multiple testing. Both adjustments may actually be combined. Specifically, the survey adjustment uses an approximate F statistic $(d-k+1)W/(kd)$, where W is the Wald test statistic, k is the dimension of the hypothesis test, and $d =$ the total number of sampled PSUs minus the total number of strata. Under the null hypothesis, $(d-k+1)F/(kd) \sim F(k, d-k+1)$, where $F(k, d-k+1)$ is an F distribution with k numerator degrees of freedom and $d-k+1$ denominator degrees of freedom. If nosvyadjust is specified, the p-value is computed using $W/k \sim F(k, d)$.

See Korn and Graubard (1990) for a detailed description of the Bonferroni adjustment technique and for a discussion of the relative merits of it and of the adjusted and unadjusted Wald tests.

Acknowledgment

The svy adjustment code was adopted from another command developed in collaboration with John L. Eltinge, Bureau of Labor Statistics.

References

Beale, E. M. L. 1960. Confidence regions in non-linear estimation. *Journal of the Royal Statistical Society, Series B* 22: 41–88.

Eltinge, J. L., and W. M. Sribney. 1996. svy5: Estimates of linear combinations and hypothesis tests for survey data. *Stata Technical Bulletin* 31: 31–42. Reprinted in *Stata Technical Bulletin Reprints*, vol. 6, pp. 246–259. College Station, TX: Stata Press.

Gourieroux, C., and A. Monfort. 1995. *Statistics and Econometric Models, Vol 1: General Concepts, Estimation, Prediction, and Algorithms*. Trans. Q. Vuong. Cambridge: Cambridge University Press.

Holm, S. 1979. A simple sequentially rejective multiple test procedure. *Scandinavian Journal of Statistics* 6: 65–70.

Judge, G. G., W. E. Griffiths, R. C. Hill, H. Lütkepohl, and T.-C. Lee. 1985. *The Theory and Practice of Econometrics*. 2nd ed. New York: Wiley.

Korn, E. L., and B. I. Graubard. 1990. Simultaneous testing of regression coefficients with complex survey data: Use of Bonferroni *t* statistics. *American Statistician* 44: 270–276.

Weesie, J. 1999. sg100: Two-stage linear constrained estimation. *Stata Technical Bulletin* 47: 24–30. Reprinted in *Stata Technical Bulletin Reprints*, vol. 8, pp. 217–225. College Station, TX: Stata Press.

Also see

[R] **anova** — Analysis of variance and covariance

[R] **anova postestimation** — Postestimation tools for anova

[R] **contrast** — Contrasts and linear hypothesis tests after estimation

[R] **lincom** — Linear combinations of estimators

[R] **lrtest** — Likelihood-ratio test after estimation

[R] **nestreg** — Nested model statistics

[R] **nlcom** — Nonlinear combinations of estimators

[R] **testnl** — Test nonlinear hypotheses after estimation

[U] **13.5 Accessing coefficients and standard errors**

[U] **20 Estimation and postestimation commands**

Title

testnl — Test nonlinear hypotheses after estimation

Syntax

> testnl *exp* = *exp* [= *exp* ...] [, *options*]

> testnl (*exp* = *exp* [= *exp* ...]) [(*exp* = *exp* [= *exp* ...]) ...] [, *options*]

options	Description
<u>m</u>test[(*opt*)]	test each condition separately
nosvyadjust	carry out the Wald test as $W/k \sim F(k, d)$; for use with svy estimation commands
<u>iterate</u>(#)	use maximum # of iterations to find the optimal step size

The second syntax means that if more than one expression is specified, each must be surrounded by parentheses.

exp is a possibly nonlinear expression containing

> _b[*coef*]
> _b[*eqno*:*coef*]
> [*eqno*]*coef*
> [*eqno*]_b[*coef*]

eqno is

> ##
> *name*

coef identifies a coefficient in the model. *coef* is typically a variable name, a level indicator, an interaction indicator, or an interaction involving continuous variables. Level indicators identify one level of a factor variable and interaction indicators identify one combination of levels of an interaction; see [U] **11.4.3 Factor variables**. *coef* may contain time-series operators; see [U] **11.4.4 Time-series varlists**.

Distinguish between [ ], which are to be typed, and [ ], which indicate optional arguments.

Menu

Statistics > Postestimation > Tests > Test nonlinear hypotheses

Description

testnl tests (linear or nonlinear) hypotheses about the estimated parameters from the most recently fit model.

2186

testnl produces Wald-type tests of smooth nonlinear (or linear) hypotheses about the estimated parameters from the most recently fit model. The p-values are based on the delta method, an approximation appropriate in large samples.

testnl can be used with svy estimation results; see [SVY] **svy postestimation**.

The format $(exp_1 = exp_2 = exp_3 \ldots)$ for a simultaneous-equality hypothesis is just a convenient shorthand for a list $(exp_1 = exp_2)$ $(exp_1 = exp_3)$, etc.

testnl may also be used to test linear hypotheses. test is faster if you want to test only linear hypotheses; see [R] **test**. testnl is the only option for testing linear and nonlinear hypotheses simultaneously.

Options

mtest $\big[(opt)\big]$ specifies that tests be performed for each condition separately. *opt* specifies the method for adjusting p-values for multiple testing. Valid values for *opt* are

<u>b</u>onferroni	Bonferroni's method
<u>h</u>olm	Holm's method
<u>s</u>idak	Šidák's method
<u>noadj</u>ust	no adjustment is to be made

Specifying mtest without an argument is equivalent to specifying mtest(noadjust).

nosvyadjust is for use with svy estimation commands; see [SVY] **svy estimation**. It specifies that the Wald test be carried out without the default adjustment for the design degrees of freedom. That is, the test is carried out as $W/k \sim F(k, d)$ rather than as $(d - k + 1)W/(kd) \sim F(k, d - k + 1)$, where $k =$ the dimension of the test and $d =$ the total number of sampled PSUs minus the total number of strata.

iterate(#) specifies the maximum number of iterations used to find the optimal step size in the calculation of numerical derivatives of the test expressions. By default, the maximum number of iterations is 100, but convergence is usually achieved after only a few iterations. You should rarely have to use this option.

Remarks

Remarks are presented under the following headings:

Introduction
Using testnl to perform linear tests
Specifying constraints
Dropped constraints
Output
Multiple constraints
Manipulability

Introduction

▷ Example 1

We have just estimated the parameters of an earnings model on cross-sectional time-series data using one of Stata's more sophisticated estimators:

```
. use http://www.stata-press.com/data/r12/earnings
(NLS Women 14-24 in 1968)

. xtgee ln_w grade age c.age#c.age, corr(exchangeable) nolog
```

GEE population-averaged model					Number of obs	=	1326
Group variable:			idcode		Number of groups	=	269
Link:			identity		Obs per group: min	=	1
Family:			Gaussian		avg	=	4.9
Correlation:			exchangeable		max	=	9
					Wald chi2(3)	=	327.33
Scale parameter:			.0976738		Prob > chi2	=	0.0000

ln_wage	Coef.	Std. Err.	z	P>|z|	[95% Conf.	Interval]
grade	.0749686	.0066111	11.34	0.000	.062011	.0879261
age	.1080806	.0235861	4.58	0.000	.0618526	.1543086
c.age#c.age	−.0016253	.0004739	−3.43	0.001	−.0025541	−.0006966
_cons	−.8788933	.2830899	−3.10	0.002	−1.433739	−.3240473

An implication of this model is that peak earnings occur at age −_b[age]/(2*_b[c.age#c.age]), which here is equal to 33.2. Say that we have a theory that peak earnings should occur at age 16 + 1/_b[grade].

```
. testnl -_b[age]/(2*_b[c.age#c.age]) = 16 + 1/_b[grade]
  (1)  -_b[age]/(2*_b[c.age#c.age]) = 16 + 1/_b[grade]
            chi2(1) =        1.71
          Prob > chi2 =      0.1914
```

These data do not reject our theory.

◁

Using testnl to perform linear tests

testnl may be used to test linear constraints, but test is faster; see [R] **test**. You could type

```
. testnl _b[x4] = _b[x1]
```

but it would take less computer time if you typed

```
. test _b[x4] = _b[x1]
```

Specifying constraints

The constraints to be tested can be formulated in many different ways. You could type

```
. testnl _b[mpg]*_b[weight] = 1
```

or

```
. testnl _b[mpg] = 1/_b[weight]
```

or you could express the constraint any other way you wished. (To say that `testnl` allows constraints to be specified in different ways does not mean that the test itself does not depend on the formulation. This point is briefly discussed later.) In formulating the constraints, you must, however, exercise one caution: users of `test` often refer to the coefficient on a variable by specifying the variable name. For example,

```
. test mpg = 0
```

More formally, they should type

```
. test _b[mpg] = 0
```

but `test` allows the `_b[]` surrounding the variable name to be omitted. `testnl` does not allow this shorthand. Typing

```
. testnl mpg=0
```

specifies the constraint that the value of variable `mpg` in the first observation is zero. If you make this mistake, sometimes `testnl` will catch it:

```
. testnl mpg=0
equation (1) contains reference to X rather than _b[X]
r(198);
```

In other cases, `testnl` may not catch the mistake; then the constraint will be dropped because it does not make sense:

```
. testnl mpg=0
  Constraint (1) dropped
```

(There are reasons other than this for constraints being dropped.) The worst case, however, is

```
. testnl _b[weight]*mpg = 1
```

when what you mean is not that `_b[weight]` equals the reciprocal of the value of `mpg` in the first observation, but rather that

```
. testnl _b[weight]*_b[mpg] = 1
```

Sometimes this mistake will be caught by the "contains reference to X rather than _b[X]" error, and sometimes it will not. Be careful.

`testnl`, like `test`, can be used after any Stata estimation command, including the survey estimators. When you use it after a multiple-equation command, such as `mlogit` or `heckman`, you refer to coefficients by using Stata's standard syntax: [*eqname*] _b[*varname*].

Stata's single-equation estimation output looks like this:

	Coef	...			
weight	12.27	...	<-	coefficient is	_b[weight]
mpg	3.21	...			

Stata's multiple-equation output looks like this:

	Coef ...
cat1	...
weight	12.27 ... <- coefficient is [cat1]_b[weight]
mpg	3.21 ...
8	...
weight	5.83 ... <- coefficient is [8]_b[weight]
mpg	7.43 ...

Dropped constraints

testnl automatically drops constraints when

- They are nonbinding, for example, _b[mpg]=_b[mpg]. More subtle cases include

      _b[mpg]*_b[weight] = 4
      _b[weight] = 2
      _b[mpg] = 2

In this example, the third constraint is nonbinding because it is implied by the first two.

- They are contradictory, for example, _b[mpg]=2 and _b[mpg]=3. More subtle cases include

      _b[mpg]*_b[weight] = 4
      _b[weight] = 2
      _b[mpg] = 3

The third constraint contradicts the first two.

Output

testnl reports the constraints being tested followed by an F or a χ^2 test:

```
. use http://www.stata-press.com/data/r12/auto
(1978 Automobile Data)
. regress price mpg weight c.weight#c.weight foreign
(output omitted)
. testnl (39*_b[mpg]^2 = _b[foreign]) (_b[mpg]/_b[weight] = 4)

 (1)  39*_b[mpg]^2 = _b[foreign]
 (2)  _b[mpg]/_b[weight] = 4

           F(2, 69) =        0.08
           Prob > F =      0.9195
. logit foreign price weight mpg
(output omitted)
. testnl (45*_b[mpg]^2 = _b[price]) (_b[mpg]/_b[weight] = 4)

 (1)  45*_b[mpg]^2 = _b[price]
 (2)  _b[mpg]/_b[weight] = 4

           chi2(2) =        2.44
           Prob > chi2 =    0.2946
```

Multiple constraints

▷ Example 2

We illustrate the simultaneous test of a series of constraints using simulated data on labor-market promotion in a given year. We fit a probit model with separate effects for education, experience, and experience-squared for men and women.

```
. use http://www.stata-press.com/data/r12/promotion
. probit promo male male#c.(yedu yexp yexp2), nolog
```

Probit regression					Number of obs	=	775
					LR chi2(7)	=	424.42
					Prob > chi2	=	0.0000
Log likelihood = -245.42768					Pseudo R2	=	0.4637

promo	Coef.	Std. Err.	z	P>|z|	[95% Conf. Interval]	
male	.6489974	.203739	3.19	0.001	.2496763	1.048318
male#c.yedu						
0	.9730237	.1056136	9.21	0.000	.7660248	1.180023
1	1.390517	.1527288	9.10	0.000	1.091174	1.68986
male#c.yexp						
0	.4559544	.0901169	5.06	0.000	.2793285	.6325803
1	1.422539	.1544255	9.21	0.000	1.11987	1.725207
male#c.yexp2						
0	-.1027149	.0573059	-1.79	0.073	-.2150325	.0096026
1	-.3749457	.1160113	-3.23	0.001	-.6023236	-.1475677
_cons	.9872018	.1148215	8.60	0.000	.7621559	1.212248

Note: 1 failure and 2 successes completely determined.

The effects of human capital seem to differ between men and women. A formal test confirms this.

```
. test (yedu#0.male = yedu#1.male) (yexp#0.male = yexp#1.male)
> (yexp2#0.male = yexp2#1.male)
 ( 1)  [promo]0b.male#c.yedu - [promo]1.male#c.yedu = 0
 ( 2)  [promo]0b.male#c.yexp - [promo]1.male#c.yexp = 0
 ( 3)  [promo]0b.male#c.yexp2 - [promo]1.male#c.yexp2 = 0
           chi2(  3) =     35.43
         Prob > chi2 =     0.0000
```

How do we interpret this gender difference? It has repeatedly been stressed (see, for example, Long [1997, 47–50]; Allison [1999]) that comparison of groups in binary response models, and similarly in other latent-variable models, is hampered by an identification problem: with β the regression coefficients for the latent variable and σ the standard deviation of the latent residual, only the β/σ are identified. In fact, in terms of the latent regression, the probit coefficients should be interpreted as β/σ, not as the β. If we cannot claim convincingly that the residual standard deviation σ does not vary between the sexes, equality of the regression coefficients β implies that the coefficients of the probit model for men and women are *proportional* but not necessarily equal. This is a nonlinear hypothesis in terms of the probit coefficients, not a linear one.

```
. testnl _b[yedu#1.male]/_b[yedu#0.male] = _b[yexp#1.male]/_b[yexp#0.male]
> = _b[yexp2#1.male]/_b[yexp2#0.male]

 (1)  _b[yedu#1.male]/_b[yedu#0.male] = _b[yexp#1.male]/_b[yexp#0.male]
 (2)  _b[yedu#1.male]/_b[yedu#0.male] = _b[yexp2#1.male]/_b[yexp2#0.male]

            chi2(2) =        9.21
         Prob > chi2 =      0.0100
```

We conclude that we find fairly strong evidence against the proportionality of the coefficients, and hence we have to conclude that success in the labor market is produced in different ways by men and women. (But remember, these were simulated data.)

◁

▷ Example 3

The syntax for specifying the equality of multiple expressions is just a convenient shorthand for specifying a series of constraints, namely, that the first expression equals the second expression, the first expression also equals the third expression, etc. The Wald test performed and the output of testnl are the same whether we use the shorthand or we specify the series of constraints. The lengthy specification as a series of constraints can be simplified using the continuation symbols ///.

```
. testnl (_b[yedu#1.male]/_b[yedu#0.male] =  ///
         _b[yexp#1.male]/_b[yexp#0.male])   ///
         (_b[yedu#1.male]/_b[yedu#0.male] =  ///
         _b[yexp2#1.male]/_b[yexp2#0.male])

 (1)  _b[yedu#1.male]/_b[yedu#0.male] = _b[yexp#1.male]/_b[yexp#0.male]
 (2)  _b[yedu#1.male]/_b[yedu#0.male] = _b[yexp2#1.male]/_b[yexp2#0.male]

            chi2(2) =        9.21
         Prob > chi2 =      0.0100
```

Having established differences between men and women, we would like to do multiple testing between the ratios. Because we did not specify hypotheses in advance, we prefer to adjust the p-values of tests using, here, Bonferroni's method.

```
. testnl _b[yedu#1.male]/_b[yedu#0.male] = ///
         _b[yexp#1.male]/_b[yexp#0.male] = ///
         _b[yexp2#1.male]/_b[yexp2#0.male], mtest(b)

 (1)  _b[yedu#1.male]/_b[yedu#0.male] = _b[yexp#1.male]/_b[yexp#0.male]
 (2)  _b[yedu#1.male]/_b[yedu#0.male] = _b[yexp2#1.male]/_b[yexp2#0.male]
```

	chi2	df	p
(1)	6.89	1	0.0173 #
(2)	0.93	1	0.6713 #
all	9.21	2	0.0100

    # Bonferroni-adjusted p-values

◁

Manipulability

Although testnl allows you to specify constraints in different ways that are mathematically equivalent, as noted above, this does not mean that the tests are the same. This difference is known as the manipulability of the Wald test for nonlinear hypotheses; also see [R] **boxcox**. The test might even be significant for one formulation but not significant for another formulation that is mathematically equivalent. Trying out different specifications to find a formulation with the desired p-value is totally inappropriate, though it may actually be fun to try. There is no variance under representation because the nonlinear Wald test is actually a standard Wald test for a linearization of the constraint, which depends on the particular specification. We note that the likelihood-ratio test is not manipulable in this sense.

From a statistical point of view, it is best to choose a specification of the constraints that is as linear is possible. Doing so usually improves the accuracy of the approximation of the null-distribution of the test by a χ^2 or an F distribution. The example above used the nonlinear Wald test to test whether the coefficients of human capital variables for men were proportional to those of women. A specification of proportionality of coefficients in terms of ratios of coefficients is fairly nonlinear if the coefficients in the denominator are close to 0. A more linear version of the test results from a bilinear formulation. Thus instead of

```
. testnl _b[yedu#1.male]/_b[yedu#0.male] = _b[yexp#1.male]/_b[yexp#0.male]

  (1)  _b[yedu#1.male]/_b[yedu#0.male] = _b[yexp#1.male]/_b[yexp#0.male]

             chi2(1) =       6.89
           Prob > chi2 =     0.0087
```

perhaps

```
. testnl _b[yedu#1.male]*_b[yexp#0.male] = _b[yedu#0.male]*_b[yexp#1.male]

  (1)  _b[yedu#1.male]*_b[yexp#0.male] = _b[yedu#0.male]*_b[yexp#1.male]

             chi2(1) =      13.95
           Prob > chi2 =     0.0002
```

is better, and in fact it has been suggested that the latter version of the test is more reliable. This assertion is confirmed by performing simulations and is in line with theoretical results of Phillips and Park (1988). There is strong evidence against the proportionality of human capital effects between men and women, implying for this example that differences in the residual variances between the sexes can be ruled out as the explanation of the sex differences in the analysis of labor market participation.

Saved results

testnl saves the following in r():

Scalars
r(df)	degrees of freedom
r(df_r)	residual degrees of freedom
r(chi2)	χ^2
r(p)	significance
r(F)	F statistic

Macros
r(mtmethod)	method specified in mtest()

Matrices
r(G)	derivatives of $R(\mathbf{b})$ with respect to $\mathbf{b}$; see *Methods and formulas* below
r(R)	$R(\mathbf{b}) - \mathbf{q}$; see *Methods and formulas* below
r(mtest)	multiple test results

Methods and formulas

`testnl` is implemented as an ado-file.

After fitting a model, define $\mathbf{b}$ as the resulting $1 \times k$ parameter vector and $\mathbf{V}$ as the $k \times k$ covariance matrix. The (linear or nonlinear) hypothesis is given by $R(\mathbf{b}) = \mathbf{q}$, where R is a function returning a $j \times 1$ vector. The Wald test formula is (Greene 2012, 528)

$$W = \left\{ R(\mathbf{b}) - \mathbf{q} \right\}' \left(\mathbf{GVG'} \right)^{-1} \left\{ R(\mathbf{b}) - \mathbf{q} \right\}$$

where $\mathbf{G}$ is the derivative matrix of $R(\mathbf{b})$ with respect to $\mathbf{b}$. W is distributed as χ^2 if $\mathbf{V}$ is an asymptotic covariance matrix. $F = W/j$ is distributed as F for linear regression.

The adjustment methods for multiple testing are described in [R] **test**. The adjustment for survey design effects is described in [SVY] **svy postestimation**.

References

Allison, P. D. 1999. Comparing logit and probit coefficients across groups. *Sociological Methods and Research* 28: 186–208.

Gould, W. W. 1996. crc43: Wald test of nonlinear hypotheses after model estimation. *Stata Technical Bulletin* 29: 2–4. Reprinted in *Stata Technical Bulletin Reprints*, vol. 5, pp. 15–18. College Station, TX: Stata Press.

Greene, W. H. 2012. *Econometric Analysis*. 7th ed. Upper Saddle River, NJ: Prentice Hall.

Long, J. S. 1997. *Regression Models for Categorical and Limited Dependent Variables*. Thousand Oaks, CA: Sage.

Phillips, P. C. B., and J. Y. Park. 1988. On the formulation of Wald tests of nonlinear restrictions. *Econometrica* 56: 1065–1083.

Also see

[R] **contrast** — Contrasts and linear hypothesis tests after estimation

[R] **lincom** — Linear combinations of estimators

[R] **lrtest** — Likelihood-ratio test after estimation

[R] **nlcom** — Nonlinear combinations of estimators

[R] **test** — Test linear hypotheses after estimation

[U] **13.5 Accessing coefficients and standard errors**

[U] **20 Estimation and postestimation commands**

Title

> **tetrachoric** — Tetrachoric correlations for binary variables

Syntax

tetrachoric *varlist* [*if*] [*in*] [*weight*] [, *options*]

options	Description
Main	
stats(*statlist*)	list of statistics; select up to 4 statistics; default is stats(rho)
edwards	use the noniterative Edwards and Edwards estimator; default is the maximum likelihood estimator
print(#)	significance level for displaying coefficients
star(#)	significance level for displaying with a star
bonferroni	use Bonferroni-adjusted significance level
sidak	use Šidák-adjusted significance level
pw	calculate all the pairwise correlation coefficients by using all available data (pairwise deletion)
zeroadjust	adjust frequencies when one cell has a zero count
matrix	display output in matrix form
notable	suppress display of correlations
posdef	modify correlation matrix to be positive semidefinite

statlist	Description
rho	tetrachoric correlation coefficient
se	standard error of rho
obs	number of observations
p	exact two-sided significance level

by is allowed; see [D] **by**.

fweights are allowed; see [U] **11.1.6 weight**.

Menu

Statistics > Summaries, tables, and tests > Summary and descriptive statistics > Tetrachoric correlations

Description

tetrachoric computes estimates of the tetrachoric correlation coefficients of the binary variables in *varlist*. All these variables should be 0, 1, or missing values.

Tetrachoric correlations assume a latent bivariate normal distribution (X_1, X_2) for each pair of variables (v_1, v_2), with a threshold model for the manifest variables, $v_i = 1$ if and only if $X_i > 0$. The means and variances of the latent variables are not identified, but the correlation, r, of X_1 and X_2 can be estimated from the joint distribution of v_1 and v_2 and is called the tetrachoric correlation coefficients.

tetrachoric computes pairwise estimates of the tetrachoric correlations by the (iterative) maximum likelihood estimator obtained from bivariate probit without explanatory variables (see [R] **biprobit**) by using the Edwards and Edwards (1984) noniterative estimator as the initial value.

The pairwise correlation matrix is returned as r(Rho) and can be used to perform a factor analysis or a principal component analysis of binary variables by using the factormat or pcamat commands; see [MV] **factor** and [MV] **pca**.

Options

─────┐ Main └──

stats(*statlist*) specifies the statistics to be displayed in the matrix of output. stats(rho) is the default. Up to four statistics may be specified. stats(rho se p obs) would display the tetrachoric correlation coefficient, its standard error, the significance level, and the number of observations. If *varlist* contains only two variables, all statistics are shown in tabular form. stats(), print(), and star() have no effect unless the matrix option is also specified.

edwards specifies that the noniterative Edwards and Edwards estimator be used. The default is the maximum likelihood estimator. If you analyze many binary variables, you may want to use the fast noniterative estimator proposed by Edwards and Edwards (1984). However, if you have skewed variables, the approximation does not perform well.

print(#) specifies the maximum significance level of correlation coefficients to be printed. Correlation coefficients with larger significance levels are left blank in the matrix. Typing tetrachoric ... , print(.10) would list only those correlation coefficients that are significant at the 10% level or lower.

star(#) specifies the maximum significance level of correlation coefficients to be marked with a star. Typing tetrachoric ... , star(.05) would "star" all correlation coefficients significant at the 5% level or lower.

bonferroni makes the Bonferroni adjustment to calculated significance levels. This option affects printed significance levels and the print() and star() options. Thus tetrachoric ... , print(.05) bonferroni prints coefficients with Bonferroni-adjusted significance levels of 0.05 or less.

sidak makes the Šidák adjustment to calculated significance levels. This option affects printed significance levels and the print() and star() options. Thus tetrachoric ... , print(.05) sidak prints coefficients with Šidák-adjusted significance levels of 0.05 or less.

pw specifies that the tetrachoric correlation be calculated by using all available data. By default, tetrachoric uses casewise deletion, where observations are ignored if any of the specified variables in *varlist* are missing.

zeroadjust specifies that when one of the cells has a zero count, a frequency adjustment be applied in such a way as to increase the zero to one-half and maintain row and column totals.

matrix forces tetrachoric to display the statistics as a matrix, even if *varlist* contains only two variables. matrix is implied if more than two variables are specified.

notable suppresses the output.

posdef modifies the correlation matrix so that it is positive semidefinite, that is, a proper correlation matrix. The modified result is the correlation matrix associated with the least-squares approximation of the tetrachoric correlation matrix by a positive-semidefinite matrix. If the correlation matrix is modified, the standard errors and significance levels are not displayed and are returned in r().

Remarks

Remarks are presented under the following headings:

Association in 2-by-2 tables
Factor analysis of dichotomous variables
Tetrachoric correlations with simulated data

Association in 2-by-2 tables

Although a wide variety of measures of association in cross tabulations have been proposed, such measures are essentially equivalent (monotonically related) in the special case of 2×2 tables—there is only 1 degree of freedom for nonindependence. Still, some measures have more desirable properties than others. Here we compare two measures: the standard Pearson correlation coefficient and the tetrachoric correlation coefficient. Given asymmetric row or column margins, Pearson correlations are limited to a range smaller than -1 to 1, although tetrachoric correlations can still span the range from -1 to 1. To illustrate, consider the following set of tables for two binary variables, X and Z:

	Z = 0	Z = 1	
X = 0	$20 - a$	$10 + a$	30
X = 1	a	$10 - a$	10
	20	20	40

For a equal to 0, 1, 2, 5, 8, 9, and 10, the Pearson and tetrachoric correlations for the above table are

a	0	1	2	5	8	9	10
Pearson	0.577	0.462	0.346	0	-0.346	-0.462	-0.577
Tetrachoric	1.000	0.792	0.607	0	-0.607	-0.792	-1.000

The restricted range for the Pearson correlation is especially unfortunate when you try to analyze the association between binary variables by using models developed for continuous data, such as factor analysis and principal component analysis.

The tetrachoric correlation of two variables (Y_1, Y_2) can be thought of as the Pearson correlation of two latent bivariate normal distributed variables (Y_1^*, Y_2^*) with threshold measurement models $Y_i = (Y_i^* > c_i)$ for unknown cutpoints c_i. Or equivalently, $Y_i = (Y_i^{**} > 0)$ where the latent bivariate normal (Y_1^{**}, Y_2^{**}) are shifted versions of (Y_1^*, Y_2^*) so that the cutpoints are zero. Obviously, you must judge whether assuming underlying latent variables is meaningful for the data. If this assumption is justified, tetrachoric correlations have two advantages. First, you have an intuitive understanding of the size of correlations that are substantively interesting in your field of research, and this intuition is based on correlations that range from -1 to 1. Second, because the tetrachoric correlation for binary variables estimates the Pearson correlation of the latent continuous variables (assumed multivariate-normal distributed), you can use the tetrachoric correlations to analyze multivariate relationships between the dichotomous variables. When doing so, remember that you must interpret the model in terms of the underlying continuous variables.

▷ Example 1

To illustrate tetrachoric correlations, we examine three binary variables from the `familyvalues` dataset (described in example 2).

```
. use http://www.stata-press.com/data/r12/familyvalues
(Attitudes on gender, relationships and family)
. tabulate RS075 RS076
```

fam att: women in charge bad	fam att: trad division of labor 0	1	Total
0	1,564	979	2,543
1	119	632	751
Total	1,683	1,611	3,294

```
. correlate RS074 RS075 RS076
(obs=3291)
```

	RS074	RS075	RS076
RS074	1.0000		
RS075	0.0396	1.0000	
RS076	0.1595	0.3830	1.0000

```
. tetrachoric RS074 RS075 RS076
(obs=3291)
```

	RS074	RS075	RS076
RS074	1.0000		
RS075	0.0689	1.0000	
RS076	0.2480	0.6427	1.0000

As usual, the tetrachoric correlation coefficients are larger (in absolute value) and more dispersed than the Pearson correlations.

◁

Factor analysis of dichotomous variables

▷ Example 2

Factor analysis is a popular model for measuring latent continuous traits. The standard estimators are appropriate only for continuous unimodal data. Because of the skewness implied by Bernoulli-distributed variables (especially when the probability is distributed unevenly), a factor analysis of a Pearson correlation matrix can be rather misleading when used in this context. A factor analysis of a matrix of tetrachoric correlations is more appropriate under these conditions (Uebersax 2000). We illustrate this with data on gender, relationship, and family attitudes of spouses using the Households in The Netherlands survey 1995 (Weesie et al. 1995). For attitude variables, it seems reasonable to assume that agreement or disagreement is just a coarse measurement of more nuanced underlying attitudes.

To demonstrate, we examine a few of the variables from the `familyvalues` dataset.

```
. use http://www.stata-press.com/data/r12/familyvalues
(Attitudes on gender, relationships and family)
. describe RS056-RS063
```

variable name	storage type	display format	value label	variable label
RS056	byte	%9.0g		fam att: should be together
RS057	byte	%9.0g		fam att: should fight for relat
RS058	byte	%9.0g		fam att: should avoid conflict
RS059	byte	%9.0g		fam att: woman better nurturer
RS060	byte	%9.0g		fam att: both spouses money goo
RS061	byte	%9.0g		fam att: woman techn school goo
RS062	byte	%9.0g		fam att: man natural breadwinne
RS063	byte	%9.0g		fam att: common leisure good

```
. summarize RS056-RS063
```

Variable	Obs	Mean	Std. Dev.	Min	Max
RS056	3298	.5630685	.4960816	0	1
RS057	3296	.5400485	.4984692	0	1
RS058	3283	.6387451	.4804374	0	1
RS059	3308	.654474	.4756114	0	1
RS060	3302	.3906723	.487975	0	1
RS061	3293	.7102946	.4536945	0	1
RS062	3307	.5857272	.4926705	0	1
RS063	3298	.5379018	.498637	0	1

```
. correlate RS056-RS063
(obs=3221)
```

	RS056	RS057	RS058	RS059	RS060	RS061	RS062
RS056	1.0000						
RS057	0.1350	1.0000					
RS058	0.2377	0.0258	1.0000				
RS059	0.1816	0.0097	0.2550	1.0000			
RS060	−0.1020	−0.0538	−0.0424	0.0126	1.0000		
RS061	−0.1137	0.0610	−0.1375	−0.2076	0.0706	1.0000	
RS062	0.2014	0.0285	0.2273	0.4098	−0.0793	−0.2873	1.0000
RS063	0.2057	0.1460	0.1049	0.0911	0.0179	−0.0233	0.0975

	RS063
RS063	1.0000

Skewness in these data is relatively modest. For comparison, here are the tetrachoric correlations:

```
. tetrachoric RS056-RS063
(obs=3221)
```

	RS056	RS057	RS058	RS059	RS060	RS061	RS062
RS056	1.0000						
RS057	0.2114	1.0000					
RS058	0.3716	0.0416	1.0000				
RS059	0.2887	0.0158	0.4007	1.0000			
RS060	-0.1620	-0.0856	-0.0688	0.0208	1.0000		
RS061	-0.1905	0.1011	-0.2382	-0.3664	0.1200	1.0000	
RS062	0.3135	0.0452	0.3563	0.6109	-0.1267	-0.4845	1.0000
RS063	0.3187	0.2278	0.1677	0.1467	0.0286	-0.0388	0.1538

	RS063
RS063	1.0000

Again we see that the tetrachoric correlations are generally larger in absolute value than the Pearson correlations. The bivariate probit and Edwards and Edwards estimators (the edwards option) implemented in tetrachoric may return a correlation matrix that is not positive semidefinite—a mathematical property of any real correlation matrix. Positive definiteness is required by commands for analyses of correlation matrices, such as factormat and pcamat; see [MV] **factor** and [MV] **pca**. The posdef option of tetrachoric tests for positive definiteness and projects the estimated correlation matrix to a positive-semidefinite matrix if needed.

```
. tetrachoric RS056-RS063, notable posdef
. matrix C = r(Rho)
```

This time, we suppressed the display of the correlations with the notable option and requested that the correlation matrix be positive semidefinite with the posdef option. Had the correlation matrix not been positive definite, tetrachoric would have displayed a warning message and then adjusted the matrix to be positive semidefinite. We placed the resulting tetrachoric correlation matrix into a matrix, C, so that we can perform a factor analysis upon it.

tetrachoric with the posdef option asserted that C was positive definite because no warning message was displayed. We can verify this by using a familiar characterization of symmetric positive-definite matrices: all eigenvalues are real and positive.

```
. matrix symeigen eigenvectors eigenvalues = C
. matrix list eigenvalues
eigenvalues[1,8]
            e1          e2          e3          e4          e5          e6          e7
r1   2.5974789   1.3544664   1.0532476   .77980391   .73462018   .57984565   .54754512
            e8
r1   .35299228
```

We can proceed with a factor analysis on the matrix C. We use factormat and select iterated principal factors as the estimation method; see [MV] **factor**.

```
. factormat C, n(3221) ipf factor(2)
(obs=3221)
```

Factor analysis/correlation	Number of obs	=	3221
Method: iterated principal factors	Retained factors	=	2
Rotation: (unrotated)	Number of params	=	15

Factor	Eigenvalue	Difference	Proportion	Cumulative
Factor1	2.06855	1.40178	0.7562	0.7562
Factor2	0.66677	0.47180	0.2438	1.0000
Factor3	0.19497	0.06432	0.0713	1.0713
Factor4	0.13065	0.10967	0.0478	1.1191
Factor5	0.02098	0.10085	0.0077	1.1267
Factor6	-0.07987	0.01037	-0.0292	1.0975
Factor7	-0.09024	0.08626	-0.0330	1.0645
Factor8	-0.17650	.	-0.0645	1.0000

LR test: independent vs. saturated: chi2(28) = 4620.01 Prob>chi2 = 0.0000

Factor loadings (pattern matrix) and unique variances

Variable	Factor1	Factor2	Uniqueness
RS056	0.5528	0.4120	0.5247
RS057	0.1124	0.4214	0.8098
RS058	0.5333	0.0718	0.7105
RS059	0.6961	-0.1704	0.4865
RS060	-0.1339	-0.0596	0.9785
RS061	-0.5126	0.2851	0.6560
RS062	0.7855	-0.2165	0.3361
RS063	0.2895	0.3919	0.7626

◁

▷ Example 3

We noted in example 2 that the matrix of estimates of the tetrachoric correlation coefficients need not be positive definite. Here is an example:

```
. use http://www.stata-press.com/data/r12/familyvalues
(Attitudes on gender, relationships and family)
. tetrachoric RS056-RS063 in 1/20, posdef
(obs=18)
```

matrix with tetrachoric correlations is not positive semidefinite;
  it has 2 negative eigenvalues
  maxdiff(corr,adj-corr) = 0.2346
  (adj-corr: tetrachoric correlations adjusted to be positive semidefinite)

adj-corr	RS056	RS057	RS058	RS059	RS060	RS061	RS062
RS056	1.0000						
RS057	0.5284	1.0000					
RS058	0.3012	0.2548	1.0000				
RS059	0.3251	0.2791	0.0550	1.0000			
RS060	-0.5197	-0.4222	-0.7163	0.0552	1.0000		
RS061	0.3448	0.4815	-0.0958	-0.1857	-0.0980	1.0000	
RS062	0.1066	-0.0375	0.0072	0.3909	-0.2333	-0.7654	1.0000
RS063	0.3830	0.4939	0.4336	0.0075	-0.8937	-0.0337	0.4934

adj-corr	RS063
RS063	1.0000

```
. mata
─────────────────────────────────────── mata (type end to exit) ───────
: C2 = st_matrix("r(Rho)")

: eigenvecs = .

: eigenvals = .

: symeigensystem(C2, eigenvecs, eigenvals)

: eigenvals
                      1              2              3              4
    ┌─────────────────────────────────────────────────────────────────┐
  1 │   3.156592567    2.065279398    1.324911199    .7554904485        │
    └─────────────────────────────────────────────────────────────────┘
                      5              6              7              8
    ┌─────────────────────────────────────────────────────────────────┐
  1 │   .4845368741    .2131895139    2.02944e-16    -1.11650e-16        │
    └─────────────────────────────────────────────────────────────────┘

: end
```

The estimated tetrachoric correlation matrix is rank-2 deficient. With this C2 matrix, we can only use models of correlation that allow for singular cases.

◁

Tetrachoric correlations with simulated data

▷ Example 4

We use drawnorm (see [D] **drawnorm**) to generate a sample of 1,000 observations from a bivariate normal distribution with means -1 and 1, unit variances, and correlation 0.4.

```
. clear
. set seed 11000
. matrix m = (1, -1)
. matrix V = (1, 0.4 \ 0.4, 1)
. drawnorm c1 c2, n(1000) means(m) cov(V)
(obs 1000)
```

Now consider the measurement model assumed by the tetrachoric correlations. We observe only whether c1 and c2 are greater than zero,

```
. generate d1 = (c1 > 0)
. generate d2 = (c2 > 0)
. tabulate d1 d2
```

	d2		
d1	0	1	Total
0	176	6	182
1	656	162	818
Total	832	168	1,000

We want to estimate the correlation of c1 and c2 from the binary variables d1 and d2. Pearson's correlation of the binary variables d1 and d2 is 0.170—a seriously biased estimate of the underlying correlation $\rho = 0.4$.

```
. correlate d1 d2
(obs=1000)
```

	d1	d2
d1	1.0000	
d2	0.1704	1.0000

The tetrachoric correlation coefficient of d1 and d2 estimates the Pearson correlation of the latent continuous variables, c1 and c2.

```
. tetrachoric d1 d2
     Number of obs =      1000
   Tetrachoric rho =       0.4790
         Std error =       0.0700

  Test of Ho: d1 and d2 are independent
   2-sided exact P =       0.0000
```

The estimate of the tetrachoric correlation of d1 and d2, 0.4790, is much closer to the underlying correlation, 0.4, between c1 and c2.

◁

Saved results

tetrachoric saves the following in r():

Scalars
r(rho)	tetrachoric correlation coefficient between variables 1 and 2
r(N)	number of observations
r(nneg)	number of negative eigenvalues (posdef only)
r(se_rho)	standard error of r(rho)
r(p)	exact two-sided significance level

Macros
r(method)	estimator used

Matrices
r(Rho)	tetrachoric correlation matrix
r(Se_Rho)	standard errors of r(Rho)
r(Nobs)	number of observations used in computing correlation
r(P)	exact two-sided significance level matrix

Methods and formulas

tetrachoric is implemented as an ado-file.

tetrachoric provides two estimators for the tetrachoric correlation ρ of two binary variables with the frequencies n_{ij}, $i, j = 0, 1$. tetrachoric defaults to the slower (iterative) maximum likelihood estimator obtained from bivariate probit without explanatory variables (see [R] **biprobit**) by using the Edwards and Edwards noniterative estimator as the initial value. A fast (noniterative) estimator is also available by specifying the edwards option (Edwards and Edwards 1984; Digby 1983)

$$\widehat{\rho} = \frac{\alpha - 1}{\alpha + 1}$$

where

$$\alpha = \left(\frac{n_{00}n_{11}}{n_{01}n_{10}}\right)^{\pi/4} \quad (\pi = 3.14\ldots)$$

if all $n_{ij} > 0$. If $n_{00} = 0$ or $n_{11} = 0$, $\widehat{\rho} = -1$; if $n_{01} = 0$ or $n_{10} = 0$, $\widehat{\rho} = 1$.

The asymptotic variance of the Edwards and Edwards estimator of the tetrachoric correlation is easily obtained by the delta method,

$$\mathrm{avar}(\widehat{\rho}) = \left(\frac{\pi\alpha}{2(1+\alpha)^2}\right)^2 \left(\frac{1}{n_{00}} + \frac{1}{n_{01}} + \frac{1}{n_{10}} + \frac{1}{n_{11}}\right)$$

provided all $n_{ij} > 0$, otherwise it is left undefined (missing). The Edwards and Edwards estimator is fast, but may be inaccurate if the margins are very skewed.

tetrachoric reports exact p-values for statistical independence, computed by the exact option of [R] **tabulate twoway**.

References

Brown, M. B. 1977. Algorithm AS 116: The tetrachoric correlation and its asymptotic standard error. *Applied Statistics* 26: 343–351.

Brown, M. B., and J. K. Benedetti. 1977. On the mean and variance of the tetrachoric correlation coefficient. *Psychometrika* 42: 347–355.

Digby, P. G. N. 1983. Approximating the tetrachoric correlation coefficient. *Biometrics* 39: 753–757.

Edwards, J. H., and A. W. F. Edwards. 1984. Approximating the tetrachoric correlation coefficient. *Biometrics* 40: 563.

Golub, G. H., and C. F. Van Loan. 1996. *Matrix Computations*. 3rd ed. Baltimore: Johns Hopkins University Press.

Uebersax, J. S. 2000. Estimating a latent trait model by factor analysis of tetrachoric correlations. http://ourworld.compuserve.com/homepages/jsuebersax/irt.htm.

Weesie, J., M. Kalmijn, W. Bernasco, and D. Giesen. 1995. *Households in The Netherlands 1995*. Utrecht, Netherlands: Datafile, ISCORE, University of Utrecht.

Also see

[MV] **factor** — Factor analysis

[MV] **pca** — Principal component analysis

[R] **tabulate twoway** — Two-way tables of frequencies

[R] **biprobit** — Bivariate probit regression

[R] **correlate** — Correlations (covariances) of variables or coefficients

[R] **spearman** — Spearman's and Kendall's correlations

Title

> **tnbreg** — Truncated negative binomial regression

Syntax

tnbreg *depvar* [*indepvars*] [*if*] [*in*] [*weight*] [, *options*]

options	Description
Model	
<u>nocon</u>stant	suppress constant term
ll(# | *varname*)	truncation point; default value is ll(0), zero truncation
<u>dis</u>persion(<u>m</u>ean)	parameterization of dispersion; the default
<u>dis</u>persion(<u>c</u>onstant)	constant dispersion for all observations
<u>exp</u>osure(*varname_e*)	include ln(*varname_e*) in model with coefficient constrained to 1
<u>off</u>set(*varname_o*)	include *varname_o* in model with coefficient constrained to 1
<u>constraints</u>(*constraints*)	apply specified linear constraints
<u>col</u>linear	keep collinear variables
SE/Robust	
vce(*vcetype*)	*vcetype* may be oim, <u>r</u>obust, <u>cl</u>uster *clustvar*, opg, <u>boot</u>strap, or <u>jack</u>knife
Reporting	
<u>level</u>(#)	set confidence level; default is level(95)
<u>nolr</u>test	suppress likelihood-ratio test
<u>irr</u>	report incidence-rate ratios
<u>nocns</u>report	do not display constraints
display_options	control column formats, row spacing, line width, and display of omitted variables and base and empty cells
Maximization	
maximize_options	control the maximization process; seldom used
<u>coefl</u>egend	display legend instead of statistics

indepvars may contain factor variables; see [U] **11.4.3 Factor variables**.
depvar and *indepvars* may contain time-series operators; see [U] **11.4.4 Time-series varlists**.
bootstrap, by, jackknife, rolling, statsby, and svy are allowed; see [U] **11.1.10 Prefix commands**.
Weights are not allowed with the bootstrap prefix; see [R] **bootstrap**.
vce() and weights are not allowed with the svy prefix; see [SVY] **svy**.
fweights, iweights, and pweights are allowed; see [U] **11.1.6 weight**.
coeflegend does not appear in the dialog box.
See [U] **20 Estimation and postestimation commands** for more capabilities of estimation commands.

Menu

Statistics > Count outcomes > Truncated negative binomial regression

Description

tnbreg estimates the parameters of a truncated negative binomial model by maximum likelihood. The dependent variable *depvar* is regressed on *indepvars*, where *depvar* is a positive count variable whose values are all above the truncation point.

Options

┌─ Model ┐

noconstant; see [R] **estimation options**.

ll(# | *varname*) specifies the truncation point, which is a nonnegative integer. The default is zero truncation, ll(0).

dispersion(mean | constant) specifies the parameterization of the model. dispersion(mean), the default, yields a model with dispersion equal to $1 + \alpha \exp(\mathbf{x}_j\boldsymbol{\beta} + \text{offset}_j)$; that is, the dispersion is a function of the expected mean: $\exp(\mathbf{x}_j\boldsymbol{\beta} + \text{offset}_j)$. dispersion(constant) has dispersion equal to $1 + \delta$; that is, it is a constant for all observations.

exposure(*varname$_e$*), offset(*varname$_o$*), constraints(*constraints*), collinear; see [R] **estimation options**.

┌─ SE/Robust ┐

vce(*vcetype*) specifies the type of standard error reported, which includes types that are derived from asymptotic theory, that are robust to some kinds of misspecification, that allow for intragroup correlation, and that use bootstrap or jackknife methods; see [R] ***vce_option***.

┌─ Reporting ┐

level(#); see [R] **estimation options**.

nolrtest suppresses fitting the Poisson model. Without this option, a comparison Poisson model is fit, and the likelihood is used in a likelihood-ratio test of the null hypothesis that the dispersion parameter is zero.

irr reports estimated coefficients transformed to incidence-rate ratios, that is, e^{β_i} rather than β_i. Standard errors and confidence intervals are similarly transformed. This option affects how results are displayed, not how they are estimated or stored. irr may be specified at estimation or when replaying previously estimated results.

nocnsreport; see [R] **estimation options**.

display_options: noomitted, vsquish, noemptycells, baselevels, allbaselevels, cformat(%*fmt*), pformat(%*fmt*), sformat(%*fmt*), and nolstretch; see [R] **estimation options**.

┌─ Maximization ┐

maximize_options: difficult, technique(*algorithm_spec*), iterate(#), [no]log, trace, gradient, showstep, hessian, showtolerance, tolerance(#), ltolerance(#), nrtolerance(#), nonrtolerance, and from(*init_specs*); see [R] **maximize**. These options are seldom used.

Setting the optimization type to technique(bhhh) resets the default *vcetype* to vce(opg).

The following option is available with tnbreg but is not shown in the dialog box:

coeflegend; see [R] **estimation options**.

Remarks

Grogger and Carson (1991) showed that overdispersion causes inconsistent estimation of the mean in the truncated Poisson model. To solve this problem, they proposed using the truncated negative binomial model as an alternative. If data are truncated but do not exhibit overdispersion, the truncated Poisson model is more appropriate; see [R] **tpoisson**. For an introduction to negative binomial regression, see Cameron and Trivedi (2005, 2010) and Long and Freese (2006). For an introduction to truncated negative binomial models, see Cameron and Trivedi (1998) and Long (1997, chap. 8).

tnbreg fits the mean-dispersion and the constant-dispersion parameterizations of truncated negative binomial models. These parameterizations extend those implemented in nbreg; see [R] **nbreg**.

▷ Example 1

We illustrate the truncated negative binomial model using the 1997 MedPar dataset (Hilbe 1999). The data are from 1,495 patients in Arizona who were assigned to a diagnostic-related group (DRG) of patients having a ventilator. Length of stay (los), the dependent variable, is a positive integer; it cannot have zero values. The data are truncated because there are no observations on individuals who stayed for zero days.

The objective of this example is to determine whether the length of stay was related to the binary variables: died, hmo, type1, type2, and type3.

The died variable was recorded as a 0 unless the patient died, in which case, it was recorded as a 1. The other variables also adopted this encoding. The hmo variable was set to 1 if the patient belonged to a health maintenance organization (HMO).

The type1–type3 variables indicated the type of admission used for the patient. The type1 variable indicated an emergency admit. The type2 variable indicated an urgent admit—that is, the first available bed. The type3 variable indicated an elective admission. Because type1–type3 were mutually exclusive, only two of the three could be used in the truncated negative binomial regression shown below.

```
. use http://www.stata-press.com/data/r12/medpar
. tnbreg los died hmo type2-type3, vce(cluster provnum) nolog
Truncated negative binomial regression
Truncation point: 0                          Number of obs   =         1495
Dispersion     = mean                        Wald chi2(4)    =        36.01
Log likelihood = -4737.535                   Prob > chi2     =       0.0000
                         (Std. Err. adjusted for 54 clusters in provnum)
```

los	Coef.	Robust Std. Err.	z	P>|z|	[95% Conf. Interval]	
died	-.2521884	.061533	-4.10	0.000	-.3727908	-.1315859
hmo	-.0754173	.0533132	-1.41	0.157	-.1799091	.0290746
type2	.2685095	.0666474	4.03	0.000	.137883	.3991359
type3	.7668101	.2183505	3.51	0.000	.338851	1.194769
_cons	2.224028	.034727	64.04	0.000	2.155964	2.292091
/lnalpha	-.630108	.0764019			-.779853	-.480363
alpha	.5325343	.0406866			.4584734	.6185588

Because observations within the same hospital (provnum) are likely to be correlated, we specified the vce(cluster provnum) option. The results show that whether the patient died in the hospital and the type of admission have significant effects on the patient's length of stay.

◁

▷ Example 2

To illustrate truncated negative binomial regression with more complex data than the previous example, similar data were created from 100 hospitals. Each hospital had its own way of tracking patient data. In particular, hospitals only recorded data from patients with a minimum length of stay, denoted by the variable minstay.

Definitions for minimum length of stay varied among hospitals, typically, from 5 to 18 days. The objective of this example is the same as before: to determine whether the length of stay, recorded in los, was related to the binary variables: died, hmo, type1, type2, and type3.

The binary variables encode the same information as in example 1 above. The minstay variable was used to allow for varying truncation points.

```
. use http://www.stata-press.com/data/r12/medproviders
. tnbreg los died hmo type2-type3, ll(minstay) vce(cluster hospital) nolog
Truncated negative binomial regression
Truncation points: minstay                  Number of obs   =       2144
Dispersion     = mean                        Wald chi2(4)    =      15.22
Log likelihood = -7864.0928                  Prob > chi2     =     0.0043
                            (Std. Err. adjusted for 100 clusters in hospital)
```

los	Coef.	Robust Std. Err.	z	P>|z|	[95% Conf. Interval]	
died	.078104	.0303603	2.57	0.010	.0185988	.1376091
hmo	-.0731132	.0368899	-1.98	0.047	-.1454162	-.0008103
type2	.0294132	.0390165	0.75	0.451	-.0470578	.1058843
type3	.0626349	.0540124	1.16	0.246	-.0432275	.1684972
_cons	3.014964	.0291045	103.59	0.000	2.95792	3.072008
/lnalpha	-.9965124	.0829428			-1.159077	-.8339475
alpha	.3691647	.0306196			.3137756	.4343314

In this analysis, two variables have a statistically significant relationship with length of stay. On average, patients who died in the hospital had longer lengths of stay ($p = 0.01$). Because the coefficient for HMO is negative, that is, $b_{\text{HMO}} = -0.073$, on average, patients who were insured by an HMO had shorter lengths of stay ($p = 0.047$). The type of admission was not statistically significant ($p > 0.05$).

◁

Saved results

tnbreg saves the following in e():

Scalars
e(N)	number of observations
e(k)	number of parameters
e(k_aux)	number of auxiliary parameters
e(k_eq)	number of equations in e(b)
e(k_eq_model)	number of equations in overall model test
e(k_dv)	number of dependent variables
e(df_m)	model degrees of freedom
e(r2_p)	pseudo-R-squared
e(ll)	log likelihood
e(ll_0)	log likelihood, constant-only model
e(ll_c)	log likelihood, comparison model
e(alpha)	value of alpha
e(N_clust)	number of clusters
e(chi2)	χ^2
e(chi2_c)	χ^2 for comparison test
e(p)	significance
e(rank)	rank of e(V)
e(rank0)	rank of e(V) for constant-only model
e(ic)	number of iterations
e(rc)	return code
e(converged)	1 if converged, 0 otherwise

Macros
e(cmd)	tnbreg
e(cmdline)	command as typed
e(depvar)	name of dependent variable
e(llopt)	contents of ll(), or 0 if not specified
e(wtype)	weight type
e(wexp)	weight expression
e(title)	title in estimation output
e(clustvar)	name of cluster variable
e(offset)	linear offset variable
e(chi2type)	Wald or LR; type of model χ^2 test
e(chi2_ct)	Wald or LR; type of model χ^2 test corresponding to e(chi2_c)
e(dispers)	mean or constant
e(vce)	*vcetype* specified in vce()
e(vcetype)	title used to label Std. Err.
e(opt)	type of optimization
e(which)	max or min; whether optimizer is to perform maximization or minimization
e(ml_method)	type of ml method
e(user)	name of likelihood-evaluator program
e(technique)	maximization technique
e(properties)	b V
e(predict)	program used to implement predict
e(asbalanced)	factor variables fvset as asbalanced
e(asobserved)	factor variables fvset as asobserved

Matrices
e(b)	coefficient vector
e(Cns)	constraints matrix
e(ilog)	iteration log (up to 20 iterations)
e(gradient)	gradient vector
e(V)	variance–covariance matrix of the estimators
e(V_modelbased)	model-based variance

Functions
e(sample)	marks estimation sample

Methods and formulas

`tnbreg` is implemented as an ado-file.

Methods and formulas are presented under the following headings:

> *Mean-dispersion model*
> *Constant-dispersion model*

Mean-dispersion model

A negative binomial distribution can be regarded as a gamma mixture of Poisson random variables. The number of times an event occurs, y_j, is distributed as $\text{Poisson}(\nu_j \mu_j)$. That is, its conditional likelihood is

$$f(y_j \mid \nu_j) = \frac{(\nu_j \mu_j)^{y_j} e^{-\nu_j \mu_j}}{\Gamma(y_j + 1)}$$

where $\mu_j = \exp(\mathbf{x}_j \boldsymbol{\beta} + \text{offset}_j)$ and ν_j is an unobserved parameter with a $\text{Gamma}(1/\alpha, \alpha)$ density:

$$g(\nu) = \frac{\nu^{(1-\alpha)/\alpha} e^{-\nu/\alpha}}{\alpha^{1/\alpha} \Gamma(1/\alpha)}$$

This gamma distribution has a mean of 1 and a variance of α, where α is our ancillary parameter.

The unconditional likelihood for the jth observation is therefore

$$f(y_j) = \int_0^\infty f(y_j \mid \nu) g(\nu)\, d\nu = \frac{\Gamma(m + y_j)}{\Gamma(y_j + 1)\Gamma(m)}\, p_j^m (1 - p_j)^{y_j}$$

where $p_j = 1/(1 + \alpha \mu_j)$ and $m = 1/\alpha$. Solutions for α are handled by searching for $\ln\alpha$ because α must be greater than zero. The conditional probability of observing y_j events given that y_j is greater than the truncation point τ_j is

$$\Pr(Y = y_j \mid y_j > \tau_j, \mathbf{x}_j) = \frac{f(y_j)}{\Pr(Y > \tau_j \mid \mathbf{x}_j)}$$

The log likelihood (with weights w_j and offsets) is given by

$$m = 1/\alpha \qquad p_j = 1/(1 + \alpha \mu_j) \qquad \mu_j = \exp(\mathbf{x}_j \boldsymbol{\beta} + \text{offset}_j)$$

$$\ln L = \sum_{j=1}^n w_j \left[\ln\{\Gamma(m + y_j)\} - \ln\{\Gamma(y_j + 1)\} \right.$$

$$\left. - \ln\{\Gamma(m)\} + m \ln(p_j) + y_j \ln(1 - p_j) - \ln\{\Pr(Y > \tau_j \mid p_j, m)\} \right]$$

Constant-dispersion model

The constant-dispersion model assumes that y_j is conditionally distributed as Poisson(μ_j^*), where $\mu_j^* \sim$ Gamma($\mu_j/\delta, \delta$) for some dispersion parameter δ [by contrast, the mean-dispersion model assumes that $\mu_j^* \sim$ Gamma($1/\alpha, \alpha\mu_j$)]. The log likelihood is given by

$$m_j = \mu_j/\delta \qquad p = 1/(1+\delta)$$

$$\ln L = \sum_{j=1}^{n} w_j \left[\ln\{\Gamma(m_j + y_j)\} - \ln\{\Gamma(y_j + 1)\} \right.$$

$$\left. - \ln\{\Gamma(m_j)\} + m_j \ln(p) + y_j \ln(1-p) - \ln\{\Pr(Y > \tau_j \mid p, m_j)\} \right]$$

with everything else defined as shown above in the calculations for the mean-dispersion model.

This command supports the Huber/White/sandwich estimator of the variance and its clustered version using `vce(robust)` and `vce(cluster` *clustvar*`)`, respectively. See [P] **_robust**, particularly *Maximum likelihood estimators* and *Methods and formulas*.

`tnbreg` also supports estimation with survey data. For details on variance–covariance estimates with survey data, see [SVY] **variance estimation**.

Acknowledgment

We gratefully acknowledge the previous work by Joseph Hilbe, Arizona State University; see Hilbe (1999).

References

Cameron, A. C., and P. K. Trivedi. 1998. *Regression Analysis of Count Data*. Cambridge: Cambridge University Press.

——. 2005. *Microeconometrics: Methods and Applications*. New York: Cambridge University Press.

——. 2010. *Microeconometrics Using Stata*. Rev. ed. College Station, TX: Stata Press.

Grogger, J. T., and R. T. Carson. 1991. Models for truncated counts. *Journal of Applied Econometrics* 6: 225–238.

Hilbe, J. M. 1998. sg91: Robust variance estimators for MLE Poisson and negative binomial regression. *Stata Technical Bulletin* 45: 26–28. Reprinted in *Stata Technical Bulletin Reprints*, vol. 8, pp. 177–180. College Station, TX: Stata Press.

——. 1999. sg102: Zero-truncated Poisson and negative binomial regression. *Stata Technical Bulletin* 47: 37–40. Reprinted in *Stata Technical Bulletin Reprints*, vol. 8, pp. 233–236. College Station, TX: Stata Press.

Long, J. S. 1997. *Regression Models for Categorical and Limited Dependent Variables*. Thousand Oaks, CA: Sage.

Long, J. S., and J. Freese. 2006. *Regression Models for Categorical Dependent Variables Using Stata*. 2nd ed. College Station, TX: Stata Press.

Simonoff, J. S. 2003. *Analyzing Categorical Data*. New York: Springer.

Also see

[R] **tnbreg postestimation** — Postestimation tools for tnbreg

[R] **nbreg** — Negative binomial regression

[R] **poisson** — Poisson regression

[R] **tpoisson** — Truncated Poisson regression

[R] **zinb** — Zero-inflated negative binomial regression

[R] **zip** — Zero-inflated Poisson regression

[SVY] **svy estimation** — Estimation commands for survey data

[XT] **xtnbreg** — Fixed-effects, random-effects, & population-averaged negative binomial models

[U] **20 Estimation and postestimation commands**

Title

tnbreg postestimation — Postestimation tools for tnbreg

Description

The following postestimation commands are available after `tnbreg`:

Command	Description
contrast	contrasts and ANOVA-style joint tests of estimates
estat	AIC, BIC, VCE, and estimation sample summary
estat (svy)	postestimation statistics for survey data
estimates	cataloging estimation results
lincom	point estimates, standard errors, testing, and inference for linear combinations of coefficients
lrtest[1]	likelihood-ratio test
margins	marginal means, predictive margins, marginal effects, and average marginal effects
marginsplot	graph the results from margins (profile plots, interaction plots, etc.)
nlcom	point estimates, standard errors, testing, and inference for nonlinear combinations of coefficients
predict	predictions, residuals, influence statistics, and other diagnostic measures
predictnl	point estimates, standard errors, testing, and inference for generalized predictions
pwcompare	pairwise comparisons of estimates
suest	seemingly unrelated estimation
test	Wald tests of simple and composite linear hypotheses
testnl	Wald tests of nonlinear hypotheses

[1] `lrtest` is not appropriate with svy estimation results.

See the corresponding entries in the *Base Reference Manual* for details, but see [SVY] **estat** for details about `estat` (svy).

Syntax for predict

> predict [*type*] *newvar* [*if*] [*in*] [, *statistic* <u>nooff</u>set]

> predict [*type*] { *stub** | *newvar*$_\text{reg}$ *newvar*$_\text{disp}$ } [*if*] [*in*] , <u>sc</u>ores

statistic	Description
Main	
n	number of events; the default
ir	incidence rate
cm	conditional mean, $E(y_j \mid y_j > \tau_j)$
pr(*n*)	probability $\Pr(y_j = n)$
pr(*a*,*b*)	probability $\Pr(a \leq y_j \leq b)$
cpr(*n*)	conditional probability $\Pr(y_j = n \mid y_j > \tau_j)$
cpr(*a*,*b*)	conditional probability $\Pr(a \leq y_j \leq b \mid y_j > \tau_j)$
xb	linear prediction
stdp	standard error of the linear prediction

These statistics are available both in and out of sample; type predict ... if e(sample) ... if wanted only for the estimation sample.

Menu

Statistics > Postestimation > Predictions, residuals, etc.

Options for predict

⌐ Main ⌐

n, the default, calculates the predicted number of events, which is $\exp(\mathbf{x}_j\boldsymbol{\beta})$ if neither offset() nor exposure() was specified when the model was fit; $\exp(\mathbf{x}_j\boldsymbol{\beta} + \text{offset}_j)$ if offset() was specified; or $\exp(\mathbf{x}_j\boldsymbol{\beta}) \times \text{exposure}_j$ if exposure() was specified.

ir calculates the incidence rate $\exp(\mathbf{x}_j\boldsymbol{\beta})$, which is the predicted number of events when exposure is 1. This is equivalent to specifying both the n and the nooffset options.

cm calculates the conditional mean,

$$E(y_j \mid y_j > \tau_j) = \frac{E(y_j)}{\Pr(y_j > \tau_j)}$$

where τ_j is the truncation point found in e(llopt).

pr(*n*) calculates the probability $\Pr(y_j = n)$, where n is a nonnegative integer that may be specified as a number or a variable.

pr(*a*,*b*) calculates the probability $\Pr(a \leq y_j \leq b)$, where a and b are nonnegative integers that may be specified as numbers or variables;

b missing ($b \geq$.) means $+\infty$;
pr(20,.) calculates $\Pr(y_j \geq 20)$;
pr(20,*b*) calculates $\Pr(y_j \geq 20)$ in observations for which $b \geq$. and calculates $\Pr(20 \leq y_j \leq b)$ elsewhere.

pr(.,b) produces a syntax error. A missing value in an observation of the variable a causes a missing value in that observation for pr(a,b).

cpr(n) calculates the conditional probability $\Pr(y_j = n \,|\, y_j > \tau_j)$, where τ_j is the truncation point found in e(llopt). n is an integer greater than the truncation point that may be specified as a number or a variable.

cpr(a,b) calculates the conditional probability $\Pr(a \le y_j \le b \,|\, y_j > \tau_j)$, where τ_j is the truncation point found in e(llopt). The syntax for this option is analogous to that used for pr(a,b) except that a must be greater than the truncation point.

xb calculates the linear prediction, which is $\mathbf{x}_j\boldsymbol{\beta}$ if neither offset() nor exposure() was specified when the model was fit; $\mathbf{x}_j\boldsymbol{\beta} + \text{offset}_j$ if offset() was specified; or $\mathbf{x}_j\boldsymbol{\beta} + \ln(\text{exposure}_j)$ if exposure() was specified; see nooffset below.

stdp calculates the standard error of the linear prediction.

nooffset is relevant only if you specified offset() or exposure() when you fit the model. It modifies the calculations made by predict so that they ignore the offset or exposure variable; the linear prediction is treated as $\mathbf{x}_j\boldsymbol{\beta}$ rather than as $\mathbf{x}_j\boldsymbol{\beta} + \text{offset}_j$ or $\mathbf{x}_j\boldsymbol{\beta} + \ln(\text{exposure}_j)$. Specifying predict ..., nooffset is equivalent to specifying predict ..., ir.

scores calculates equation-level score variables.

The first new variable will contain $\partial \ln L / \partial(\mathbf{x}_j\boldsymbol{\beta})$.

The second new variable will contain $\partial \ln L / \partial(\ln\alpha)$ for dispersion(mean).

The second new variable will contain $\partial \ln L / \partial(\ln\delta)$ for dispersion(constant).

Methods and formulas

All postestimation commands listed above are implemented as ado-files.

In the following formulas, we use the same notation as in [R] **tnbreg**.

Methods and formulas are presented under the following headings:

> Mean-dispersion model
> Constant-dispersion model

Mean-dispersion model

The equation-level scores are given by

$$\text{score}(\mathbf{x}\boldsymbol{\beta})_j = p_j(y_j - \mu_j) - \frac{p_j^{(m+1)}\mu_j}{\Pr(Y > \tau_j \,|\, p_j, m)}$$

$$\text{score}(\omega)_j = -m\left\{\frac{\alpha(\mu_j - y_j)}{1 + \alpha\mu_j} - \ln(1 + \alpha\mu_j) + \psi(y_j + m) - \psi(m)\right\}$$

$$- \frac{p_j^m}{\Pr(Y > \tau_j \,|\, p_j, m)}\{m\ln(p_j) + \mu_j p_j\}$$

where $\omega_j = \ln\alpha_j$, $\psi(z)$ is the digamma function, and τ_j is the truncation point found in e(llopt).

Constant-dispersion model

The equation-level scores are given by

$$\text{score}(\mathbf{x}\boldsymbol{\beta})_j = m_j \left\{ \psi(y_j + m_j) - \psi(m_j) + \ln(p) + \frac{p^{m_j} \ln(p)}{\Pr(Y > \tau_j \mid p, m_j)} \right\}$$

$$\text{score}(\omega)_j = y_j - (y_j + m_j)(1 - p) - \text{score}(\mathbf{x}\boldsymbol{\beta})_j - \frac{\mu_j p}{\Pr(Y > \tau_j \mid p, m_j)}$$

where $\omega_j = \ln\delta_j$ and τ_j is the truncation point found in e(llopt).

Also see

[R] **tnbreg** — Truncated negative binomial regression

[U] **20 Estimation and postestimation commands**

Title

> **tobit** — Tobit regression

Syntax

$$\underline{\text{tobit}} \; depvar \; \big[indepvars\big] \; \big[if\big] \; \big[in\big] \; \big[weight\big] \text{, } \text{ll}\big[(\#)\big] \; \text{ul}\big[(\#)\big] \; \big[options\big]$$

options	Description
Model	
<u>nocon</u>stant	suppress constant term
*ll$\big[(\#)\big]$	left-censoring limit
*ul$\big[(\#)\big]$	right-censoring limit
<u>off</u>set(*varname*)	include *varname* in model with coefficient constrained to 1
SE/Robust	
vce(*vcetype*)	*vcetype* may be oim, <u>r</u>obust, <u>cl</u>uster *clustvar*, <u>boot</u>strap, or jackknife
Reporting	
<u>l</u>evel(#)	set confidence level; default is level(95)
display_options	control column formats, row spacing, line width, and display of omitted variables and base and empty cells
Maximization	
maximize_options	control the maximization process; seldom used
<u>coefl</u>egend	display legend instead of statistics

*You must specify at least one of ll$\big[(\#)\big]$ or ul$\big[(\#)\big]$.

indepvars may contain factor variables; see [U] **11.4.3 Factor variables**.

depvar and *indepvars* may contain time-series operators; see [U] **11.4.4 Time-series varlists**.

bootstrap, by, jackknife, nestreg, rolling, statsby, stepwise, and svy are allowed; see
[U] **11.1.10 Prefix commands**.

Weights are not allowed with the bootstrap prefix; see [R] **bootstrap**.

aweights are not allowed with the jackknife prefix; see [R] **jackknife**.

vce() and weights are not allowed with the svy prefix; see [SVY] **svy**.

aweights, fweights, pweights, and iweights are allowed; see [U] **11.1.6 weight**.

coeflegend does not appear in the dialog box.

See [U] **20 Estimation and postestimation commands** for more capabilities of estimation commands.

Menu

Statistics > Linear models and related > Censored regression > Tobit regression

Description

tobit fits a model of *depvar* on *indepvars* where the censoring values are fixed.

Options

Model

noconstant; see [R] **estimation options**.

ll[(#)] and ul[(#)] indicate the lower and upper limits for censoring, respectively. You may specify one or both. Observations with *depvar* ≤ ll() are left-censored; observations with *depvar* ≥ ul() are right-censored; and remaining observations are not censored. You do not have to specify the censoring values at all. It is enough to type ll, ul, or both. When you do not specify a censoring value, tobit assumes that the lower limit is the minimum observed in the data (if ll is specified) and the upper limit is the maximum (if ul is specified).

offset(*varname*); see [R] **estimation options**.

SE/Robust

vce(*vcetype*) specifies the type of standard error reported, which includes types that are derived from asymptotic theory, that are robust to some kinds of misspecification, that allow for intragroup correlation, and that use bootstrap or jackknife methods; see [R] ***vce_option***.

Reporting

level(#); see [R] **estimation options**.

display_options: noomitted, vsquish, noemptycells, baselevels, allbaselevels, cformat(%*fmt*), pformat(%*fmt*), sformat(%*fmt*), and nolstretch; see [R] **estimation options**.

Maximization

maximize_options: iterate(#), [no]log, trace, tolerance(#), ltolerance(#), nrtolerance(#), and nonrtolerance; see [R] **maximize**. These options are seldom used.

Unlike most maximum likelihood commands, tobit defaults to nolog—it suppresses the iteration log.

The following option is available with tobit but is not shown in the dialog box:

coeflegend; see [R] **estimation options**.

Remarks

Tobit estimation was originally developed by Tobin (1958). A consumer durable was purchased if a consumer's desire was high enough, where desire was measured by the dollar amount spent by the purchaser. If no purchase was made, the measure of desire was censored at zero.

▷ Example 1

We will demonstrate tobit with an artificial example, which in the process will allow us to emphasize the assumptions underlying the estimation. We have a dataset containing the mileage ratings and weights of 74 cars. There are no censored variables in this dataset, but we are going to create one. Before that, however, the relationship between mileage and weight in our complete data is

```
. use http://www.stata-press.com/data/r12/auto
(1978 Automobile Data)

. generate wgt = weight/1000

. regress mpg wgt
```

Source	SS	df	MS		Number of obs =	74
					F(1, 72) =	134.62
Model	1591.99024	1	1591.99024		Prob > F =	0.0000
Residual	851.469221	72	11.8259614		R-squared =	0.6515
					Adj R-squared =	0.6467
Total	2443.45946	73	33.4720474		Root MSE =	3.4389

mpg	Coef.	Std. Err.	t	P>|t|	[95% Conf. Interval]	
wgt	-6.008687	.5178782	-11.60	0.000	-7.041058	-4.976316
_cons	39.44028	1.614003	24.44	0.000	36.22283	42.65774

(We divided weight by 1,000 simply to make discussing the resulting coefficients easier. We find that each additional 1,000 pounds of weight reduces mileage by 6 mpg.)

mpg in our data ranges from 12 to 41. Let us now pretend that our data were censored in the sense that we could not observe a mileage rating below 17 mpg. If the true mpg is 17 or less, all we know is that the mpg is less than or equal to 17:

```
. replace mpg=17 if mpg<=17
(14 real changes made)

. tobit mpg wgt, ll
```

Tobit regression				Number of obs	=	74
				LR chi2(1)	=	72.85
				Prob > chi2	=	0.0000
Log likelihood = -164.25438				Pseudo R2	=	0.1815

mpg	Coef.	Std. Err.	t	P>|t|	[95% Conf. Interval]	
wgt	-6.87305	.7002559	-9.82	0.000	-8.268658	-5.477442
_cons	41.49856	2.05838	20.16	0.000	37.39621	45.6009
/sigma	3.845701	.3663309			3.115605	4.575797

```
Obs. summary:        18  left-censored observations at mpg<=17
                     56      uncensored observations
                      0 right-censored observations
```

The replace before estimation was not really necessary—we remapped all the mileage ratings below 17 to 17 merely to reassure you that tobit was not somehow using uncensored data. We typed ll after tobit to inform tobit that the data were left-censored. tobit found the minimum of mpg in our data and assumed that was the censoring point. We could also have dispensed with replace and typed ll(17), informing tobit that all values of the dependent variable 17 and below are really censored at 17. In either case, at the bottom of the table, we are informed that there are, as a result, 18 left-censored observations.

On these data, our estimate is now a reduction of 6.9 mpg per 1,000 extra pounds of weight as opposed to 6.0. The parameter reported as /sigma is the estimated standard error of the regression; the resulting 3.8 is comparable with the estimated root mean squared error reported by regress of 3.4.

◁

❏ Technical note

You would never want to throw away information by purposefully censoring variables. The regress estimates are in every way preferable to those of tobit. Our example is designed solely to illustrate the relationship between tobit and regress. If you have uncensored data, use regress. If your data are censored, you have no choice but to use tobit.

❏

▷ Example 2

tobit can also fit models that are censored from above. This time, let's assume that we do not observe the actual mileage rating of cars yielding 24 mpg or better—we know only that it is at least 24. (Also assume that we have undone the change to mpg we made in the previous example.)

```
. use http://www.stata-press.com/data/r12/auto, clear
(1978 Automobile Data)
. generate wgt = weight/1000
. regress mpg wgt
  (output omitted)
. tobit mpg wgt, ul(24)
```

Tobit regression				Number of obs	=	74
				LR chi2(1)	=	90.72
				Prob > chi2	=	0.0000
Log likelihood = -129.8279				Pseudo R2	=	0.2589

mpg	Coef.	Std. Err.	t	P>|t|	[95% Conf. Interval]	
wgt	-5.080645	.43493	-11.68	0.000	-5.947459	-4.213831
_cons	36.08037	1.432056	25.19	0.000	33.22628	38.93445
/sigma	2.385357	.2444604			1.898148	2.872566

```
Obs. summary:          0  left-censored observations
                      51      uncensored observations
                      23 right-censored observations at mpg>=24
```

◁

▷ Example 3

tobit can also fit models that are censored from both sides (the so-called two-limit tobit):

```
. tobit mpg wgt, ll(17) ul(24)
```

Tobit regression				Number of obs	=	74
				LR chi2(1)	=	77.60
				Prob > chi2	=	0.0000
Log likelihood = -104.25976				Pseudo R2	=	0.2712

mpg	Coef.	Std. Err.	t	P>|t|	[95% Conf. Interval]	
wgt	-5.764448	.7245417	-7.96	0.000	-7.208457	-4.320438
_cons	38.07469	2.255917	16.88	0.000	33.57865	42.57072
/sigma	2.886337	.3952143			2.098676	3.673998

```
Obs. summary:         18  left-censored observations at mpg<=17
                      33      uncensored observations
                      23 right-censored observations at mpg>=24
```

◁

Saved results

tobit saves the following in e():

Scalars
e(N)	number of observations
e(N_unc)	number of uncensored observations
e(N_lc)	number of left-censored observations
e(N_rc)	number of right-censored observations
e(llopt)	contents of ll(), if specified
e(ulopt)	contents of ul(), if specified
e(k_aux)	number of auxiliary parameters
e(df_m)	model degrees of freedom
e(df_r)	residual degrees of freedom
e(r2_p)	pseudo-R-squared
e(chi2)	χ^2
e(ll)	log likelihood
e(ll_0)	log likelihood, constant-only model
e(N_clust)	number of clusters
e(F)	F statistic
e(p)	significance
e(rank)	rank of e(V)
e(converged)	1 if converged, 0 otherwise

Macros
e(cmd)	tobit
e(cmdline)	command as typed
e(depvar)	name of dependent variable
e(wtype)	weight type
e(wexp)	weight expression
e(title)	title in estimation output
e(clustvar)	name of cluster variable
e(offset)	linear offset variable
e(chi2type)	LR; type of model χ^2 test
e(vce)	*vcetype* specified in vce()
e(vcetype)	title used to label Std. Err.
e(properties)	b V
e(predict)	program used to implement predict
e(footnote)	program and arguments to display footnote
e(asbalanced)	factor variables fvset as asbalanced
e(asobserved)	factor variables fvset as asobserved

Matrices
e(b)	coefficient vector
e(V)	variance–covariance matrix of the estimators
e(V_modelbased)	model-based variance

Functions
e(sample)	marks estimation sample

James Tobin (1918–2002) was an American economist who after education and research at Harvard moved to Yale, where he was on the faculty from 1950 to 1988. He made many outstanding contributions to economics and was awarded the Nobel Prize in 1981 "for his analysis of financial markets and their relations to expenditure decisions, employment, production and prices". He trained in the U.S. Navy with the writer in Herman Wouk, who later fashioned a character after Tobin in the novel *The Caine Mutiny* (1951): "A mandarin-like midshipman named Tobit, with a domed forehead, measured quiet speech, and a mind like a sponge, was ahead of the field by a spacious percentage."

Methods and formulas

tobit is implemented as an ado-file.

See *Methods and formulas* in [R] **intreg**.

See Tobin (1958) for the original derivation of the tobit model. An introductory description of the tobit model can be found in, for instance, Wooldridge (2009, 587–595), Davidson and MacKinnon (2004, 484–486), Long (1997, 196–210), and Maddala and Lahiri (2006, 333–336). Cameron and Trivedi (2010, chap. 16) discuss the tobit model using Stata examples.

This command supports the Huber/White/sandwich estimator of the variance and its clustered version using vce(robust) and vce(cluster *clustvar*), respectively. See [P] **_robust**, particularly *Maximum likelihood estimators* and *Methods and formulas*.

tobit also supports estimation with survey data. For details on VCEs with survey data, see [SVY] **variance estimation**.

References

Amemiya, T. 1973. Regression analysis when the dependent variable is truncated normal. *Econometrica* 41: 997–1016.

——. 1984. Tobit models: A survey. *Journal of Econometrics* 24: 3–61.

Burke, W. J. 2009. Fitting and interpreting Cragg's tobit alternative using Stata. *Stata Journal* 9: 584–592.

Cameron, A. C., and P. K. Trivedi. 2010. *Microeconometrics Using Stata*. Rev. ed. College Station, TX: Stata Press.

Cong, R. 2000. sg144: Marginal effects of the tobit model. *Stata Technical Bulletin* 56: 27–34. Reprinted in *Stata Technical Bulletin Reprints*, vol. 10, pp. 189–197. College Station, TX: Stata Press.

Davidson, R., and J. G. MacKinnon. 2004. *Econometric Theory and Methods*. New York: Oxford University Press.

Drukker, D. M. 2002. Bootstrapping a conditional moments test for normality after tobit estimation. *Stata Journal* 2: 125–139.

Goldberger, A. S. 1983. Abnormal selection bias. In *Studies in Econometrics, Time Series, and Multivariate Statistics*, ed. S. Karlin, T. Amemiya, and L. A. Goodman, 67–84. New York: Academic Press.

Hurd, M. 1979. Estimation in truncated samples when there is heteroscedasticity. *Journal of Econometrics* 11: 247–258.

Kendall, M. G., and A. Stuart. 1973. *The Advanced Theory of Statistics, Vol. 2: Inference and Relationship*. New York: Hafner.

Long, J. S. 1997. *Regression Models for Categorical and Limited Dependent Variables*. Thousand Oaks, CA: Sage.

Maddala, G. S., and K. Lahiri. 2006. *Introduction to Econometrics*. 4th ed. New York: Wiley.

McDonald, J. F., and R. A. Moffitt. 1980. The use of tobit analysis. *Review of Economics and Statistics* 62: 318–321.

Shiller, R. J. 1999. The ET interview: Professor James Tobin. *Econometric Theory* 15: 867–900.

Stewart, M. B. 1983. On least squares estimation when the dependent variable is grouped. *Review of Economic Studies* 50: 737–753.

Tobin, J. 1958. Estimation of relationships for limited dependent variables. *Econometrica* 26: 24–36.

Wooldridge, J. M. 2009. *Introductory Econometrics: A Modern Approach*. 4th ed. Cincinnati, OH: South-Western.

Also see

[R] **tobit postestimation** — Postestimation tools for tobit

[R] **intreg** — Interval regression

[R] **heckman** — Heckman selection model

[R] **ivtobit** — Tobit model with continuous endogenous regressors

[R] **regress** — Linear regression

[R] **truncreg** — Truncated regression

[SVY] **svy estimation** — Estimation commands for survey data

[XT] **xtintreg** — Random-effects interval-data regression models

[XT] **xttobit** — Random-effects tobit models

[U] **20 Estimation and postestimation commands**

Title

> **tobit postestimation** — Postestimation tools for tobit

Description

The following postestimation commands are available after `tobit`:

Command	Description
contrast	contrasts and ANOVA-style joint tests of estimates
estat	AIC, BIC, VCE, and estimation sample summary
estat (svy)	postestimation statistics for survey data
estimates	cataloging estimation results
hausman	Hausman's specification test
lincom	point estimates, standard errors, testing, and inference for linear combinations of coefficients
linktest	link test for model specification
lrtest[1]	likelihood-ratio test
margins	marginal means, predictive margins, marginal effects, and average marginal effects
marginsplot	graph the results from margins (profile plots, interaction plots, etc.)
nlcom	point estimates, standard errors, testing, and inference for nonlinear combinations of coefficients
predict	predictions, residuals, influence statistics, and other diagnostic measures
predictnl	point estimates, standard errors, testing, and inference for generalized predictions
pwcompare	pairwise comparisons of estimates
suest	seemingly unrelated estimation
test	Wald tests of simple and composite linear hypotheses
testnl	Wald tests of nonlinear hypotheses

[1] `lrtest` is not appropriate with svy estimation results.

See the corresponding entries in the *Base Reference Manual* for details, but see [SVY] **estat** for details about `estat` (svy).

Syntax for predict

predict [*type*] *newvar* [*if*] [*in*] [, *statistic* <u>nooff</u>set]

predict [*type*] { *stub** | *newvar*_{reg} *newvar*_{sigma} } [*if*] [*in*] , <u>sc</u>ores

statistic	Description
Main	
xb	linear prediction; the default
stdp	standard error of the linear prediction
stdf	standard error of the forecast
<u>pr</u>(*a*,*b*)	$\Pr(a < y_j < b)$
e(*a*,*b*)	$E(y_j \mid a < y_j < b)$
<u>y</u>star(*a*,*b*)	$E(y_j^*)$, $y_j^* = \max\{a, \min(y_j, b)\}$

These statistics are available both in and out of sample; type predict ... if e(sample) ... if wanted only for the estimation sample.

stdf is not allowed with svy estimation results.

where *a* and *b* may be numbers or variables; *a* missing ($a \geq .$) means $-\infty$, and *b* missing ($b \geq .$) means $+\infty$; see [U] **12.2.1 Missing values**.

Menu

Statistics > Postestimation > Predictions, residuals, etc.

Options for predict

> ⌐ Main ⌐

xb, the default, calculates the linear prediction.

stdp calculates the standard error of the prediction, which can be thought of as the standard error of the predicted expected value or mean for the observation's covariate pattern. The standard error of the prediction is also referred to as the standard error of the fitted value.

stdf calculates the standard error of the forecast, which is the standard error of the point prediction for 1 observation. It is commonly referred to as the standard error of the future or forecast value. By construction, the standard errors produced by stdf are always larger than those produced by stdp; see *Methods and formulas* in [R] **regress postestimation**.

pr(*a*,*b*) calculates $\Pr(a < \mathbf{x}_j \mathbf{b} + u_j < b)$, the probability that $y_j \mid \mathbf{x}_j$ would be observed in the interval (a, b).

a and *b* may be specified as numbers or variable names; *lb* and *ub* are variable names;
pr(20,30) calculates $\Pr(20 < \mathbf{x}_j \mathbf{b} + u_j < 30)$;
pr(*lb*,*ub*) calculates $\Pr(lb < \mathbf{x}_j \mathbf{b} + u_j < ub)$; and
pr(20,*ub*) calculates $\Pr(20 < \mathbf{x}_j \mathbf{b} + u_j < ub)$.

a missing ($a \geq .$) means $-\infty$; pr(.,30) calculates $\Pr(-\infty < \mathbf{x}_j \mathbf{b} + u_j < 30)$;
pr(*lb*,30) calculates $\Pr(-\infty < \mathbf{x}_j \mathbf{b} + u_j < 30)$ in observations for which *lb* $\geq .$
and calculates $\Pr(lb < \mathbf{x}_j \mathbf{b} + u_j < 30)$ elsewhere.

b missing ($b \geq .$) means $+\infty$; $\text{pr}(20,.)$ calculates $\Pr(+\infty > \mathbf{x}_j \mathbf{b} + u_j > 20)$;
$\text{pr}(20,ub)$ calculates $\Pr(+\infty > \mathbf{x}_j \mathbf{b} + u_j > 20)$ in observations for which $ub \geq .$
and calculates $\Pr(20 < \mathbf{x}_j \mathbf{b} + u_j < ub)$ elsewhere.

$\text{e}(a,b)$ calculates $E(\mathbf{x}_j \mathbf{b} + u_j \mid a < \mathbf{x}_j \mathbf{b} + u_j < b)$, the expected value of $y_j | \mathbf{x}_j$ conditional on $y_j | \mathbf{x}_j$ being in the interval (a, b), meaning that $y_j | \mathbf{x}_j$ is truncated.
a and b are specified as they are for $\text{pr}()$.

$\text{ystar}(a,b)$ calculates $E(y_j^*)$, where $y_j^* = a$ if $\mathbf{x}_j \mathbf{b} + u_j \leq a$, $y_j^* = b$ if $\mathbf{x}_j \mathbf{b} + u_j \geq b$, and $y_j^* = \mathbf{x}_j \mathbf{b} + u_j$ otherwise, meaning that y_j^* is censored. a and b are specified as they are for $\text{pr}()$.

nooffset is relevant only if you specified offset(*varname*). It modifies the calculations made by predict so that they ignore the offset variable; the linear prediction is treated as $\mathbf{x}_j \mathbf{b}$ rather than as $\mathbf{x}_j \mathbf{b} + \text{offset}_j$.

scores calculates equation-level score variables.

The first new variable will contain $\partial \ln L / \partial (\mathbf{x}_j \beta)$.

The second new variable will contain $\partial \ln L / \partial \sigma$.

Remarks

Following Cong (2000), write the tobit model as

$$
y_i^* = \begin{cases} y_i, & \text{if } a < y_i < b \\ a, & \text{if } y_i \leq a \\ b, & \text{if } y_i \geq b \end{cases}
$$

y_i is a latent variable; instead, we observe y_i^*, which is bounded between a and b if y_i is outside those bounds.

There are four types of marginal effects that may be of interest in the tobit model, depending on the application:

1. The β coefficients themselves measure how the unobserved variable y_i changes with respect to changes in the regressors.

2. The marginal effects of the truncated expected value $E(y_i^* | a < y_i^* < b)$ measure the changes in y_i with respect to changes in the regressors among the subpopulation for which y_i is not at a boundary.

3. The marginal effects of the censored expected value $E(y_i^*)$ describe how the observed variable y_i^* changes with respect to the regressors.

4. The marginal effects of $\Pr(a < y_i^* < b)$ describe how the probability of being uncensored changes with respect to the regressors.

In the next example, we show how to obtain each of these.

▷ Example 1

In example 3 of [R] **tobit**, we fit a two-limit tobit model of mpg on wgt. The marginal effects of y_i are simply the β's reported by tobit:

```
. use http://www.stata-press.com/data/r12/auto
(1978 Automobile Data)

. generate wgt = weight/1000
```

```
. tobit mpg wgt, ll(17) ul(24)
```

Tobit regression

```
                                                      Number of obs   =        74
                                                      LR chi2(1)      =     77.60
                                                      Prob > chi2     =    0.0000
Log likelihood = -104.25976                           Pseudo R2       =    0.2712
```

mpg	Coef.	Std. Err.	t	P>|t|	[95% Conf. Interval]	
wgt	-5.764448	.7245417	-7.96	0.000	-7.208457	-4.320438
_cons	38.07469	2.255917	16.88	0.000	33.57865	42.57072
/sigma	2.886337	.3952143			2.098676	3.673998

```
Obs. summary:          18  left-censored observations at mpg<=17
                       33      uncensored observations
                       23 right-censored observations at mpg>=24
```

If mpg were not censored at 17 and 24, then a 1,000-pound increase in a car's weight (which is a 1-unit increase in wgt) would lower fuel economy by 5.8 mpg.

To get the marginal effect on the truncated expected value conditional on weight being at each of five values (1,000; 2,000; 3,000; 4,000; and 5,000 pounds), we type

```
. margins, dydx(wgt) predict(e(17,24)) at(wgt=(1 2 3 4 5))
Conditional marginal effects                          Number of obs   =        74
Model VCE      : OIM

Expression     : E(mpg|17<mpg<24), predict(e(17,24))
dy/dx w.r.t.   : wgt

1._at          : wgt             =            1
2._at          : wgt             =            2
3._at          : wgt             =            3
4._at          : wgt             =            4
5._at          : wgt             =            5
```

		dy/dx	Delta-method Std. Err.	z	P>|z|	[95% Conf. Interval]	
wgt	_at						
	1	-.4286109	.067281	-6.37	0.000	-.5604793	-.2967425
	2	-1.166572	.0827549	-14.10	0.000	-1.328768	-1.004375
	3	-2.308842	.4273727	-5.40	0.000	-3.146477	-1.471207
	4	-1.288896	.0889259	-14.49	0.000	-1.463188	-1.114604
	5	-.4685559	.0740565	-6.33	0.000	-.613704	-.3234079

For cars whose fuel economy lies between 17 and 24 mpg, a 1,000-pound increase in the car's weight lowers fuel economy most for a 3,000 pound car, 2.3 mpg, and least for a 5,000 pound car, 0.5 mpg.

To get the marginal effect on the censored expected value at the same levels of `wgt`, we type

```
. margins, dydx(wgt) predict(ystar(17,24)) at(wgt=(1 2 3 4 5))
Conditional marginal effects                     Number of obs   =         74
Model VCE     : OIM

Expression    : E(mpg*|17<mpg<24), predict(ystar(17,24))
dy/dx w.r.t.  : wgt

1._at         : wgt             =           1
2._at         : wgt             =           2
3._at         : wgt             =           3
4._at         : wgt             =           4
5._at         : wgt             =           5
```

		Delta-method				
	dy/dx	Std. Err.	z	P>|z|	[95% Conf. Interval]	
wgt						
_at						
1	-.0117778	.0204876	-0.57	0.565	-.0519328	.0283771
2	-1.0861	.311273	-3.49	0.000	-1.696184	-.4760162
3	-4.45315	.4772541	-9.33	0.000	-5.388551	-3.51775
4	-1.412822	.3289702	-4.29	0.000	-2.057591	-.768052
5	-.0424155	.0390014	-1.09	0.277	-.1188568	.0340258

The decrease in observed fuel economy (which is bounded between 17 and 24) for a 1,000 pound increase in weight is about 4.5 mpg for a 3,000 pound car. For a 5,000 pound car, the decrease is almost negligible because of the binding truncation at 17 mpg.

◁

Methods and formulas

All postestimation commands listed above are implemented as ado-files.

References

Cong, R. 2000. sg144: Marginal effects of the tobit model. *Stata Technical Bulletin* 56: 27–34. Reprinted in *Stata Technical Bulletin Reprints*, vol. 10, pp. 189–197. College Station, TX: Stata Press.

McDonald, J. F., and R. A. Moffitt. 1980. The use of tobit analysis. *Review of Economics and Statistics* 62: 318–321.

Also see

[R] **tobit** — Tobit regression

[U] **20 Estimation and postestimation commands**

Title

> **total** — Estimate totals

Syntax

> total *varlist* [*if*] [*in*] [*weight*] [, *options*]

options	Description
if/in/over	
over(*varlist*[, <u>nolabel</u>])	group over subpopulations defined by *varlist*; optionally, suppress group labels
SE/Cluster	
vce(*vcetype*)	*vcetype* may be analytic, <u>c</u>luster *clustvar*, <u>boot</u>strap, or <u>jackknife</u>
Reporting	
<u>level</u>(*#*)	set confidence level; default is level(95)
<u>nohe</u>ader	suppress table header
<u>nol</u>egend	suppress table legend
display_options	control column formats and line width
<u>coefl</u>egend	display legend instead of statistics

bootstrap, jackknife, mi estimate, rolling, statsby, and svy are allowed; see [U] **11.1.10 Prefix commands**.

vce(bootstrap) and vce(jackknife) are not allowed with the mi estimate prefix.

Weights are not allowed with the bootstrap prefix; see [R] **bootstrap**.

vce() and weights are not allowed with the svy prefix; see [SVY] **svy**.

fweights, pweights, and iweights are allowed; see [U] **11.1.6 weight**.

coeflegend does not appear in the dialog box.

See [U] **20 Estimation and postestimation commands** for more capabilities of estimation commands.

Menu

Statistics > Summaries, tables, and tests > Summary and descriptive statistics > Totals

Description

total produces estimates of totals, along with standard errors.

Options

    ⌐ if/in/over ⌐

over(*varlist* [, nolabel]) specifies that estimates be computed for multiple subpopulations, which are identified by the different values of the variables in *varlist*.

When this option is supplied with one variable name, such as over(*varname*), the value labels of *varname* are used to identify the subpopulations. If *varname* does not have labeled values (or there are unlabeled values), the values themselves are used, provided that they are nonnegative integers. Noninteger values, negative values, and labels that are not valid Stata names are substituted with a default identifier.

When over() is supplied with multiple variable names, each subpopulation is assigned a unique default identifier.

nolabel specifies that value labels attached to the variables identifying the subpopulations be ignored.

⌐ SE/Cluster ⌐

vce(*vcetype*) specifies the type of standard error reported, which includes types that are derived from asymptotic theory, that allow for intragroup correlation, and that use bootstrap or jackknife methods; see [R] *vce_option*.

vce(analytic), the default, uses the analytically derived variance estimator associated with the sample total.

⌐ Reporting ⌐

level(#); see [R] **estimation options**.

noheader prevents the table header from being displayed. This option implies nolegend.

nolegend prevents the table legend identifying the subpopulations from being displayed.

display_options: cformat(%*fmt*) and nolstretch; see [R] **estimation options**.

The following option is available with total but is not shown in the dialog box:

coeflegend; see [R] **estimation options**.

Remarks

▷ Example 1

Suppose that we collected data on incidence of heart attacks. The variable heartatk indicates whether a person ever had a heart attack (1 means yes; 0 means no). We can then estimate the total number of persons who have had heart attacks for each sex in the population represented by the data we collected.

```
. use http://www.stata-press.com/data/r12/total
. total heartatk [pw=swgt], over(sex)
Total estimation                  Number of obs   =   4946
        Male: sex = Male
      Female: sex = Female
```

Over	Total	Std. Err.	[95% Conf. Interval]	
heartatk				
Male	944559	104372.3	739943	1149175
Female	581590	82855.59	419156.3	744023.7

◁

Saved results

total saves the following in e():

Scalars
e(N)	number of observations
e(N_over)	number of subpopulations
e(N_clust)	number of clusters
e(k_eq)	number of equations in e(b)
e(df_r)	sample degrees of freedom
e(rank)	rank of e(V)

Macros
e(cmd)	total
e(cmdline)	command as typed
e(varlist)	*varlist*
e(wtype)	weight type
e(wexp)	weight expression
e(title)	title in estimation output
e(cluster)	name of cluster variable
e(over)	*varlist* from over()
e(over_labels)	labels from over() variables
e(over_namelist)	names from e(over_labels)
e(vce)	*vcetype* specified in vce()
e(vcetype)	title used to label Std. Err.
e(properties)	b V
e(estat_cmd)	program used to implement estat
e(marginsnotok)	predictions disallowed by margins

Matrices
e(b)	vector of total estimates
e(V)	(co)variance estimates
e(_N)	vector of numbers of nonmissing observations
e(error)	error code corresponding to e(b)

Functions
e(sample)	marks estimation sample

Methods and formulas

total is implemented as an ado-file.

Methods and formulas are presented under the following headings:

> *The total estimator*
> *Survey data*
> *The survey total estimator*
> *The poststratified total estimator*
> *Subpopulation estimation*

The total estimator

Let y denote the variable on which to calculate the total and $y_j, j = 1, \ldots, n$, denote an individual observation on y. Let w_j be the frequency weight (or iweight), and if no weight is specified, define $w_j = 1$ for all j. See the next section for pweighted data. The sum of the weights is an estimate of the population size:

$$\widehat{N} = \sum_{j=1}^{n} w_j$$

If the population values of y are denoted by $Y_j, j = 1, \ldots, N$, the associated population total is

$$Y = \sum_{j=1}^{N} Y_j = N\overline{y}$$

where $\overline{y}$ is the population mean. The total is estimated as

$$\widehat{Y} = \widehat{N}\overline{y}$$

The variance estimator for the total is

$$\widehat{V}(\widehat{Y}) = \widehat{N}^2 \widehat{V}(\overline{y})$$

where $\widehat{V}(\overline{y})$ is the variance estimator for the mean; see [R] **mean**. The standard error of the total is the square root of the variance.

If x, x_j, $\overline{x}$, and $\widehat{X}$ are similarly defined for another variable (observed jointly with y), the covariance estimator between $\widehat{X}$ and $\widehat{Y}$ is

$$\widehat{\mathrm{Cov}}(\widehat{X}, \widehat{Y}) = \widehat{N}^2 \widehat{\mathrm{Cov}}(\overline{x}, \overline{y})$$

where $\widehat{\mathrm{Cov}}(\overline{x}, \overline{y})$ is the covariance estimator between two means; see [R] **mean**.

Survey data

See [SVY] **variance estimation** and [SVY] **poststratification** for discussions that provide background information for the following formulas.

The survey total estimator

Let Y_j be a survey item for the jth individual in the population, where $j = 1, \ldots, M$ and M is the size of the population. The associated population total for the item of interest is

$$Y = \sum_{j=1}^{M} Y_j$$

Let y_j be the survey item for the jth sampled individual from the population, where $j = 1, \ldots, m$ and m is the number of observations in the sample.

The estimator $\widehat{Y}$ for the population total Y is

$$\widehat{Y} = \sum_{j=1}^{m} w_j y_j$$

where w_j is a sampling weight. The estimator for the number of individuals in the population is

$$\widehat{M} = \sum_{j=1}^{m} w_j$$

The score variable for the total estimator is the variable itself,

$$z_j(\widehat{Y}) = y_j$$

The poststratified total estimator

Let P_k denote the set of sampled observations that belong to poststratum k, and define $I_{P_k}(j)$ to indicate if the jth observation is a member of poststratum k, where $k = 1, \ldots, L_P$ and L_P is the number of poststrata. Also, let M_k denote the population size for poststratum k. P_k and M_k are identified by specifying the `poststrata()` and `postweight()` options on svyset; see [SVY] **svyset**.

The estimator for the poststratified total is

$$\widehat{Y}^P = \sum_{k=1}^{L_P} \frac{M_k}{\widehat{M}_k} \widehat{Y}_k = \sum_{k=1}^{L_P} \frac{M_k}{\widehat{M}_k} \sum_{j=1}^{m} I_{P_k}(j)\, w_j y_j$$

where

$$\widehat{M}_k = \sum_{j=1}^{m} I_{P_k}(j) w_j$$

The score variable for the poststratified total is

$$z_j(\widehat{Y}^P) = \sum_{k=1}^{L_P} I_{P_k}(j) \frac{M_k}{\widehat{M}_k} \left(y_j - \frac{\widehat{Y}_k}{\widehat{M}_k} \right)$$

Subpopulation estimation

Let S denote the set of sampled observations that belong to the subpopulation of interest, and define $I_S(j)$ to indicate if the jth observation falls within the subpopulation.

The estimator for the subpopulation total is

$$\widehat{Y}^S = \sum_{j=1}^{m} I_S(j)\, w_j y_j$$

and its score variable is

$$z_j(\widehat{Y}^S) = I_S(j)\, y_j$$

The estimator for the poststratified subpopulation total is

$$\widehat{Y}^{PS} = \sum_{k=1}^{L_P} \frac{M_k}{\widehat{M}_k} \widehat{Y}_k^S = \sum_{k=1}^{L_P} \frac{M_k}{\widehat{M}_k} \sum_{j=1}^{m} I_{P_k}(j) I_S(j)\, w_j y_j$$

and its score variable is

$$z_j(\widehat{Y}^{PS}) = \sum_{k=1}^{L_P} I_{P_k}(j) \frac{M_k}{\widehat{M}_k} \left\{ I_S(j)\, y_j - \frac{\widehat{Y}_k^S}{\widehat{M}_k} \right\}$$

References

Cochran, W. G. 1977. *Sampling Techniques*. 3rd ed. New York: Wiley.

Stuart, A., and J. K. Ord. 1994. *Kendall's Advanced Theory of Statistics: Distribution Theory, Vol I*. 6th ed. London: Arnold.

Also see

[R] **total postestimation** — Postestimation tools for total

[R] **mean** — Estimate means

[R] **proportion** — Estimate proportions

[R] **ratio** — Estimate ratios

[MI] **estimation** — Estimation commands for use with mi estimate

[SVY] **direct standardization** — Direct standardization of means, proportions, and ratios

[SVY] **poststratification** — Poststratification for survey data

[SVY] **subpopulation estimation** — Subpopulation estimation for survey data

[SVY] **svy estimation** — Estimation commands for survey data

[SVY] **variance estimation** — Variance estimation for survey data

[U] **20 Estimation and postestimation commands**

Title

> **total postestimation** — Postestimation tools for total

Description

The following postestimation commands are available after `total`:

Command	Description
estat	VCE
estat (svy)	postestimation statistics for survey data
estimates	cataloging estimation results
lincom	point estimates, standard errors, testing, and inference for linear combinations of coefficients
nlcom	point estimates, standard errors, testing, and inference for nonlinear combinations of coefficients
test	Wald tests of simple and composite linear hypotheses
testnl	Wald tests of nonlinear hypotheses

See the corresponding entries in the *Base Reference Manual* for details, but see [SVY] **estat** for details about estat (svy).

Remarks

> Example 1

Continuing with our data on incidence of heart attacks from example 1 in [R] **total**, we want to test whether there are twice as many heart attacks among men than women in the population.

```
. use http://www.stata-press.com/data/r12/total
. total heartatk [pw=swgt], over(sex)
(output omitted)
. test _b[Male] = 2*_b[Female]
 ( 1)  [heartatk]Male - 2 [heartatk]Female = 0

       F(  1,  4945) =     1.25
            Prob > F =    0.2643
```

Thus we do not reject our hypothesis that the total number of heart attacks for men is twice that for women in the population.

◁

Methods and formulas

All postestimation commands listed above are implemented as ado-files.

Also see

[R] **total** — Estimate totals

[U] **20 Estimation and postestimation commands**

Title

tpoisson — Truncated Poisson regression

Syntax

tpoisson *depvar* [*indepvars*] [*if*] [*in*] [*weight*] [, *options*]

options	Description
Model	
<u>nocon</u>stant	suppress constant term
ll(# | *varname*)	truncation point; default value is ll(0), zero truncation
<u>exp</u>osure(*varname_e*)	include ln(*varname_e*) in model with coefficient constrained to 1
<u>off</u>set(*varname_o*)	include *varname_o* in model with coefficient constrained to 1
<u>constr</u>aints(*constraints*)	apply specified linear constraints
<u>col</u>linear	keep collinear variables
SE/Robust	
vce(*vcetype*)	*vcetype* may be oim, <u>r</u>obust, <u>cl</u>uster *clustvar*, opg, <u>boot</u>strap, or <u>jack</u>knife
Reporting	
<u>l</u>evel(#)	set confidence level; default is level(95)
<u>irr</u>	report incidence-rate ratios
<u>nocns</u>report	do not display constraints
display_options	control column formats, row spacing, line width, and display of omitted variables and base and empty cells
Maximization	
maximize_options	control the maximization process; seldom used
<u>coefl</u>egend	display legend instead of statistics

indepvars may contain factor variables; see [U] **11.4.3 Factor variables**.

depvar and *indepvars* may contain time-series operators; see [U] **11.4.4 Time-series varlists**.

bootstrap, by, jackknife, rolling, statsby, and svy are allowed; see [U] **11.1.10 Prefix commands**.

Weights are not allowed with the bootstrap prefix; see [R] **bootstrap**.

vce() and weights are not allowed with the svy prefix; see [SVY] **svy**.

fweights, iweights, and pweights are allowed; see [U] **11.1.6 weight**.

coeflegend does not appear in the dialog box.

See [U] **20 Estimation and postestimation commands** for more capabilities of estimation commands.

Menu

Statistics > Count outcomes > Truncated Poisson regression

Description

tpoisson estimates the parameters of a truncated Poisson model by maximum likelihood. The dependent variable *depvar* is regressed on *indepvars*, where *depvar* is a positive count variable whose values are all above the truncation point.

Options

   ◻ Model ◻

noconstant; see [R] **estimation options**.

ll(# | *varname*) specifies the truncation point, which is a nonnegative integer. The default is zero truncation, ll(0).

exposure(*varname_e*), offset(*varname_o*), constraints(*constraints*), collinear; see [R] **estimation options**.

   ◻ SE/Robust ◻

vce(*vcetype*) specifies the type of standard error reported, which includes types that are derived from asymptotic theory, that are robust to some kinds of misspecification, that allow for intragroup correlation, and that use bootstrap or jackknife methods; see [R] **vce_option**.

   ◻ Reporting ◻

level(*#*); see [R] **estimation options**.

irr reports estimated coefficients transformed to incidence-rate ratios, that is, e^{β_i} rather than β_i. Standard errors and confidence intervals are similarly transformed. This option affects how results are displayed, not how they are estimated. irr may be specified at estimation or when replaying previously estimated results.

nocnsreport; see [R] **estimation options**.

display_options: noomitted, vsquish, noemptycells, baselevels, allbaselevels, cformat(*%fmt*), pformat(*%fmt*), sformat(*%fmt*), and nolstretch; see [R] **estimation options**.

   ◻ Maximization ◻

maximize_options: difficult, technique(*algorithm_spec*), iterate(*#*), [no]log, trace, gradient, showstep, hessian, showtolerance, tolerance(*#*), ltolerance(*#*), nrtolerance(*#*), nonrtolerance, and from(*init_specs*); see [R] **maximize**. These options are seldom used.

Setting the optimization type to technique(bhhh) resets the default *vcetype* to vce(opg).

The following option is available with tpoisson but is not shown in the dialog box:

coeflegend; see [R] **estimation options**.

Remarks

Truncated Poisson regression is used to model the number of occurrences of an event when that number is restricted to be above the truncation point. If the dependent variable is not truncated, standard Poisson regression may be more appropriate; see [R] **poisson**. Truncated Poisson regression was first proposed by Grogger and Carson (1991). For an introduction to Poisson regression, see Cameron and Trivedi (2005, 2010) and Long and Freese (2006). For an introduction to truncated Poisson models, see Cameron and Trivedi (1998) and Long (1997, chap. 8).

Suppose that the patients admitted to a hospital for a given condition form a random sample from a population of interest and that each admitted patient stays at least one day. You are interested in modeling the length of stay of patients in days. The sample is truncated at zero because you only have data on individuals who stayed at least one day. tpoisson accounts for the truncated sample, whereas poisson does not.

Truncation is not the same as censoring. Right-censored Poisson regression was implemented in Stata by Raciborski (2011).

▷ Example 1

Consider the Simonoff (2003) dataset of running shoes for a sample of runners who registered an online running log. A running-shoe marketing executive is interested in knowing how the number of running shoes purchased relates to other factors such as gender, marital status, age, education, income, typical number of runs per week, average miles run per week, and the preferred type of running. These data are naturally truncated at zero. A truncated Poisson model is fit to the number of shoes owned on runs per week, miles run per week, gender, age, and marital status.

No options are needed because zero truncation is the default for tpoisson.

```
. use http://www.stata-press.com/data/r12/runshoes

. tpoisson shoes rpweek mpweek male age married
Iteration 0:   log likelihood = -88.328151
Iteration 1:   log likelihood = -86.272639
Iteration 2:   log likelihood = -86.257999
Iteration 3:   log likelihood = -86.257994
```

Truncated Poisson regression					Number of obs	=	60
Truncation point: 0					LR chi2(5)	=	22.75
					Prob > chi2	=	0.0004
Log likelihood = -86.257994					Pseudo R2	=	0.1165

shoes	Coef.	Std. Err.	z	P>|z|	[95% Conf. Interval]	
rpweek	.1575811	.1097893	1.44	0.151	-.057602	.3727641
mpweek	.0210673	.0091113	2.31	0.021	.0032094	.0389252
male	.0446134	.2444626	0.18	0.855	-.4345246	.5237513
age	.0185565	.0137786	1.35	0.178	-.008449	.045562
married	-.1283912	.2785044	-0.46	0.645	-.6742498	.4174674
_cons	-1.205844	.6619774	-1.82	0.069	-2.503296	.0916078

Using the zero-truncated Poisson regression with these data, only the coefficient on average miles per week is statistically significant at the 5% level.

◁

▷ Example 2

Semiconductor manufacturing requires that silicon wafers be coated with a layer of metal oxide. The depth of this layer is strictly controlled. In this example, a critical oxide layer is designed for 300 ± 20 angstroms (Å).

After the oxide layer is coated onto a wafer, the wafer enters a photolithography step in which the lines representing the electrical connections are printed on the oxide and later etched and filled with metal. The widths of these lines are measured. In this example, they are controlled to 90 ± 5 micrometers (μm).

After these and other steps, each wafer is electrically tested at probe. If too many failures are discovered, the wafer is rejected and sent for engineering analysis. In this example, the maximum number of probe failures tolerated for this product is 10.

A major failure at probe has been encountered—88 wafers had more than 10 failures each. The 88 wafers that failed were tested using 4 probe machines. The engineer suspects that the failures were a result of faulty probe machines, poor depth control, or poor line widths. The line widths and depths in these data are the actual measurement minus its specification target, 300 Å for the oxide depths and 90 μm for the line widths.

The following table tabulates the average failure rate for each probe using Stata's mean command; see [R] **mean**.

```
. use http://www.stata-press.com/data/r12/probe

. mean failures, over(probe) nolegend
Mean estimation                      Number of obs   =        88
```

Over	Mean	Std. Err.	[95% Conf. Interval]	
failures				
1	15.875	1.186293	13.51711	18.23289
2	14.95833	.5912379	13.78318	16.13348
3	16.47059	.9279866	14.62611	18.31506
4	23.09677	.9451117	21.21826	24.97529

The 95% confidence intervals in this table suggest that there are about 5–11 additional failures per wafer on probe 4. These are unadjusted for varying line widths and oxide depths. Possibly, probe 4 received the wafers with larger line widths or extreme oxide depths.

Truncated Poisson regression more clearly identifies the root causes for the increased failures by estimating the differences between probes adjusted for the line widths and oxide depths. It also allows us to determine whether the deviations from specifications in line widths or oxide depths might be contributing to the problem.

```
. tpoisson failures i.probe depth width, ll(10) nolog
Truncated Poisson regression                  Number of obs   =        88
Truncation point: 10                          LR chi2(5)      =     73.70
                                              Prob > chi2     =    0.0000
Log likelihood = -239.35746                   Pseudo R2       =    0.1334
```

failures	Coef.	Std. Err.	z	P>|z|	[95% Conf. Interval]	
probe						
2	-.1113037	.1019786	-1.09	0.275	-.3111781	.0885707
3	.0114339	.1036032	0.11	0.912	-.1916245	.2144924
4	.4254115	.0841277	5.06	0.000	.2605242	.5902989
depth	-.0005034	.0033375	-0.15	0.880	-.0070447	.006038
width	.0330225	.015573	2.12	0.034	.0025001	.063545
_cons	2.714025	.0752617	36.06	0.000	2.566515	2.861536

The coefficients listed for the probes are testing the null hypothesis: H_0: $probe_i = probe_1$, where i equals 2, 3, and 4. Because the only coefficient that is statistically significant is the one for testing for H_0: $probe_4 = probe_1$, $p < 0.001$, and because the p-values for the other probes are not statistically significant, that is, $p \geq 0.275$, the implication is that there is a difference between probe 4 and the other machines. Because the coefficient for this test is positive, 0.425, the conclusion is that the average failure rate for probe 4, after adjusting for line widths and oxide depths, is higher than the other probes. Possibly, probe 4 needs calibration or the head used with this machine is defective.

Line-width control is statistically significant, $p = 0.034$, but variation in oxide depths is not causing the increased failure rate. The engineer concluded that the sudden increase in failures is the result of two problems. First, probe 4 is malfunctioning, and second, there is a possible lithography or etching problem.

◁

Saved results

tpoisson saves the following in e():

Scalars

e(N)	number of observations
e(k)	number of parameters
e(k_eq)	number of equations in e(b)
e(k_eq_model)	number of equations in overall model test
e(k_dv)	number of dependent variables
e(df_m)	model degrees of freedom
e(r2_p)	pseudo-R-squared
e(ll)	log likelihood
e(ll_0)	log likelihood, constant-only model
e(N_clust)	number of clusters
e(chi2)	χ^2
e(p)	significance
e(rank)	rank of e(V)
e(ic)	number of iterations
e(rc)	return code
e(converged)	1 if converged, 0 otherwise

Macros

e(cmd)	tpoisson
e(cmdline)	command as typed
e(depvar)	name of dependent variable
e(llopt)	contents of ll(), or 0 if not specified
e(wtype)	weight type
e(wexp)	weight expression
e(title)	title in estimation output
e(clustvar)	name of cluster variable
e(offset)	linear offset variable
e(chi2type)	Wald or LR; type of model χ^2 test
e(vce)	vcetype specified in vce()
e(vcetype)	title used to label Std. Err.
e(opt)	type of optimization
e(which)	max or min; whether optimizer is to perform maximization or minimization
e(ml_method)	type of ml method
e(user)	name of likelihood-evaluator program
e(technique)	maximization technique
e(properties)	b V
e(predict)	program used to implement predict
e(asbalanced)	factor variables fvset as asbalanced
e(asobserved)	factor variables fvset as asobserved

Matrices
    e(b)   coefficient vector
    e(Cns)   constraints matrix
    e(ilog)   iteration log (up to 20 iterations)
    e(gradient)   gradient vector
    e(V)   variance–covariance matrix of the estimators
    e(V_modelbased)   model-based variance
Functions
    e(sample)   marks estimation sample

Methods and formulas

tpoisson is implemented as an ado-file.

The conditional probability of observing y_j events given that $y_j > \tau_j$, where τ_j is the truncation point, is given by

$$\Pr(Y = y_j \mid y_j > \tau_j, \mathbf{x}_j) = \frac{\exp(-\lambda)\lambda^{y_j}}{y_j!\Pr(Y > \tau_j \mid \mathbf{x}_j)}$$

The log likelihood (with weights w_j and offsets) is given by

$$\xi_j = \mathbf{x}_j\boldsymbol{\beta} + \text{offset}_j$$

$$f(y_j) = \frac{\exp\{-\exp(\xi_j)\}\exp(\xi_j y_j)}{y_j!\Pr(Y > \tau_j \mid \xi_j)}$$

$$\ln L = \sum_{j=1}^{n} w_j\left[-\exp(\xi_j) + \xi_j y_j - \ln(y_j!) - \ln\{\Pr(Y > \tau_j \mid \xi_j)\}\right]$$

This command supports the Huber/White/sandwich estimator of the variance and its clustered version using vce(robust) and vce(cluster *clustvar*), respectively. See [P] _robust, particularly *Maximum likelihood estimators* and *Methods and formulas*.

tpoisson also supports estimation with survey data. For details on variance–covariance estimates with survey data, see [SVY] **variance estimation**.

Acknowledgment

We gratefully acknowledge the previous work by Joseph Hilbe, Arizona State University; see Hilbe (1999).

References

Cameron, A. C., and P. K. Trivedi. 1998. *Regression Analysis of Count Data*. Cambridge: Cambridge University Press.

———. 2005. *Microeconometrics: Methods and Applications*. New York: Cambridge University Press.

———. 2010. *Microeconometrics Using Stata*. Rev. ed. College Station, TX: Stata Press.

Farbmacher, H. 2011. Estimation of hurdle models for overdispersed count data. *Stata Journal* 11: 82–94.

Grogger, J. T., and R. T. Carson. 1991. Models for truncated counts. *Journal of Applied Econometrics* 6: 225–238.

Hilbe, J. M. 1998. sg91: Robust variance estimators for MLE Poisson and negative binomial regression. *Stata Technical Bulletin* 45: 26–28. Reprinted in *Stata Technical Bulletin Reprints*, vol. 8, pp. 177–180. College Station, TX: Stata Press.

———. 1999. sg102: Zero-truncated Poisson and negative binomial regression. *Stata Technical Bulletin* 47: 37–40. Reprinted in *Stata Technical Bulletin Reprints*, vol. 8, pp. 233–236. College Station, TX: Stata Press.

Hilbe, J. M., and D. H. Judson. 1998. sg94: Right, left, and uncensored Poisson regression. *Stata Technical Bulletin* 46: 18–20. Reprinted in *Stata Technical Bulletin Reprints*, vol. 8, pp. 186–189. College Station, TX: Stata Press.

Long, J. S. 1997. *Regression Models for Categorical and Limited Dependent Variables*. Thousand Oaks, CA: Sage.

Long, J. S., and J. Freese. 2006. *Regression Models for Categorical Dependent Variables Using Stata*. 2nd ed. College Station, TX: Stata Press.

Raciborski, R. 2011. Right-censored Poisson regression model. *Stata Journal* 11: 95–105.

Simonoff, J. S. 2003. *Analyzing Categorical Data*. New York: Springer.

Also see

[R] **tpoisson postestimation** — Postestimation tools for tpoisson

[R] **poisson** — Poisson regression

[R] **tnbreg** — Truncated negative binomial regression

[R] **nbreg** — Negative binomial regression

[R] **zinb** — Zero-inflated negative binomial regression

[R] **zip** — Zero-inflated Poisson regression

[SVY] **svy estimation** — Estimation commands for survey data

[XT] **xtpoisson** — Fixed-effects, random-effects, and population-averaged Poisson models

[U] **20 Estimation and postestimation commands**

Title

> **tpoisson postestimation** — Postestimation tools for tpoisson

Description

The following postestimation commands are available after `tpoisson`:

Command	Description
contrast	contrasts and ANOVA-style joint tests of estimates
estat	AIC, BIC, VCE, and estimation sample summary
estat (svy)	postestimation statistics for survey data
estimates	cataloging estimation results
lincom	point estimates, standard errors, testing, and inference for linear combinations of coefficients
lrtest[1]	likelihood-ratio test
margins	marginal means, predictive margins, marginal effects, and average marginal effects
marginsplot	graph the results from margins (profile plots, interaction plots, etc.)
nlcom	point estimates, standard errors, testing, and inference for nonlinear combinations of coefficients
predict	predictions, residuals, influence statistics, and other diagnostic measures
predictnl	point estimates, standard errors, testing, and inference for generalized predictions
pwcompare	pairwise comparisons of estimates
suest	seemingly unrelated estimation
test	Wald tests of simple and composite linear hypotheses
testnl	Wald tests of nonlinear hypotheses

[1] `lrtest` is not appropriate with svy estimation results.

See the corresponding entries in the *Base Reference Manual* for details, but see [SVY] **estat** for details about `estat` (svy).

Syntax for predict

predict [*type*] *newvar* [*if*] [*in*] [, *statistic* <u>nooff</u>set]

statistic	Description
Main	
n	number of events; the default
ir	incidence rate
cm	conditional mean, $E(y_j \mid y_j > \tau_j)$
pr(*n*)	probability $\Pr(y_j = n)$
pr(*a*,*b*)	probability $\Pr(a \le y_j \le b)$
cpr(*n*)	conditional probability $\Pr(y_j = n \mid y_j > \tau_j)$
cpr(*a*,*b*)	conditional probability $\Pr(a \le y_j \le b \mid y_j > \tau_j)$
xb	linear prediction
stdp	standard error of the linear prediction
<u>score</u>	first derivative of the log likelihood with respect to $\mathbf{x}_j \boldsymbol{\beta}$

These statistics are available both in and out of sample; type predict ... if e(sample) ... if wanted only for the estimation sample.

Menu

Statistics > Postestimation > Predictions, residuals, etc.

Options for predict

Main

n, the default, calculates the predicted number of events, which is $\exp(\mathbf{x}_j \boldsymbol{\beta})$ if neither offset() nor exposure() was specified when the model was fit; $\exp(\mathbf{x}_j \boldsymbol{\beta} + \text{offset}_j)$ if offset() was specified; or $\exp(\mathbf{x}_j \boldsymbol{\beta}) \times \text{exposure}_j$ if exposure() was specified.

ir calculates the incidence rate $\exp(\mathbf{x}_j \boldsymbol{\beta})$, which is the predicted number of events when exposure is 1. This is equivalent to specifying both the n and the nooffset options.

cm calculates the conditional mean,

$$E(y_j \mid y_j > \tau_j) = \frac{E(y_j)}{\Pr(y_j > \tau_j)}$$

where τ_j is the truncation point found in e(llopt).

pr(*n*) calculates the probability $\Pr(y_j = n)$, where n is a nonnegative integer that may be specified as a number or a variable.

pr(*a*,*b*) calculates the probability $\Pr(a \le y_j \le b)$, where a and b are nonnegative integers that may be specified as numbers or variables;

b missing ($b \ge .$) means $+\infty$;
pr(20,.) calculates $\Pr(y_j \ge 20)$;
pr(20,*b*) calculates $\Pr(y_j \ge 20)$ in observations for which $b \ge .$ and calculates $\Pr(20 \le y_j \le b)$ elsewhere.

pr(.,*b*) produces a syntax error. A missing value in an observation of the variable *a* causes a missing value in that observation for pr(*a*,*b*).

cpr(*n*) calculates the conditional probability $\Pr(y_j = n \mid y_j > \tau_j)$, where τ_j is the truncation point found in e(llopt). *n* is an integer greater than the truncation point that may be specified as a number or a variable.

cpr(*a*,*b*) calculates the conditional probability $\Pr(a \le y_j \le b \mid y_j > \tau_j)$, where τ_j is the truncation point found in e(llopt). The syntax for this option is analogous to that used for pr(*a*,*b*) except that *a* must be greater than the truncation point.

xb calculates the linear prediction, which is $\mathbf{x}_j\boldsymbol{\beta}$ if neither offset() nor exposure() was specified when the model was fit; $\mathbf{x}_j\boldsymbol{\beta} + \text{offset}_j$ if offset() was specified; or $\mathbf{x}_j\boldsymbol{\beta} + \ln(\text{exposure}_j)$ if exposure() was specified; see nooffset below.

stdp calculates the standard error of the linear prediction.

score calculates the equation-level score, $\partial \ln L / \partial(\mathbf{x}_j\boldsymbol{\beta})$.

nooffset is relevant only if you specified offset() or exposure() when you fit the model. It modifies the calculations made by predict so that they ignore the offset or exposure variable; the linear prediction is treated as $\mathbf{x}_j\boldsymbol{\beta}$ rather than as $\mathbf{x}_j\boldsymbol{\beta} + \text{offset}_j$ or $\mathbf{x}_j\boldsymbol{\beta} + \ln(\text{exposure}_j)$. Specifying predict ... , nooffset is equivalent to specifying predict ... , ir.

Methods and formulas

All postestimation commands listed above are implemented as ado-files.

In the following formula, we use the same notation as in [R] **tpoisson**.

The equation-level scores are given by

$$\text{score}(\mathbf{x}\boldsymbol{\beta})_j = y_j - e^{\xi_j} - \frac{e^{-e^{\xi_j}} e^{\xi_j}}{\Pr(Y > \tau_j \mid \xi_j)}$$

where τ_j is the truncation point found in e(llopt).

Also see

[R] **tpoisson** — Truncated Poisson regression

[U] **20 Estimation and postestimation commands**

Title

translate — Print and translate logs

Syntax

Print log and SMCL files

> print *filename* [, like(*ext*) name(*windowname*) *override_options*]

Translate log files to SMCL files and vice versa

> translate *filename*$_{\text{in}}$ *filename*$_{\text{out}}$ [, <u>t</u>ranslator(*tname*) name(*windowname*)
>
> *override_options* replace]

View translator parameter settings

> translator <u>q</u>uery [*tname*]

Change translator parameter settings

> translator set [*tname setopt setval*]

Return translator parameter settings to default values

> translator reset *tname*

List current mappings from one extension to another

> transmap <u>q</u>uery [.*ext*]

Specify that files with one extension be treated the same as files with another extension

> transmap <u>def</u>ine .*ext*$_{\text{new}}$.*ext*$_{\text{old}}$

filename in print, in addition to being a filename to be printed, may be specified as @Results to mean the Results window and @Viewer to mean the Viewer window.

filename$_{\text{in}}$ in translate may be specified just as *filename* in print.

tname in translator specifies the name of a translator; see the translator() option under *Options for translate*.

Description

print prints log, SMCL, and text files. Although there is considerable flexibility in how print (and translate, which print uses) can be set to work, they have already been set up and should just work:

```
. print mylog.smcl
. print mylog.log
```

Unix users may discover that they need to do a bit of setup before `print` works; see *Printing files, Unix* below. International Unix users may also wish to modify the default paper size. All users can tailor `print` and `translate` to their needs.

`print` may also be used to print the current contents of the Results window or the Viewer. For instance, the current contents of the Results window could be printed by typing

> . print @Results

`translate` translates log and SMCL files from one format to another, the other typically being suitable for printing. `translate` can also translate SMCL logs (logs created by typing, say, `log using mylog`) to plain text:

> . translate mylog.smcl mylog.log

You can use `translate` to recover a log when you have forgotten to start one. You may type

> . translate @Results mylog.txt

to capture as plain text what is currently shown in the Results window.

This entry provides a general overview of `print` and `translate` and covers in detail the printing and translation of text (nongraphic) files.

`translator query`, `translator set`, and `translator reset` show, change, and restore the default values of the settings for each translator.

`transmap define` and `transmap query` create and show mappings from one file extension to another for use with `print` and `translate`.

For example, `print myfile.txt` knows to use a translator appropriate for printing text files because of the `.txt` extension. However, it does not know what to do with `.xyz` files. If you have `.xyz` files and always wish to treat them as `.txt` files, you can type `transmap define .xyz .txt`.

Options for print

`like(`*ext*`)` specifies how the file should be translated to a form suitable for printing. The default is to determine the translation method from the extension of *filename*. Thus `mylog.smcl` is translated according to the rule for translating `smcl` files, `myfile.txt` is translated according to the rule for translating `txt` files, and so on. (These rules are, in fact, `translate`'s `smcl2prn` and `txt2prn` translators, but put that aside for the moment.)

Rules for the following extensions are predefined:

.txt	assume input file contains plain text
.log	assume input file contains Stata log text
.smcl	assume input file contains SMCL

To print a file that has an extension different from those listed above, you can define a new extension, but you do not have to do that. Assume that you wish to print the file `read.me`, which you know to contain plain text. If you were just to type `print read.me`, you would be told that Stata cannot translate `.me` files. (You would actually be told that the translator for `me2prn` was not found.) You could type `print read.me, like(txt)` to tell `print` to print `read.me` like a `.txt` file.

On the other hand, you could type

> . transmap define .me .txt

to tell Stata that .me files are always to be treated like .txt files. If you did that, Stata would remember the new rule, even in future sessions.

When you specify the like() option, you override the recorded rules. So, if you were to type print mylog.smcl, like(txt), the file would be printed as plain text (meaning that all the SMCL commands would show).

name(*windowname*) specifies which window to print when printing a Viewer. The default is for Stata to print the topmost Viewer [Unix(GUI) users: See the second technical note in *Printing files, Unix*]. The name() option is ignored when printing the Results window.

The window name is located inside parentheses in the window title. For example, if the title for a Viewer window is *Viewer (#1) [help print]*, the name for the window is *#1*.

override_options refers to translate's options for overriding default values. print uses translate to translate the file into a format suitable for sending to the printer, and thus translate's *override_options* may also be used with print. The settings available vary between each translator (for example, smcl2ps will have different settings than smcl2txt) and may also differ across operating systems (for example, Windows may have different printing options than Mac OS X). To find out what you can override when printing .smcl files, type

```
. translator query smcl2prn
  (output omitted)
```

In the omitted output, you might learn that there is an rmargin # tunable value, which specifies the right margin in inches. You could specify the *override_option* rmargin(#) to temporarily override the default value, or you could type translator set smcl2prn rmargin # beforehand to permanently reset the value.

Alternatively, on some computers with some translators, you might discover that nothing can be set.

Options for translate

translator(*tname*) specifies the name of the translator to be used to translate the file. The available translators are

tname	Input	Output
smcl2ps	SMCL	PostScript
log2ps	Stata text log	PostScript
txt2ps	generic text file	PostScript
Viewer2ps	Viewer window	PostScript
Results2ps	Results window	PostScript
smcl2prn	SMCL	default printer format
log2prn	Stata text log	default printer format
txt2prn	generic text log	default printer format
Results2prn	Results window	default printer format
Viewer2prn	Viewer window	default printer format
smcl2txt	SMCL	generic text file
smcl2log	SMCL	Stata text log
Results2txt	Results window	generic text file
Viewer2txt	Viewer window	generic text file
smcl2pdf	SMCL	PDF
log2pdf	Stata text log	PDF
txt2pdf	generic text log	PDF
Results2pdf	Results window	PDF
Viewer2pdf	Viewer window	PDF

If `translator()` is not specified, `translate` determines which translator to use from extensions of the filenames specified. Typing `translate myfile.smcl myfile.ps` would use the `smcl2ps` translator. Typing `translate myfile.smcl myfile.ps, translate(smcl2prn)` would override the default and use the `smcl2prn` translator.

Actually, when you type `translate` *a.b c.d*, `translate` looks up *.b* in the `transmap` extension-synonym table. If *.b* is not found, the translator *b2d* is used. If *.b* is found in the table, the mapped extension is used (call it *b′*), and then the translator *b′2d* is used. For example,

Command	Translator used
. translate myfile.smcl myfile.ps	smcl2ps
. translate myfile.odd myfile.ps	odd2ps, which does not exist, so error
. transmap define .odd .txt	
. translate myfile.odd myfile.ps	txt2ps

You can list the mappings that `translate` uses by typing `transmap query`.

`name(`*windowname*`)` specifies which window to translate when translating a Viewer. The default is for Stata to translate the topmost Viewer. The `name()` option is ignored when translating the Results window.

The window name is located inside parentheses in the window title. For example, if the title for a Viewer window is *Viewer (#1) [help print]*, the name for the window is *#1*.

override_options override any of the default options of the specified or implied translator. To find out what you can override for, say, `log2ps`, type

> . `translator query log2ps`
> (*output omitted*)

In the omitted output, you might learn that there is an `rmargin #` tunable value, which, for `log2ps`, specifies the right margin in inches. You could specify the *override_option* `rmargin(#)` to temporarily override the default value or type `translator set log2ps rmargin #` beforehand to permanently reset the value.

`replace` specifies that *filename*out be replaced if it already exists.

Remarks

Remarks are presented under the following headings:

> *Printing files*
> *Printing files, Mac and Windows*
> *Printing files, Unix*
> *Translating files from one format to another*

Printing files

Printing should be easy; just type

> . `print mylog.smcl`
> . `print mylog.log`

You can use `print` to print SMCL files, plain text files, and even the contents of the Results and Viewer windows:

> . `print @Results`

```
. print @Viewer
. print @Viewer, name(#2)
```

For information about printing and translating graph files, see [G-2] **graph print** and see [G-2] **graph export**.

Printing files, Mac and Windows

When you type `print`, you are using the same facility that you would be using if you had selected **Print** from the **File** menu. If you try to print a file that Stata does not know about, Stata will complain:

```
. print read.me
translator me2prn not found
(perhaps you need to specify the like() option)
r(111);
```

Then you could type

```
. print read.me, like(txt)
```

to indicate that you wanted `read.me` sent to the printer in the same fashion as if the file were named `readme.txt`, or you could type

```
. transmap define .me .txt
. print read.me
```

Here you are telling Stata once and for all that you want files ending in `.me` to be treated in the same way as files ending in `.txt`. Stata will remember this mapping, even across sessions. To clear the `.me` mapping, type

```
. transmap define .me
```

To see all the mappings, type

```
. transmap query
```

To print to a file, use the `translate` command, not `print`:

```
. translate mylog.smcl mylog.prn
```

`translate` prints to a file by using the Windows print driver when the new filename ends in `.prn`. Under Mac, the `prn` translators are the same as the `pdf` translators. We suggest that you simply use the `.pdf` file extension when printing to a file.

Printing files, Unix

Stata assumes that you have a PostScript printer attached to your Unix computer and that the Unix command `lpr(1)` can be used to send PostScript files to it, but you can change this. On your Unix system, typing

```
mycomputer$ lpr < filename
```

may not be sufficient to print PostScript files. For instance, perhaps on your system you would need to type

```
mycomputer$ lpr -Plexmark < filename
```

or

```
mycomputer$ lpr -Plexmark filename
```

or something else. To set the print command to be `lpr -Plexmark` *filename* and to state that the printer expects to receive PostScript files, type

```
. printer define prn ps "lpr -Plexmark @"
```

To set the print command to `lpr -Plexmark` < *filename* and to state that the printer expects to receive plain text files, type

```
. printer define prn txt "lpr -Plexmark < @"
```

That is, just type the command necessary to send files to your printer and include an @ sign where the filename should be substituted. Two file formats are available: `ps` and `txt`. The default setting, as shipped from the factory, is

```
. printer define prn ps "lpr < @"
```

We will return to the `printer` command in the technical note that follows because it has some other capabilities you should know about.

In any case, after you redefine the default printer, the following should just work:

```
. print mylog.smcl
. print mylog.log
```

If you try to print a file that Stata does not know about, it will complain:

```
. print read.me
translator me2prn not found
r(111);
```

Here you could type

```
. print read.me, like(txt)
```

to indicate that you wanted `read.me` sent to the printer in the same fashion as if the file were named `readme.txt`, or you could type

```
. transmap define .me .txt
. print read.me
```

Here you are telling Stata once and for all that you want files ending in `.me` to be treated in the same way as files ending in `.txt`. Stata will remember this setting for `.me`, even across sessions.

If you want to clear the `.me` setting, type

```
. transmap define .me
```

If you want to see all your settings, type

```
. transmap query
```

❏ Technical note

The syntax of the `printer` command is

> `printer` <u>def</u>ine *printername* [{ `ps` | `txt` } "*Unix command with @*"]

> `printer` <u>q</u>uery [*printername*]

You may define multiple printers. By default, `print` uses the printer named `prn`, but `print` has the syntax

> `print` *filename* [, `like`(*ext*) `printer`(*printername*) *override_options*]

so, if you define multiple printers, you may route your output to them.

For instance, if you have a second printer on your system, you might type

```
. printer define lexmark ps "lpr -Plexmark < @"
```

After doing that, you could type

```
. print myfile.smcl, printer(lexmark)
```

Any printers that you set will be remembered even across sessions. You can delete printers:

```
. printer define lexmark
```

You can list all the defined printers by typing `printer query`, and you can list the definition of a particular printer, say, prn, by typing `printer query prn`.

The default printer `prn` we have predefined for you is

```
. printer define prn ps "lpr < @"
```

meaning that we assume that it is a PostScript printer and that the Unix command `lpr(1)`, without options, is sufficient to cause files to print. Feel free to change the default definition. If you change it, the change will be remembered across sessions.   ❏

❏ Technical note

Unix(GUI) users should note that X-Windows does not have the concept of a window z-order, which prevents Stata from determining which window is the topmost window. Instead, Stata determines which window is topmost based on which window has the focus. However, some window managers will set the focus to a window without bringing the window to the top. What Stata considers the topmost window may not appear topmost visually. For this reason, you should always use the `name()` option to ensure that the correct window is printed.   ❏

❏ Technical note

When you select the Results window to print from the **Print** menu or toolbar button, the result is the same as if you were to issue the `print` command. When you select a Viewer window to print from the **Print** menu or toolbar button, the result is the same as if you were to issue the `print` command with a `name()` option.   ❏

The translation to PostScript format is done by `translate` and, in particular, is performed by the translators `smcl2ps`, `log2ps`, and `txt2ps`. There are many tunable parameters in each of these translators. You can display the current values of these tunable parameters for, say, `smcl2ps` by typing

```
. translator query smcl2ps
  (output omitted )
```

and you can set any of the tunable parameters (for instance, setting `smcl2ps`'s `rmargin` value to 1) by typing

```
. translator set smcl2ps rmargin 1
  (output omitted )
```

Any settings you make will be remembered across sessions. You can reset `smcl2ps` to be as it was when Stata was shipped by typing

```
. translator reset smcl2ps
```

Translating files from one format to another

If you have a SMCL log, which you might have created by previously typing `log using mylog`, you can translate it to an text log by typing

```
. translate myfile.smcl myfile.log
```

and you can translate it to a PostScript file by typing

```
. translate myfile.smcl myfile.ps
```

`translate` translates files from one format to another, and, in fact, `print` uses `translate` to produce a file suitable for sending to the printer.

When you type

```
. translate a.b c.d
```

`translate` looks for the predefined translator $b2d$ and uses that to perform the translation. If there is a `transmap` synonym for b, however, the mapped value b' is used: $b'2d$.

Only certain translators exist, and they are listed under the description of the `translate()` option in *Options for translate* above, or you can type

```
. translator query
```

for a complete (and perhaps more up-to-date) list.

Anyway, `translate` forms the name $b2d$ or $b'2d$, and if the translator does not exist, `translate` issues an error message. With the `translator()` option, you can specify exactly which translator to use, and then it does not matter how your files are named.

The only other thing to know is that some translators have tunable parameters that affect how they perform their translation. You can type

```
. translator query translator_name
```

to find out what those parameters are. Some translators have no tunable parameters, and some have many:

```
. translator query smcl2ps
```

header	on							
headertext								
logo	on							
user								
projecttext								
cmdnumber	on							

fontsize	9	lmargin	1.00	
pagesize	letter	rmargin	1.00	
pagewidth	8.50	tmargin	1.00	
pageheight	11.00	bmargin	1.00	

scheme	monochrome

cust1_result_color	0	0	0	cust2_result_color	0	0	0
cust1_standard_color	0	0	0	cust2_standard_color	0	0	0
cust1_error_color	0	0	0	cust2_error_color	255	0	0
cust1_input_color	0	0	0	cust2_input_color	0	0	0
cust1_link_color	0	0	0	cust2_link_color	0	0	255
cust1_hilite_color	0	0	0	cust2_hilite_color	0	0	0
cust1_result_bold	on			cust2_result_bold	on		
cust1_standard_bold	off			cust2_standard_bold	off		
cust1_error_bold	on			cust2_error_bold	on		
cust1_input_bold	off			cust2_input_bold	off		
cust1_link_bold	off			cust2_link_bold	off		
cust1_hilite_bold	on			cust2_hilite_bold	on		
cust1_link_underline	on			cust2_link_underline	on		
cust1_hilite_underline	off			cust2_hilite_underline	off		

You can temporarily override any setting by specifying the *setopt*(*setval*) option on the translate (or print) command. For instance, you can type

```
. translate ..., ... cmdnumber(off)
```

or you can reset the value permanently by typing

```
. translator set smcl2ps setopt setval
```

For instance,

```
. translator set smcl2ps cmdnumber off
```

If you reset a value, Stata will remember the change, even in future sessions.

Mac and Windows users: The smcl2ps (and the other *2ps translators) are not used by print, even when you have a PostScript printer attached to your computer. Instead, the Mac or Windows print driver is used. Resetting smcl2ps values will not affect printing; instead, you change the defaults in the Printers Control Panel in Windows and by selecting **Page Setup...** from the **File** menu in Mac. You can, however, translate files yourself using the smcl2ps translator and the other *2ps translators.

Saved results

transmap query .*ext* saves in macro r(suffix) the mapped extension (without the leading period) or saves *ext* if the *ext* is not mapped.

translator query *translatorname* saves *setval* in macro r(*setopt*) for every *setopt*, *setval* pair.

`printer query` *printername* (Unix only) saves in macro `r(suffix)` the "filetype" of the input that the printer expects (currently "ps" or "txt") and, in macro `r(command)`, the command to send output to the printer.

Methods and formulas

`print` is implemented as an ado-file.

Also see

[R] **log** — Echo copy of session to file

[G-2] **graph export** — Export current graph

[G-2] **graph print** — Print a graph

[G-2] **graph set** — Set graphics options

[P] **smcl** — Stata Markup and Control Language

[U] **15 Saving and printing output—log files**

Title

> **treatreg** — Treatment-effects model

Syntax

Basic syntax

> treatreg *depvar* $\left[\,indepvars\,\right]$, <u>tr</u>eat(*depvar*$_t$ = *indepvars*$_t$) $\left[\,\underline{\text{two}}\text{step}\,\right]$

Full syntax for maximum likelihood estimates only

> treatreg *depvar* $\left[\,indepvars\,\right]$ $\left[\,if\,\right]$ $\left[\,in\,\right]$ $\left[\,weight\,\right]$,
>
> $\quad$ <u>tr</u>eat(*depvar*$_t$ = *indepvars*$_t$ $\left[\,,\ \underline{\text{no}}\text{constant}\,\right]$) $\left[\,treatreg\_ml\_options\,\right]$

Full syntax for two-step consistent estimates only

> treatreg *depvar* $\left[\,indepvars\,\right]$ $\left[\,if\,\right]$ $\left[\,in\,\right]$,
>
> $\quad$ <u>tr</u>eat(*depvar*$_t$ = *indepvars*$_t$ $\left[\,,\ \underline{\text{no}}\text{constant}\,\right]$) <u>two</u>step $\left[\,treatreg\_ts\_options\,\right]$

treatreg_ml_options	Description
Model	
*<u>tr</u>eat()	equation for treatment effects
<u>no</u>constant	suppress constant term
<u>constr</u>aints(*constraints*)	apply specified linear constraints
<u>col</u>linear	keep collinear variables
SE/Robust	
vce(*vcetype*)	*vcetype* may be oim, <u>r</u>obust, <u>c</u>luster *clustvar*, opg, <u>boot</u>strap, or jackknife
Reporting	
<u>l</u>evel(#)	set confidence level; default is level(95)
<u>fir</u>st	report first-step probit estimates
noskip	perform likelihood-ratio test
<u>haz</u>ard(*newvar*)	create *newvar* containing hazard from treatment equation
<u>nocns</u>report	do not display constraints
display_options	control column formats, row spacing, line width, and display of omitted variables and base and empty cells
Maximization	
maximize_options	control the maximization process; seldom used
<u>coefl</u>egend	display legend instead of statistics

*<u>tr</u>eat(*depvar*$_t$ = *indepvars*$_t$ $\left[\,,\ \underline{\text{no}}\text{constant}\,\right]$) is required.

treatreg_ts_options	Description
Model	
* <u>treat</u>()	equation for treatment effects
* <u>twostep</u>	produce two-step consistent estimate
<u>no</u>constant	suppress constant term
SE	
vce(*vcetype*)	*vcetype* may be <u>conventional</u>, <u>boot</u>strap, or <u>jack</u>knife
Reporting	
<u>level</u>(#)	set confidence level; default is level(95)
<u>first</u>	report first-step probit estimates
<u>hazard</u>(*newvar*)	create *newvar* containing hazard from treatment equation
display_options	control column formats, row spacing, line width, and display of omitted variables and base and empty cells
<u>coefl</u>egend	display legend instead of statistics

*treat(*depvar*$_t$ = *indepvars*$_t$ [, <u>no</u>constant]) and twostep are required.

indepvars and *indepvars*$_t$ may contain factor variables; see [U] **11.4.3 Factor variables**.
depvar, *indepvars*, *depvar*$_t$, and *indepvars*$_t$ may contain time-series operators; see [U] **11.4.4 Time-series varlists**.
bootstrap, by, jackknife, rolling, statsby, and svy are allowed; see [U] **11.1.10 Prefix commands**.
Weights are not allowed with the bootstrap prefix; see [R] **bootstrap**.
aweights are not allowed with the jackknife prefix; see [R] **jackknife**.
twostep, vce(), first, noskip, hazard(), and weights are not allowed with the svy prefix; see [SVY] **svy**.
pweights, aweights, fweights, and iweights are allowed with maximum likelihood estimation;
   see [U] **11.1.6 weight**. No weights are allowed if twostep is specified.
coeflegend does not appear in the dialog box.
See [U] **20 Estimation and postestimation commands** for more capabilities of estimation commands.

Menu

treatreg for maximum likelihood estimates

Statistics > Sample-selection models > Treatment-effects model (ML)

treatreg for two-step consistent estimates

Statistics > Sample-selection models > Treatment-effects model (two-step)

Description

treatreg fits a treatment-effects model by using either a two-step consistent estimator or full maximum likelihood. The treatment-effects model considers the effect of an endogenously chosen binary treatment on another endogenous continuous variable, conditional on two sets of independent variables.

Options for maximum likelihood estimates

> **⌐ Model ⌐**

treat(*depvar*$_t$ = *indepvars*$_t$ [, noconstant]) specifies the variables and options for the treatment equation. It is an integral part of specifying a treatment-effects model and is required.

noconstant, constraints(*constraints*), collinear; see [R] **estimation options**.

> **⌐ SE/Robust ⌐**

vce(*vcetype*) specifies the type of standard error reported, which includes types that are derived from asymptotic theory, that are robust to some kinds of misspecification, that allow for intragroup correlation, and that use bootstrap or jackknife methods; see [R] ***vce_option***.

> **⌐ Reporting ⌐**

level(*#*); see [R] **estimation options**.

first specifies that the first-step probit estimates of the treatment equation be displayed before estimation.

noskip specifies that a full maximum-likelihood model with only a constant for the regression equation be fit. This model is not displayed but is used as the base model to compute a likelihood-ratio test for the model test statistic displayed in the estimation header. By default, the overall model test statistic is an asymptotically equivalent Wald test that all the parameters in the regression equation are zero (except the constant). For many models, this option can substantially increase estimation time.

hazard(*newvar*) will create a new variable containing the hazard from the treatment equation. The hazard is computed from the estimated parameters of the treatment equation.

nocnsreport; see [R] **estimation options**.

display_options: noomitted, vsquish, noemptycells, baselevels, allbaselevels, cformat(%*fmt*), pformat(%*fmt*), sformat(%*fmt*), and nolstretch; see [R] **estimation options**.

> **⌐ Maximization ⌐**

maximize_options: difficult, technique(*algorithm_spec*), iterate(*#*), [no]log, trace, gradient, showstep, hessian, showtolerance, tolerance(*#*), ltolerance(*#*), nrtolerance(*#*), nonrtolerance, and from(*init_specs*); see [R] **maximize**. These options are seldom used.

Setting the optimization type to technique(bhhh) resets the default *vcetype* to vce(opg).

The following option is available with treatreg but is not shown in the dialog box:

coeflegend; see [R] **estimation options**.

Options for two-step consistent estimates

> **⌐ Model ⌐**

treat(*depvar*$_t$ = *indepvars*$_t$ [, noconstant]) specifies the variables and options for the treatment equation. It is an integral part of specifying a treatment-effects model and is required.

twostep specifies that two-step consistent estimates of the parameters, standard errors, and covariance matrix be produced, instead of the default maximum likelihood estimates.

noconstant; see [R] **estimation options**.

⌐SE⌐

vce(*vcetype*) specifies the type of standard error reported, which includes types that are derived from asymptotic theory and that use bootstrap or jackknife methods; see [R] *vce_option*.

vce(conventional), the default, uses the conventionally derived variance estimator for the two-step estimator of the treatment-effects model.

⌐Reporting⌐

level(*#*); see [R] **estimation options**.

first specifies that the first-step probit estimates of the treatment equation be displayed before estimation.

hazard(*newvar*) will create a new variable containing the hazard from the treatment equation. The hazard is computed from the estimated parameters of the treatment equation.

display_options: <u>noomit</u>ted, vsquish, <u>noempty</u>cells, <u>base</u>levels, <u>allbase</u>levels, cformat(%*fmt*), pformat(%*fmt*), sformat(%*fmt*), and nolstretch; see [R] **estimation options**.

The following option is available with treatreg but is not shown in the dialog box:

coeflegend; see [R] **estimation options**.

Remarks

The treatment-effects model estimates the effect of an endogenous binary treatment, z_j, on a continuous, fully observed variable y_j, conditional on the independent variables x_j and w_j. The primary interest is in the regression function

$$y_j = \mathbf{x}_j\boldsymbol{\beta} + \delta z_j + \epsilon_j$$

where z_j is an endogenous dummy variable indicating whether the treatment is assigned or not. The binary decision to obtain the treatment z_j is modeled as the outcome of an unobserved latent variable, z_j^*. It is assumed that z_j^* is a linear function of the exogenous covariates $\mathbf{w}_j$ and a random component u_j. Specifically,

$$z_j^* = \mathbf{w}_j\boldsymbol{\gamma} + u_j$$

and the observed decision is

$$z_j = \begin{cases} 1, & \text{if } z_j^* > 0 \\ 0, & \text{otherwise} \end{cases}$$

where ϵ and u are bivariate normal with mean zero and covariance matrix

$$\begin{bmatrix} \sigma^2 & \rho\sigma \\ \rho\sigma & 1 \end{bmatrix}$$

There are many variations of this model in the literature. Maddala (1983) derives the maximum likelihood and two-step estimators of the version implemented here and also gives a brief review of several empirical applications of this model. Barnow, Cain, and Goldberger (1981) provide another useful derivation of this model. Barnow, Cain, and Goldberger (1981) concentrate on deriving the conditions for which the self-selection bias of the simple OLS estimator of the treatment effect, δ, is nonzero and of a specific sign.

▷ Example 1

We will illustrate `treatreg` with part of the Mroz dataset distributed with Berndt (1996). This dataset contains 753 observations on women's labor supply. Our subsample is of 250 observations, with 150 market laborers and 100 nonmarket laborers.

```
. use http://www.stata-press.com/data/r12/labor

. describe
Contains data from http://www.stata-press.com/data/r12/labor.dta
  obs:           250
  vars:           15                          18 Apr 2011 05:01
  size:        15,000
```

variable name	storage type	display format	value label	variable label
lfp	float	%9.0g		1 if woman worked in 1975
whrs	float	%9.0g		wife's hours of work
kl6	float	%9.0g		# of children younger than 6
k618	float	%9.0g		# of children between 6 and 18
wa	float	%9.0g		wife's age
we	float	%9.0g		wife's education attainment
ww	float	%9.0g		wife's wage
hhrs	float	%9.0g		husband's hours worked in 1975
ha	float	%9.0g		husband's age
he	float	%9.0g		husband's educational attainment
hw	float	%9.0g		husband's wage
faminc	float	%9.0g		family income
wmed	float	%9.0g		wife's mother's educational attainment
wfed	float	%9.0g		wife's father's educational attainment
cit	float	%9.0g		1 if live in large city

```
Sorted by:

. summarize
```

Variable	Obs	Mean	Std. Dev.	Min	Max
lfp	250	.6	.4908807	0	1
whrs	250	799.84	915.6035	0	4950
kl6	250	.236	.5112234	0	3
k618	250	1.364	1.370774	0	8
wa	250	42.92	8.426483	30	60
we	250	12.352	2.164912	5	17
ww	250	2.27523	2.59775	0	14.631
hhrs	250	2234.832	600.6702	768	5010
ha	250	45.024	8.171322	30	60
he	250	12.536	3.106009	3	17
hw	250	7.494435	4.636192	1.0898	40.509
faminc	250	23062.54	12923.98	3305	91044
wmed	250	9.136	3.536031	0	17
wfed	250	8.608	3.751082	0	17
cit	250	.624	.4853517	0	1

We will assume that the wife went to college if her educational attainment is more than 12 years. Let `wc` be the dummy variable indicating whether the individual went to college. With this definition, our sample contains the following distribution of college education:

```
. generate wc = 0
. replace wc = 1 if we > 12
(69 real changes made)
. tab wc
```

wc	Freq.	Percent	Cum.
0	181	72.40	72.40
1	69	27.60	100.00
Total	250	100.00	

We will model the wife's wage as a function of her age, whether the family was living in a big city, and whether she went to college. An ordinary least-squares estimation produces the following results:

```
. regress ww wa cit wc
```

Source	SS	df	MS		Number of obs =	250
					F(3, 246) =	4.82
Model	93.2398568	3	31.0799523		Prob > F =	0.0028
Residual	1587.08776	246	6.45157627		R-squared =	0.0555
					Adj R-squared =	0.0440
Total	1680.32762	249	6.74830369		Root MSE =	2.54

ww	Coef.	Std. Err.	t	P>|t|	[95% Conf. Interval]	
wa	-.0104985	.0192667	-0.54	0.586	-.0484472	.0274502
cit	.1278922	.3389058	0.38	0.706	-.5396351	.7954195
wc	1.332192	.3644344	3.66	0.000	.6143819	2.050001
_cons	2.278337	.8432385	2.70	0.007	.6174488	3.939225

Is 1.332 a consistent estimate of the marginal effect of a college education on wages? If individuals choose whether to attend college and the error term of the model that gives rise to this choice is correlated with the error term in the wage equation, then the answer is no. (See Barnow, Cain, and Goldberger [1981] for a good discussion of the existence and sign of selectivity bias.) We might suspect that individuals with higher abilities, either innate or due to the circumstances of their birth, would be more likely to go to college and to earn higher wages. Such ability is, of course, unobserved. Furthermore, if the error term in our model for going to college is correlated with ability, and the error term in our wage equation is correlated with ability, the two terms should be positively correlated. These conditions make the problem of signing the selectivity bias equivalent to an omitted-variable problem. In the case at hand, because we would expect the correlation between the omitted variable and a college education to be positive, we suspect that OLS is biased upward.

To account for the bias, we fit the treatment-effects model. We model the wife's college decision as a function of her mother's and her father's educational attainment. Thus we are interested in fitting the model

$$ww = \beta_0 + \beta_1 wa + \beta_2 cit + \delta wc + \epsilon$$
$$wc^* = \gamma_0 + \gamma_1 wmed + \gamma_2 wfed + u$$

where

$$wc = \begin{cases} 1, & wc^* > 0, \text{ that is, wife went to college} \\ 0, & \text{otherwise} \end{cases}$$

and where ϵ and u have a bivariate normal distribution with zero mean and covariance matrix

$$\begin{bmatrix} \sigma^2 & \rho\sigma \\ \rho\sigma & 1 \end{bmatrix}$$

The following output gives the maximum likelihood estimates of the parameters of this model:

```
. treatreg ww wa cit, treat(wc=wmed wfed)
Iteration 0:    log likelihood = -707.07237
Iteration 1:    log likelihood = -707.07215
Iteration 2:    log likelihood = -707.07215

Treatment-effects model -- MLE                    Number of obs   =        250
                                                  Wald chi2(3)    =       4.11
Log likelihood = -707.07215                       Prob > chi2     =     0.2501
```

	Coef.	Std. Err.	z	P>|z|	[95% Conf. Interval]	
ww						
wa	-.0110424	.0199652	-0.55	0.580	-.0501735	.0280887
cit	.127636	.3361938	0.38	0.704	-.5312917	.7865638
wc	1.271327	.7412951	1.72	0.086	-.1815842	2.724239
_cons	2.318638	.9397573	2.47	0.014	.4767478	4.160529
wc						
wmed	.1198055	.0320056	3.74	0.000	.0570757	.1825352
wfed	.0961886	.0290868	3.31	0.001	.0391795	.1531977
_cons	-2.631876	.3309128	-7.95	0.000	-3.280453	-1.983299
/athrho	.0178668	.1899898	0.09	0.925	-.3545063	.3902399
/lnsigma	.9241584	.0447455	20.65	0.000	.8364588	1.011858
rho	.0178649	.1899291			-.3403659	.371567
sigma	2.519747	.1127473			2.308179	2.750707
lambda	.0450149	.4786442			-.8931105	.9831404

```
LR test of indep. eqns. (rho = 0):   chi2(1) =      0.01   Prob > chi2 = 0.9251
```

In the input, we specified that the continuous dependent variable, ww (wife's wage), is a linear function of cit and wa. Note the syntax for the treatment variable. The treatment wc is not included in the first variable list; it is specified in the treat() option. In this example, wmed and wfed are specified as the exogenous covariates in the treatment equation.

The output has the form of many two-equation estimators in Stata. We note that our conjecture that the OLS estimate was biased upward is verified. But it is perhaps more interesting that the size of the bias is negligible, and the likelihood-ratio test at the bottom of the output indicates that we cannot reject the null hypothesis that the two error terms are uncorrelated. This result might be due to several specification errors. We ignored the selectivity bias due to the endogeneity of entering the labor market. We have also written both the wage equation and the college-education equation in linear form, ignoring any higher power terms or interactions.

The results for the two ancillary parameters require explanation. For numerical stability during optimization, treatreg does not directly estimate ρ or σ. Instead, treatreg estimates the inverse hyperbolic tangent of ρ,

$$\operatorname{atanh} \rho = \frac{1}{2} \ln \left(\frac{1 + \rho}{1 - \rho} \right)$$

and $\ln\sigma$. Also, treatreg reports $\lambda = \rho\sigma$, along with an estimate of the standard error of the estimate and confidence interval.

◁

▷ Example 2

Stata also produces a two-step estimator of the model with the `twostep` option. Maximum likelihood estimation of the parameters can be time consuming with large datasets, and the two-step estimates may provide a good alternative in such cases. Continuing with the women's wage model, we can obtain the two-step estimates with consistent covariance estimates by typing

```
. treatreg ww wa cit, treat(wc=wmed wfed) twostep
Treatment-effects model -- two-step estimates     Number of obs    =       250
                                                  Wald chi2(3)     =      3.67
                                                  Prob > chi2      =    0.2998
```

	Coef.	Std. Err.	z	P>|z|	[95% Conf. Interval]	
ww						
wa	-.0111623	.020152	-0.55	0.580	-.0506594	.0283348
cit	.1276102	.33619	0.38	0.704	-.53131	.7865305
wc	1.257995	.8007428	1.57	0.116	-.3114319	2.827422
_cons	2.327482	.9610271	2.42	0.015	.4439031	4.21106
wc						
wmed	.1198888	.0319862	3.75	0.000	.0571971	.1825806
wfed	.0960764	.0290583	3.31	0.001	.0391233	.1530295
_cons	-2.631496	.3308389	-7.95	0.000	-3.279928	-1.983063
hazard						
lambda	.0548738	.5283928	0.10	0.917	-.9807571	1.090505
rho	0.02178					
sigma	2.5198211					

The reported `lambda` (λ) is the parameter estimate on the hazard from the augmented regression, which is derived in Maddala (1983) and presented in *Methods and formulas* below.

◁

❑ Technical note

The difference in expected earnings between participants and nonparticipants is

$$E\left(y_j \mid z_j = 1\right) - E\left(y_j \mid z_j = 0\right) = \delta + \rho\sigma \left[\frac{\phi(\mathbf{w}_j\boldsymbol{\gamma})}{\Phi(\mathbf{w}_j\boldsymbol{\gamma})\{1 - \Phi(\mathbf{w}_j\boldsymbol{\gamma})\}} \right]$$

where ϕ is the standard normal density and Φ is the standard normal cumulative distribution function. If the correlation between the error terms, ρ, is zero, the problem reduces to one estimable by OLS and the difference is simply δ. Because ρ is positive in the example, least squares overestimates the treatment effect, δ.

❑

Saved results

treatreg (maximum likelihood) saves the following in e():

Scalars

e(N)	number of observations
e(k)	number of parameters
e(k_eq)	number of equations in e(b)
e(k_eq_model)	number of equations in overall model test
e(k_aux)	number of auxiliary parameters
e(k_dv)	number of dependent variables
e(df_m)	model degrees of freedom
e(ll)	log likelihood
e(N_clust)	number of clusters
e(lambda)	λ
e(selambda)	standard error of λ
e(sigma)	estimate of sigma
e(chi2)	χ^2
e(chi2_c)	χ^2 for comparison test
e(p_c)	p-value for comparison test
e(p)	significance
e(rho)	ρ
e(rank)	rank of e(V)
e(ic)	number of iterations
e(rc)	return code
e(converged)	1 if converged, 0 otherwise

Macros

e(cmd)	treatreg
e(cmdline)	command as typed
e(depvar)	name of dependent variable
e(hazard)	variable containing hazard
e(wtype)	weight type
e(wexp)	weight expression
e(title)	title in estimation output
e(clustvar)	name of cluster variable
e(chi2type)	Wald or LR; type of model χ^2 test
e(chi2_ct)	Wald or LR; type of model χ^2 test corresponding to e(chi2_c)
e(vce)	*vcetype* specified in vce()
e(vcetype)	title used to label Std. Err.
e(opt)	type of optimization
e(which)	max or min; whether optimizer is to perform maximization or minimization
e(method)	ml
e(ml_method)	type of ml method
e(user)	name of likelihood-evaluator program
e(technique)	maximization technique
e(properties)	b V
e(predict)	program used to implement predict
e(footnote)	program used to implement the footnote display
e(marginsok)	predictions allowed by margins
e(asbalanced)	factor variables fvset as asbalanced
e(asobserved)	factor variables fvset as asobserved

Matrices

e(b)	coefficient vector
e(Cns)	constraints matrix
e(ilog)	iteration log (up to 20 iterations)
e(gradient)	gradient vector
e(V)	variance–covariance matrix of the estimators
e(V_modelbased)	model-based variance

Functions

e(sample)	marks estimation sample

`treatreg` (two-step) saves the following in `e()`:

Scalars
e(N)	number of observations
e(df_m)	model degrees of freedom
e(lambda)	λ
e(selambda)	standard error of λ
e(sigma)	estimate of sigma
e(chi2)	χ^2
e(p)	significance
e(rho)	ρ
e(rank)	rank of e(V)

Macros
e(cmd)	treatreg
e(cmdline)	command as typed
e(depvar)	name of dependent variable
e(title)	title in estimation output
e(chi2type)	Wald or LR; type of model χ^2 test
e(vce)	*vcetype* specified in vce()
e(vcetype)	title used to label Std. Err.
e(hazard)	variable specified in hazard()
e(method)	ml or twostep
e(properties)	b V
e(predict)	program used to implement predict
e(footnote)	program used to implement the footnote display
e(marginsok)	predictions allowed by margins
e(marginsnotok)	predictions disallowed by margins
e(asbalanced)	factor variables fvset as asbalanced
e(asobserved)	factor variables fvset as asobserved

Matrices
e(b)	coefficient vector
e(V)	variance–covariance matrix of the estimators

Functions
e(sample)	marks estimation sample

Methods and formulas

`treatreg` is implemented as an ado-file. Maddala (1983, 117–122) derives both the maximum likelihood and the two-step estimator implemented here. Greene (2012, 890–894) also provides an introduction to the treatment-effects model.

The primary regression equation of interest is

$$y_j = \mathbf{x}_j\boldsymbol{\beta} + \delta z_j + \epsilon_j$$

where z_j is a binary decision variable that is assumed to stem from an unobservable latent variable:

$$z_j^* = \mathbf{w}_j\boldsymbol{\gamma} + u_j$$

The decision to obtain the treatment is made according to the rule

$$z_j = \begin{cases} 1, & \text{if } z_j^* > 0 \\ 0, & \text{otherwise} \end{cases}$$

where ϵ and u are bivariate normal with mean zero and covariance matrix

$$\begin{bmatrix} \sigma^2 & \rho\sigma \\ \rho\sigma & 1 \end{bmatrix}$$

The likelihood function for this model is given in Maddala (1983, 122). Greene (2000, 180) discusses the standard method of reducing a bivariate normal to a function of a univariate normal and the correlation ρ. The following is the log likelihood for observation j,

$$\ln L_j = \begin{cases} \ln\Phi\left\{\dfrac{\mathbf{w}_j\boldsymbol{\gamma} + (y_j - \mathbf{x}_j\boldsymbol{\beta} - \delta)\rho/\sigma}{\sqrt{1-\rho^2}}\right\} - \dfrac{1}{2}\left(\dfrac{y_j - \mathbf{x}_j\boldsymbol{\beta} - \delta}{\sigma}\right)^2 - \ln(\sqrt{2\pi}\sigma), & z_j = 1 \\[4mm] \ln\Phi\left\{\dfrac{-\mathbf{w}_j\boldsymbol{\gamma} - (y_j - \mathbf{x}_j\boldsymbol{\beta})\rho/\sigma}{\sqrt{1-\rho^2}}\right\} - \dfrac{1}{2}\left(\dfrac{y_j - \mathbf{x}_j\boldsymbol{\beta}}{\sigma}\right)^2 - \ln(\sqrt{2\pi}\sigma), & z_j = 0 \end{cases}$$

where $\Phi(\cdot)$ is the cumulative distribution function of the standard normal distribution.

In the maximum likelihood estimation, σ and ρ are not directly estimated. Rather $\ln\sigma$ and $\mathrm{atanh}\,\rho$ are directly estimated, where

$$\mathrm{atanh}\,\rho = \frac{1}{2}\ln\left(\frac{1+\rho}{1-\rho}\right)$$

The standard error of $\lambda = \rho\sigma$ is approximated through the delta method, which is given by

$$\mathrm{Var}(\lambda) \approx \mathbf{D}\,\mathrm{Var}\{(\mathrm{atanh}\,\rho \ \ \ln\sigma)\}\,\mathbf{D}'$$

where $\mathbf{D}$ is the Jacobian of λ with respect to $\mathrm{atanh}\,\rho$ and $\ln\sigma$.

With maximum likelihood estimation, this command supports the Huber/White/sandwich estimator of the variance and its clustered version using vce(robust) and vce(cluster *clustvar*), respectively. See [P] _robust, particularly *Maximum likelihood estimators* and *Methods and formulas*.

The maximum likelihood version of treatreg also supports estimation with survey data. For details on VCEs with survey data, see [SVY] **variance estimation**.

Maddala (1983, 120–122) also derives the two-step estimator. In the first stage, probit estimates are obtained of the treatment equation

$$\mathrm{Pr}(z_j = 1 \mid \mathbf{w}_j) = \Phi(\mathbf{w}_j\boldsymbol{\gamma})$$

From these estimates, the hazard, h_j, for each observation j is computed as

$$h_j = \begin{cases} \phi(\mathbf{w}_j\widehat{\boldsymbol{\gamma}})\big/\Phi(\mathbf{w}_j\widehat{\boldsymbol{\gamma}}), & z_j = 1 \\[3mm] -\phi(\mathbf{w}_j\widehat{\boldsymbol{\gamma}})\big/\{1 - \Phi(\mathbf{w}_j\widehat{\boldsymbol{\gamma}})\}, & z_j = 0 \end{cases}$$

where ϕ is the standard normal density function. If

$$d_j = h_j(h_j + \widehat{\boldsymbol{\gamma}}\mathbf{w}_j)$$

then

$$E\left(y_j \mid z_j\right) = \mathbf{x}_j \boldsymbol{\beta} + \delta \mathbf{z}_j + \rho \sigma h_j$$

$$\mathrm{Var}\left(y_j \mid z_j\right) = \sigma^2 \left(1 - \rho^2 d_j\right)$$

The two-step parameter estimates of $\boldsymbol{\beta}$ and δ are obtained by augmenting the regression equation with the hazard h. Thus the regressors become $[\,\mathbf{x}\ \mathbf{z}\ h\,]$, and the additional parameter estimate β_h is obtained on the variable containing the hazard. A consistent estimate of the regression disturbance variance is obtained using the residuals from the augmented regression and the parameter estimate on the hazard

$$\widehat{\sigma}^{\,2} = \frac{\mathbf{e}'\mathbf{e} + \beta_h^2 \sum_{j=1}^{N} d_j}{N}$$

The two-step estimate of ρ is then

$$\widehat{\rho} = \frac{\beta_h}{\widehat{\sigma}}$$

To understand how the consistent estimates of the coefficient covariance matrix based on the augmented regression are derived, let $\mathbf{A} = [\,\mathbf{x}\ \mathbf{z}\ h\,]$ and $\mathbf{D}$ be a square diagonal matrix of size N with $(1 - \widehat{\rho}^{\,2} d_j)$ on the diagonal elements. The conventional VCE is

$$\mathbf{V}_{\mathrm{twostep}} = \widehat{\sigma}^{\,2}(\mathbf{A}'\mathbf{A})^{-1}(\mathbf{A}'\mathbf{D}\mathbf{A} + \mathbf{Q})(\mathbf{A}'\mathbf{A})^{-1}$$

where

$$\mathbf{Q} = \widehat{\rho}^{\,2}(\mathbf{A}'\mathbf{D}\mathbf{A})\mathbf{V_p}(\mathbf{A}'\mathbf{D}\mathbf{A})$$

and $\mathbf{V_p}$ is the variance–covariance estimate from the probit estimation of the treatment equation.

References

Barnow, B. S., G. G. Cain, and A. S. Goldberger. 1981. Issues in the analysis of selectivity bias. In Vol. 5 of *Evaluation Studies Review Annual*, ed. E. W. Stromsdorfer and G. Farkas, 123–126. Beverly Hills: Sage.

Berndt, E. R. 1996. *The Practice of Econometrics: Classic and Contemporary*. New York: Addison–Wesley.

Cong, R., and D. M. Drukker. 2000. sg141: Treatment effects model. *Stata Technical Bulletin* 55: 25–33. Reprinted in *Stata Technical Bulletin Reprints*, vol. 10, pp. 159–169. College Station, TX: Stata Press.

Greene, W. H. 2000. *Econometric Analysis*. 4th ed. Upper Saddle River, NJ: Prentice Hall.

———. 2012. *Econometric Analysis*. 7th ed. Upper Saddle River, NJ: Prentice Hall.

Maddala, G. S. 1983. *Limited-Dependent and Qualitative Variables in Econometrics*. Cambridge: Cambridge University Press.

Nannicini, T. 2007. Simulation-based sensitivity analysis for matching estimators. *Stata Journal* 7: 334–350.

Nichols, A. 2007. Causal inference with observational data. *Stata Journal* 7: 507–541.

Also see

[R] **treatreg postestimation** — Postestimation tools for treatreg

[R] **heckman** — Heckman selection model

[R] **probit** — Probit regression

[R] **regress** — Linear regression

[SVY] **svy estimation** — Estimation commands for survey data

[U] **20 Estimation and postestimation commands**

Title

treatreg postestimation — Postestimation tools for treatreg

Description

The following postestimation commands are available after `treatreg`:

Command	Description
contrast	contrasts and ANOVA-style joint tests of estimates
estat[1]	AIC, BIC, VCE, and estimation sample summary
estat (svy)	postestimation statistics for survey data
estimates	cataloging estimation results
lincom	point estimates, standard errors, testing, and inference for linear combinations of coefficients
lrtest[2]	likelihood-ratio test
margins	marginal means, predictive margins, marginal effects, and average marginal effects
marginsplot	graph the results from margins (profile plots, interaction plots, etc.)
nlcom	point estimates, standard errors, testing, and inference for nonlinear combinations of coefficients
predict	predictions, residuals, influence statistics, and other diagnostic measures
predictnl	point estimates, standard errors, testing, and inference for generalized predictions
pwcompare	pairwise comparisons of estimates
suest[1]	seemingly unrelated estimation
test	Wald tests of simple and composite linear hypotheses
testnl	Wald tests of nonlinear hypotheses

[1] `estat ic` and `suest` are not appropriate after `treatreg, twostep`.

[2] `lrtest` is not appropriate with svy estimation results.

See the corresponding entries in the *Base Reference Manual* for details, but see [SVY] **estat** for details about `estat` (svy).

Syntax for predict

After ML or twostep

> predict [*type*] *newvar* [*if*] [*in*] [, *statistic*]

After ML

> predict [*type*] { *stub** | *newvar*$_\text{reg}$ *newvar*$_\text{treat}$ *newvar*$_\text{athrho}$ *newvar*$_\text{lnsigma}$ }
>
> [*if*] [*in*] , <u>sc</u>ores

statistic	Description
Main	
xb	linear prediction; the default
stdp	standard error of the prediction
stdf	standard error of the forecast
<u>yc</u>trt	$E(y_j \mid \text{treatment} = 1)$
<u>ycn</u>trt	$E(y_j \mid \text{treatment} = 0)$
<u>p</u>trt	$\Pr(\text{treatment} = 1)$
<u>xb</u>trt	linear prediction for treatment equation
<u>stdp</u>trt	standard error of the linear prediction for treatment equation

These statistics are available both in and out of sample; type predict ... if e(sample) ... if wanted only for the estimation sample.

stdf is not allowed with svy estimation results.

Menu

Statistics > Postestimation > Predictions, residuals, etc.

Options for predict

    Main

xb, the default, calculates the linear prediction, $\mathbf{x}_j\mathbf{b}$.

stdp calculates the standard error of the prediction, which can be thought of as the standard error of the predicted expected value or mean for the observation's covariate pattern. The standard error of the prediction is also referred to as the standard error of the fitted value.

stdf calculates the standard error of the forecast, which is the standard error of the point prediction for one observation. It is commonly referred to as the standard error of the future or forecast value. By construction, the standard errors produced by stdf are always larger than those produced by stdp; see *Methods and formulas* in [R] **regress postestimation**.

yctrt calculates the expected value of the dependent variable conditional on the presence of the treatment: $E(y_j \mid \text{treatment} = 1)$.

ycntrt calculates the expected value of the dependent variable conditional on the absence of the treatment: $E(y_j \mid \text{treatment} = 0)$.

ptrt calculates the probability of the presence of the treatment:
$\Pr(\text{treatment} = 1) = \Pr(\mathbf{w}_j\boldsymbol{\gamma} + u_j > 0)$.

xbtrt calculates the linear prediction for the treatment equation.

stdptrt calculates the standard error of the linear prediction for the treatment equation.

scores, not available with twostep, calculates equation-level score variables.

> The first new variable will contain $\partial \ln L / \partial(\boldsymbol{x}_j \boldsymbol{\beta})$.

> The second new variable will contain $\partial \ln L / \partial(\boldsymbol{w}_j \boldsymbol{\gamma})$.

> The third new variable will contain $\partial \ln L / \partial \operatorname{atanh} \rho$.

> The fourth new variable will contain $\partial \ln L / \partial \ln \sigma$.

Remarks

▷ Example 1

The default statistic produced by predict after treatreg is the expected value of the dependent variable from the underlying distribution of the regression model. For example 1 in [R] **treatreg**, this model is

$$ \text{ww} = \beta_0 + \beta_1 \text{wa} + \beta_2 \text{cit} + \delta \text{wc} + \epsilon $$

Several other interesting aspects of the treatment-effects model can be explored with predict. We continue with our wage model, the wife's expected wage, conditional on attending college, can be obtained with the yctrt option. The wife's expected wages, conditional on not attending college, can be obtained with the ycntrt option. Thus the difference in expected wages between participants and nonparticipants is the difference between yctrt and ycntrt. For the case at hand, we have the following calculation:

```
. predict wwctrt, yctrt
. predict wwcntrt, ycntrt
. generate diff = wwctrt - wwcntrt
. summarize diff
```

Variable	Obs	Mean	Std. Dev.	Min	Max
diff	250	1.356912	.0134202	1.34558	1.420173

◁

Methods and formulas

All postestimation commands listed above are implemented as ado-files.

Also see

[R] **treatreg** — Treatment-effects model

[U] **20 Estimation and postestimation commands**

Title

> **truncreg** — Truncated regression

Syntax

> truncreg *depvar* [*indepvars*] [*if*] [*in*] [*weight*] [, *options*]

options	Description
Model	
<u>nocon</u>stant	suppress constant term
<u>ll</u>(*varname* | #)	lower limit for left-truncation
<u>ul</u>(*varname* | #)	upper limit for right-truncation
<u>off</u>set(*varname*)	include *varname* in model with coefficient constrained to 1
<u>constraints</u>(*constraints*)	apply specified linear constraints
<u>col</u>linear	keep collinear variables
SE/Robust	
vce(*vcetype*)	*vcetype* may be oim, <u>r</u>obust, <u>cl</u>uster *clustvar*, opg, <u>boot</u>strap, or <u>jack</u>knife
Reporting	
<u>level</u>(#)	set confidence level; default is level(95)
noskip	perform likelihood-ratio test
<u>nocns</u>report	do not display constraints
display_options	control column formats, row spacing, line width, and display of omitted variables and base and empty cells
Maximization	
maximize_options	control the maximization process; seldom used
<u>coefl</u>egend	display legend instead of statistics

indepvars may contain factor variables; see [U] **11.4.3 Factor variables**.

depvar and *indepvars* may contain time-series operators; see [U] **11.4.4 Time-series varlists**.

bootstrap, by, jackknife, mi estimate, rolling, statsby, and svy are allowed; see
   [U] **11.1.10 Prefix commands**.

vce(bootstrap) and vce(jackknife) are not allowed with the mi estimate prefix; see [MI] **mi estimate**.

Weights are not allowed with the bootstrap prefix; see [R] **bootstrap**.

aweights are not allowed with the jackknife prefix; see [R] **jackknife**.

vce(), noskip, and weights are not allowed with the svy prefix; see [SVY] **svy**.

aweights, fweights, iweights, and pweights are allowed; see [U] **11.1.6 weight**.

coeflegend does not appear in the dialog box.

See [U] **20 Estimation and postestimation commands** for more capabilities of estimation commands.

Menu

Statistics > Linear models and related > Truncated regression

Description

truncreg fits a regression model of *depvar* on *indepvars* from a sample drawn from a restricted part of the population. Under the normality assumption for the whole population, the error terms in the truncated regression model have a truncated normal distribution, which is a normal distribution that has been scaled upward so that the distribution integrates to one over the restricted range.

Options

   ⌐ Model ⌐

noconstant; see [R] **estimation options**.

ll(*varname* | #) and ul(*varname* | #) indicate the lower and upper limits for truncation, respectively. You may specify one or both. Observations with *depvar* ≤ ll() are left-truncated, observations with *depvar* ≥ ul() are right-truncated, and the remaining observations are not truncated. See [R] **tobit** for a more detailed description.

offset(*varname*), constraints(*constraints*), collinear; see [R] **estimation options**.

   ⌐ SE/Robust ⌐

vce(*vcetype*) specifies the type of standard error reported, which includes types that are derived from asymptotic theory, that are robust to some kinds of misspecification, that allow for intragroup correlation, and that use bootstrap or jackknife methods; see [R] *vce_option*.

   ⌐ Reporting ⌐

level(#); see [R] **estimation options**.

noskip specifies that a full maximum-likelihood model with only a constant for the regression equation be fit. This model is not displayed but is used as the base model to compute a likelihood-ratio test for the model test statistic displayed in the estimation header. By default, the overall model test statistic is an asymptotically equivalent Wald test of all the parameters in the regression equation being zero (except the constant). For many models, this option can substantially increase estimation time.

nocnsreport; see [R] **estimation options**.

display_options: noomitted, vsquish, noemptycells, baselevels, allbaselevels, cformat(%*fmt*), pformat(%*fmt*), sformat(%*fmt*), and nolstretch; see [R] **estimation options**.

   ⌐ Maximization ⌐

maximize_options: difficult, technique(*algorithm_spec*), iterate(#), [no]log, trace, gradient, showstep, hessian, showtolerance, tolerance(#), ltolerance(#), nrtolerance(#), nonrtolerance, and from(*init_specs*); see [R] **maximize**. These options are seldom used, but you may use the ltol(#) option to relax the convergence criterion; the default is 1e-6 during specification searches.

Setting the optimization type to technique(bhhh) resets the default *vcetype* to vce(opg).

The following option is available with truncreg but is not shown in the dialog box:

coeflegend; see [R] **estimation options**.

Remarks

Truncated regression fits a model of a dependent variable on independent variables from a restricted part of a population. Truncation is essentially a characteristic of the distribution from which the sample data are drawn. If x has a normal distribution with mean μ and standard deviation σ, the density of the truncated normal distribution is

$$f\left(x \mid a < x < b\right) = \frac{f(x)}{\Phi\left(\frac{b-\mu}{\sigma}\right) - \Phi\left(\frac{a-\mu}{\sigma}\right)}$$

$$= \frac{\frac{1}{\sigma}\phi\left(\frac{x-\mu}{\sigma}\right)}{\Phi\left(\frac{b-\mu}{\sigma}\right) - \Phi\left(\frac{a-\mu}{\sigma}\right)}$$

where ϕ and Φ are the density and distribution functions of the standard normal distribution.

Compared with the mean of the untruncated variable, the mean of the truncated variable is greater if the truncation is from below, and the mean of the truncated variable is smaller if the truncation is from above. Moreover, truncation reduces the variance compared with the variance in the untruncated distribution.

▷ Example 1

We will demonstrate `truncreg` with part of the Mroz dataset distributed with Berndt (1996). This dataset contains 753 observations on women's labor supply. Our subsample is of 250 observations, with 150 market laborers and 100 nonmarket laborers.

```
. use http://www.stata-press.com/data/r12/laborsub

. describe

Contains data from http://www.stata-press.com/data/r12/laborsub.dta
  obs:           250
  vars:            6                          25 Sep 2010 18:36
  size:        1,750
```

variable name	storage type	display format	value label	variable label
lfp	byte	%9.0g		1 if woman worked in 1975
whrs	int	%9.0g		Wife's hours of work
kl6	byte	%9.0g		# of children younger than 6
k618	byte	%9.0g		# of children between 6 and 18
wa	byte	%9.0g		Wife's age
we	byte	%9.0g		Wife's educational attainment

```
Sorted by:

. summarize, sep(0)
```

Variable	Obs	Mean	Std. Dev.	Min	Max
lfp	250	.6	.4908807	0	1
whrs	250	799.84	915.6035	0	4950
kl6	250	.236	.5112234	0	3
k618	250	1.364	1.370774	0	8
wa	250	42.92	8.426483	30	60
we	250	12.352	2.164912	5	17

We first perform ordinary least-squares estimation on the market laborers.

```
. regress whrs kl6 k618 wa we if whrs > 0
```

Source	SS	df	MS
Model	7326995.15	4	1831748.79
Residual	94793104.2	145	653745.546
Total	102120099	149	685369.794

```
                                    Number of obs =     150
                                    F(  4,   145) =    2.80
                                    Prob > F      =  0.0281
                                    R-squared     =  0.0717
                                    Adj R-squared =  0.0461
                                    Root MSE      =  808.55
```

whrs	Coef.	Std. Err.	t	P>|t|	[95% Conf. Interval]	
kl6	-421.4822	167.9734	-2.51	0.013	-753.4748	-89.48953
k618	-104.4571	54.18616	-1.93	0.056	-211.5538	2.639668
wa	-4.784917	9.690502	-0.49	0.622	-23.9378	14.36797
we	9.353195	31.23793	0.30	0.765	-52.38731	71.0937
_cons	1629.817	615.1301	2.65	0.009	414.0371	2845.597

Now we use truncreg to perform truncated regression with truncation from below zero.

```
. truncreg whrs kl6 k618 wa we, ll(0)
(note: 100 obs. truncated)

Fitting full model:

Iteration 0:   log likelihood = -1205.6992
Iteration 1:   log likelihood = -1200.9873
Iteration 2:   log likelihood = -1200.9159
Iteration 3:   log likelihood = -1200.9157
Iteration 4:   log likelihood = -1200.9157

Truncated regression
Limit:   lower =           0               Number of obs =     150
         upper =        +inf               Wald chi2(4)  =   10.05
Log likelihood = -1200.9157               Prob > chi2   =  0.0395
```

whrs	Coef.	Std. Err.	z	P>|z|	[95% Conf. Interval]	
kl6	-803.0042	321.3614	-2.50	0.012	-1432.861	-173.1474
k618	-172.875	88.72898	-1.95	0.051	-346.7806	1.030579
wa	-8.821123	14.36848	-0.61	0.539	-36.98283	19.34059
we	16.52873	46.50375	0.36	0.722	-74.61695	107.6744
_cons	1586.26	912.355	1.74	0.082	-201.9233	3374.442
/sigma	983.7262	94.44303	10.42	0.000	798.6213	1168.831

If we assume that our data were censored, the tobit model is

```
. tobit whrs kl6 k618 wa we, ll(0)
```

Tobit regression

		Number of obs	=	250
		LR chi2(4)	=	23.03
		Prob > chi2	=	0.0001
Log likelihood = -1367.0903		Pseudo R2	=	0.0084

whrs	Coef.	Std. Err.	t	P>|t|	[95% Conf. Interval]	
kl6	-827.7657	214.7407	-3.85	0.000	-1250.731	-404.8008
k618	-140.0192	74.22303	-1.89	0.060	-286.2129	6.174547
wa	-24.97919	13.25639	-1.88	0.061	-51.08969	1.131317
we	103.6896	41.82393	2.48	0.014	21.31093	186.0683
_cons	589.0001	841.5467	0.70	0.485	-1068.556	2246.556
/sigma	1309.909	82.73335			1146.953	1472.865

```
Obs. summary:      100  left-censored observations at whrs<=0
                   150      uncensored observations
                     0  right-censored observations
```
◁

❏ Technical note

Whether truncated regression is more appropriate than the ordinary least-squares estimation depends on the purpose of that estimation. If we are interested in the mean of wife's working hours conditional on the subsample of market laborers, least-squares estimation is appropriate. However if we are interested in the mean of wife's working hours regardless of market or nonmarket labor status, least-squares estimates could be seriously misleading.

Truncation and censoring are different concepts. A sample has been censored if no observations have been systematically excluded but some of the information contained in them has been suppressed. In a truncated distribution, only the part of the distribution above (or below, or between) the truncation points is relevant to our computations. We need to scale it up by the probability that an observation falls in the range that interests us to make the distribution integrate to one. The censored distribution used by tobit, however, is a mixture of discrete and continuous distributions. Instead of rescaling over the observable range, we simply assign the full probability from the censored regions to the censoring points. The truncated regression model is sometimes less well behaved than the tobit model. Davidson and MacKinnon (1993) provide an example where truncation results in more inconsistency than censoring.

❏

Saved results

truncreg saves the following in e():

Scalars
e(N)	number of observations
e(N_bf)	number of obs. before truncation
e(chi2)	model χ^2
e(k_eq)	number of equations in e(b)
e(k_eq_model)	number of equations in overall model test
e(k_aux)	number of auxiliary parameters
e(df_m)	model degrees of freedom
e(ll)	log likelihood
e(ll_0)	log likelihood, constant-only model
e(N_clust)	number of clusters
e(sigma)	estimate of sigma
e(p)	significance
e(rank)	rank of e(V)
e(ic)	number of iterations
e(rc)	return code
e(converged)	1 if converged, 0 otherwise

Macros
e(cmd)	truncreg
e(cmdline)	command as typed
e(llopt)	contents of ll(), if specified
e(ulopt)	contents of ul(), if specified
e(depvar)	name of dependent variable
e(wtype)	weight type
e(wexp)	weight expression
e(title)	title in estimation output
e(clustvar)	name of cluster variable
e(offset1)	offset
e(chi2type)	Wald or LR; type of model χ^2 test
e(vce)	*vcetype* specified in vce()
e(vcetype)	title used to label Std. Err.
e(opt)	type of optimization
e(which)	max or min; whether optimizer is to perform maximization or minimization
e(ml_method)	type of ml method
e(user)	name of likelihood-evaluator program
e(technique)	maximization technique
e(properties)	b V
e(predict)	program used to implement predict
e(asbalanced)	factor variables fvset as asbalanced
e(asobserved)	factor variables fvset as asobserved

Matrices
e(b)	coefficient vector
e(Cns)	constraints matrix
e(ilog)	iteration log (up to 20 iterations)
e(gradient)	gradient vector
e(V)	variance–covariance matrix of the estimators
e(V_modelbased)	model-based variance
e(means)	means of independent variables
e(dummy)	indicator for dummy variables

Functions
e(sample)	marks estimation sample

Methods and formulas

truncreg is implemented as an ado-file. Greene (2012, 833–839) and Davidson and MacKinnon (1993, 534–537) provide introductions to the truncated regression model.

Let $\mathbf{y} = \mathbf{X}\boldsymbol{\beta} + \boldsymbol{\epsilon}$ be the model. $\mathbf{y}$ represents continuous outcomes either observed or not observed. Our model assumes that $\boldsymbol{\epsilon} \sim N(\mathbf{0}, \sigma^2 \mathbf{I})$.

Let a be the lower limit and b be the upper limit. The log likelihood is

$$\ln L = -\frac{n}{2}\log(2\pi\sigma^2) - \frac{1}{2\sigma^2}\sum_{j=1}^{n}(y_j - \mathbf{x}_j\boldsymbol{\beta})^2 - \sum_{j=1}^{n}\log\left\{\Phi\left(\frac{b - \mathbf{x}_j\boldsymbol{\beta}}{\sigma}\right) - \Phi\left(\frac{a - \mathbf{x}_j\boldsymbol{\beta}}{\sigma}\right)\right\}$$

This command supports the Huber/White/sandwich estimator of the variance and its clustered version using vce(robust) and vce(cluster *clustvar*), respectively. See [P] _robust, particularly *Maximum likelihood estimators* and *Methods and formulas*.

truncreg also supports estimation with survey data. For details on VCEs with survey data, see [SVY] **variance estimation**.

References

Berndt, E. R. 1996. *The Practice of Econometrics: Classic and Contemporary*. New York: Addison–Wesley.

Cong, R. 1999. sg122: Truncated regression. *Stata Technical Bulletin* 52: 47–52. Reprinted in *Stata Technical Bulletin Reprints*, vol. 9, pp. 248–255. College Station, TX: Stata Press.

Davidson, R., and J. G. MacKinnon. 1993. *Estimation and Inference in Econometrics*. New York: Oxford University Press.

Greene, W. H. 2012. *Econometric Analysis*. 7th ed. Upper Saddle River, NJ: Prentice Hall.

Also see

[R] **truncreg postestimation** — Postestimation tools for truncreg

[R] **regress** — Linear regression

[R] **tobit** — Tobit regression

[MI] **estimation** — Estimation commands for use with mi estimate

[SVY] **svy estimation** — Estimation commands for survey data

[U] **20 Estimation and postestimation commands**

Title

truncreg postestimation — Postestimation tools for truncreg

Description

The following postestimation commands are available after `truncreg`:

Command	Description
contrast	contrasts and ANOVA-style joint tests of estimates
estat	AIC, BIC, VCE, and estimation sample summary
estat (svy)	postestimation statistics for survey data
estimates	cataloging estimation results
lincom	point estimates, standard errors, testing, and inference for linear combinations of coefficients
lrtest[1]	likelihood-ratio test
margins	marginal means, predictive margins, marginal effects, and average marginal effects
marginsplot	graph the results from margins (profile plots, interaction plots, etc.)
nlcom	point estimates, standard errors, testing, and inference for nonlinear combinations of coefficients
predict	predictions, residuals, influence statistics, and other diagnostic measures
predictnl	point estimates, standard errors, testing, and inference for generalized predictions
pwcompare	pairwise comparisons of estimates
suest	seemingly unrelated estimation
test	Wald tests of simple and composite linear hypotheses
testnl	Wald tests of nonlinear hypotheses

[1] `lrtest` is not appropriate with svy estimation results.

See the corresponding entries in the *Base Reference Manual* for details, but see [SVY] **estat** for details about `estat` (svy).

Syntax for predict

> predict [*type*] *newvar* [*if*] [*in*] [, *statistic* <u>nooff</u>set]

> predict [*type*] { *stub** | *newvar*$_{\text{reg}}$ *newvar*$_{\text{lnsigma}}$ } [*if*] [*in*] , <u>sc</u>ores

statistic	Description
Main	
xb	linear prediction; the default
stdp	standard error of the prediction
stdf	standard error of the forecast
<u>pr</u>(*a*,*b*)	$\Pr(a < y_j < b)$
e(*a*,*b*)	$E(y_j \mid a < y_j < b)$
<u>y</u>star(*a*,*b*)	$E(y_j^*), \; y_j^* = \max\{a, \min(y_j, b)\}$

These statistics are available both in and out of sample; type predict ... if e(sample) ... if wanted only for the estimation sample.

stdf is not allowed with svy estimation results.

where *a* and *b* may be numbers or variables; *a* missing ($a \geq .$) means $-\infty$, and *b* missing ($b \geq .$) means $+\infty$; see [U] **12.2.1 Missing values**.

Menu

Statistics > Postestimation > Predictions, residuals, etc.

Options for predict

⌐ Main ⌐

xb, the default, calculates the linear prediction.

stdp calculates the standard error of the prediction, which can be thought of as the standard error of the predicted expected value or mean for the observation's covariate pattern. The standard error of the prediction is also referred to as the standard error of the fitted value.

stdf calculates the standard error of the forecast, which is the standard error of the point prediction for 1 observation. It is commonly referred to as the standard error of the future or forecast value. By construction, the standard errors produced by stdf are always larger than those produced by stdp; see *Methods and formulas* in [R] **regress postestimation**.

pr(*a*,*b*) calculates $\Pr(a < \mathbf{x}_j\mathbf{b} + u_j < b)$, the probability that $y_j \mid \mathbf{x}_j$ would be observed in the interval (a, b).

a and *b* may be specified as numbers or variable names; *lb* and *ub* are variable names;
pr(20,30) calculates $\Pr(20 < \mathbf{x}_j\mathbf{b} + u_j < 30)$;
pr(*lb*,*ub*) calculates $\Pr(lb < \mathbf{x}_j\mathbf{b} + u_j < ub)$; and
pr(20,*ub*) calculates $\Pr(20 < \mathbf{x}_j\mathbf{b} + u_j < ub)$.

a missing ($a \geq .$) means $-\infty$; pr(.,30) calculates $\Pr(-\infty < \mathbf{x}_j\mathbf{b} + u_j < 30)$;
pr(*lb*,30) calculates $\Pr(-\infty < \mathbf{x}_j\mathbf{b} + u_j < 30)$ in observations for which *lb* $\geq$.
and calculates $\Pr(lb < \mathbf{x}_j\mathbf{b} + u_j < 30)$ elsewhere.

b missing ($b \geq .$) means $+\infty$; $pr(20,.)$ calculates $\Pr(+\infty > \mathbf{x}_j\mathbf{b} + u_j > 20)$; $pr(20,ub)$ calculates $\Pr(+\infty > \mathbf{x}_j\mathbf{b} + u_j > 20)$ in observations for which $ub \geq .$ and calculates $\Pr(20 < \mathbf{x}_j\mathbf{b} + u_j < ub)$ elsewhere.

$e(a,b)$ calculates $E(\mathbf{x}_j\mathbf{b} + u_j \mid a < \mathbf{x}_j\mathbf{b} + u_j < b)$, the expected value of $y_j|\mathbf{x}_j$ conditional on $y_j|\mathbf{x}_j$ being in the interval (a,b), meaning that $y_j|\mathbf{x}_j$ is truncated. a and b are specified as they are for $pr()$.

$ystar(a,b)$ calculates $E(y_j^*)$, where $y_j^* = a$ if $\mathbf{x}_j\mathbf{b} + u_j \leq a$, $y_j^* = b$ if $\mathbf{x}_j\mathbf{b} + u_j \geq b$, and $y_j^* = \mathbf{x}_j\mathbf{b} + u_j$ otherwise, meaning that y_j^* is censored. a and b are specified as they are for $pr()$.

nooffset is relevant only if you specified offset(*varname*). It modifies the calculations made by predict so that they ignore the offset variable; the linear prediction is treated as $\mathbf{x}_j\mathbf{b}$ rather than as $\mathbf{x}_j\mathbf{b} + \text{offset}_j$.

scores calculates equation-level score variables.

The first new variable will contain $\partial \ln L/\partial(\mathbf{x}_j\boldsymbol{\beta})$.

The second new variable will contain $\partial \ln L/\partial \sigma$.

Methods and formulas

All postestimation commands listed above are implemented as ado-files.

Also see

[R] **truncreg** — Truncated regression

[U] **20 Estimation and postestimation commands**

Title

> **ttest** — Mean-comparison tests

Syntax

One-sample mean-comparison test

   ttest *varname* == # $\big[$ *if* $\big]$ $\big[$ *in* $\big]$ $\big[$, $\underline{\text{l}}$evel(#) $\big]$

Two-sample mean-comparison test (unpaired)

   ttest *varname*$_1$ == *varname*$_2$ $\big[$ *if* $\big]$ $\big[$ *in* $\big]$, $\underline{\text{un}}$paired $\big[$ $\underline{\text{une}}$qual $\underline{\text{w}}$elch $\underline{\text{l}}$evel(#) $\big]$

Two-sample mean-comparison test (paired)

   ttest *varname*$_1$ == *varname*$_2$ $\big[$ *if* $\big]$ $\big[$ *in* $\big]$ $\big[$, $\underline{\text{l}}$evel(#) $\big]$

Two-group mean-comparison test

   ttest *varname* $\big[$ *if* $\big]$ $\big[$ *in* $\big]$, by(*groupvar*) $\big[$ *options*$_1$ $\big]$

Immediate form of one-sample mean-comparison test

   ttesti #$_{\text{obs}}$ #$_{\text{mean}}$ #$_{\text{sd}}$ #$_{\text{val}}$ $\big[$, $\underline{\text{l}}$evel(#) $\big]$

Immediate form of two-sample mean-comparison test

   ttesti #$_{\text{obs1}}$ #$_{\text{mean1}}$ #$_{\text{sd1}}$ #$_{\text{obs2}}$ #$_{\text{mean2}}$ #$_{\text{sd2}}$ $\big[$, *options*$_2$ $\big]$

options$_1$	Description
Main	
*by(*groupvar*)	variable defining the groups
$\underline{\text{une}}$qual	unpaired data have unequal variances
$\underline{\text{w}}$elch	use Welch's approximation
$\underline{\text{l}}$evel(#)	set confidence level; default is level(95)

*by(*groupvar*) is required.

options$_2$	Description
Main	
$\underline{\text{une}}$qual	unpaired data have unequal variances
$\underline{\text{w}}$elch	use Welch's approximation
$\underline{\text{l}}$evel(#)	set confidence level; default is level(95)

by is allowed with test; see [D] **by**.

Menu

one-sample

Statistics > Summaries, tables, and tests > Classical tests of hypotheses > One-sample mean-comparison test

two-sample, unpaired

Statistics > Summaries, tables, and tests > Classical tests of hypotheses > Two-sample mean-comparison test

two-sample, paired

Statistics > Summaries, tables, and tests > Classical tests of hypotheses > Mean-comparison test, paired data

two-group

Statistics > Summaries, tables, and tests > Classical tests of hypotheses > Two-group mean-comparison test

immediate command: one-sample

Statistics > Summaries, tables, and tests > Classical tests of hypotheses > One-sample mean-comparison calculator

immediate command: two-sample

Statistics > Summaries, tables, and tests > Classical tests of hypotheses > Two-sample mean-comparison calculator

Description

ttest performs t tests on the equality of means. In the first form, ttest tests that *varname* has a mean of #. In the second form, ttest tests that *varname*$_1$ and *varname*$_2$ have the same mean, assuming *unpaired* data. In the third form, ttest tests that *varname*$_1$ and *varname*$_2$ have the same mean, assuming *paired* data. In the fourth form, ttest tests that *varname* has the same mean within the two groups defined by *groupvar*.

ttesti is the immediate form of ttest; see [U] **19 Immediate commands**.

For the equivalent of a two-sample t test with sampling weights (pweights), use the svy: mean command with the over() option, and then use lincom; see [R] **mean** and [SVY] **svy postestimation**.

Options

⌐ Main ⌐

by(*groupvar*) specifies the *groupvar* that defines the two groups that ttest will use to test the hypothesis that their means are equal. Specifying by(*groupvar*) implies an unpaired (two sample) t test. Do not confuse the by() option with the by prefix; you can specify both.

unpaired specifies that the data be treated as unpaired. The unpaired option is used when the two sets of values to be compared are in different variables.

unequal specifies that the unpaired data not be assumed to have equal variances.

welch specifies that the approximate degrees of freedom for the test be obtained from Welch's formula (1947) rather than from Satterthwaite's approximation formula (1946), which is the default when unequal is specified. Specifying welch implies unequal.

level(#) specifies the confidence level, as a percentage, for confidence intervals. The default is level(95) or as set by set level; see [U] **20.7 Specifying the width of confidence intervals**.

Remarks

▷ Example 1: One-sample mean-comparison test

In the first form, ttest tests whether the mean of the sample is equal to a known constant under the assumption of unknown variance. Assume that we have a sample of 74 automobiles. We know each automobile's average mileage rating and wish to test whether the overall average for the sample is 20 miles per gallon.

```
. use http://www.stata-press.com/data/r12/auto
(1978 Automobile Data)
. ttest mpg==20
One-sample t test
```

Variable	Obs	Mean	Std. Err.	Std. Dev.	[95% Conf. Interval]
mpg	74	21.2973	.6725511	5.785503	19.9569 22.63769

```
        mean = mean(mpg)                                        t =    1.9289
Ho: mean = 20                                    degrees of freedom =        73

    Ha: mean < 20                 Ha: mean != 20                 Ha: mean > 20
 Pr(T < t) = 0.9712         Pr(|T| > |t|) = 0.0576          Pr(T > t) = 0.0288
```

The test indicates that the underlying mean is not 20 with a significance level of 5.8%.

◁

▷ Example 2: Two-sample mean-comparison test

We are testing the effectiveness of a new fuel additive. We run an experiment with 12 cars. We run the cars without and with the fuel treatment. The results of the experiment are as follows:

Without treatment	With treatment	Without treatment	With treatment
20	24	18	17
23	25	24	28
21	21	20	24
25	22	24	27
18	23	23	21
17	18	19	23

By creating two variables called mpg1 and mpg2 representing mileage without and with the treatment, respectively, we can test the equality of means by typing

```
. use http://www.stata-press.com/data/r12/fuel
. ttest mpg1==mpg2
Paired t test
```

Variable	Obs	Mean	Std. Err.	Std. Dev.	[95% Conf. Interval]
mpg1	12	21	.7881701	2.730301	19.26525 22.73475
mpg2	12	22.75	.9384465	3.250874	20.68449 24.81551
diff	12	-1.75	.7797144	2.70101	-3.46614 -.0338602

```
    mean(diff) = mean(mpg1 - mpg2)                              t =   -2.2444
Ho: mean(diff) = 0                               degrees of freedom =        11

Ha: mean(diff) < 0            Ha: mean(diff) != 0            Ha: mean(diff) > 0
 Pr(T < t) = 0.0232         Pr(|T| > |t|) = 0.0463          Pr(T > t) = 0.9768
```

We find that the means are statistically different from each other at any level greater than 4.6%.

◁

▷ Example 3: Group mean-comparison test

Let's pretend that the preceding data were collected by running 24 cars: 12 cars with the additive and 12 without. Although we might be tempted to enter the data in the same way, we should resist (see the technical note below). Instead, we enter the data as 24 observations on mpg with an additional variable, treated, taking on 1 if the car received the fuel treatment and 0 otherwise:

```
. use http://www.stata-press.com/data/r12/fuel3
. ttest mpg, by(treated)
Two-sample t test with equal variances
```

Group	Obs	Mean	Std. Err.	Std. Dev.	[95% Conf. Interval]	
0	12	21	.7881701	2.730301	19.26525	22.73475
1	12	22.75	.9384465	3.250874	20.68449	24.81551
combined	24	21.875	.6264476	3.068954	20.57909	23.17091
diff		-1.75	1.225518		-4.291568	.7915684

```
    diff = mean(0) - mean(1)                                    t =  -1.4280
Ho: diff = 0                                 degrees of freedom =        22

    Ha: diff < 0                 Ha: diff != 0                  Ha: diff > 0
 Pr(T < t) = 0.0837      Pr(|T| > |t|) = 0.1673           Pr(T > t) = 0.9163
```

This time we do not find a statistically significant difference.

If we were not willing to assume that the variances were equal and wanted to use Welch's formula, we could type

```
. ttest mpg, by(treated) welch
Two-sample t test with unequal variances
```

Group	Obs	Mean	Std. Err.	Std. Dev.	[95% Conf. Interval]	
0	12	21	.7881701	2.730301	19.26525	22.73475
1	12	22.75	.9384465	3.250874	20.68449	24.81551
combined	24	21.875	.6264476	3.068954	20.57909	23.17091
diff		-1.75	1.225518		-4.28369	.7836902

```
    diff = mean(0) - mean(1)                                    t =  -1.4280
Ho: diff = 0                     Welch's degrees of freedom =   23.2465

    Ha: diff < 0                 Ha: diff != 0                  Ha: diff > 0
 Pr(T < t) = 0.0833      Pr(|T| > |t|) = 0.1666           Pr(T > t) = 0.9167
```

◁

❑ Technical note

In two-group randomized designs, subjects will sometimes refuse the assigned treatment but still be measured for an outcome. In this case, take care to specify the group properly. You might be tempted to let *varname* contain missing where the subject refused and thus let ttest drop such observations from the analysis. Zelen (1979) argues that it would be better to specify that the subject belongs to the group in which he or she was randomized, even though such inclusion will dilute the measured effect.

❑

❏ Technical note

There is a second, inferior way to organize the data in the preceding example. Remember, we ran a test on 24 cars, 12 without the additive and 12 with. Nevertheless, we could have entered the data in the same way as we did when we had 12 cars, each run without and with the additive, by creating two variables—mpg1 and mpg2.

This method is inferior because it suggests a connection that is not there. For the 12-car experiment, there was most certainly a connection—it was the same car. In the 24-car experiment, however, it is arbitrary which mpg results appear next to which. Nevertheless, if our data are organized like this, ttest can accommodate us.

```
. use http://www.stata-press.com/data/r12/fuel
. ttest mpg1==mpg2, unpaired
Two-sample t test with equal variances
```

Variable	Obs	Mean	Std. Err.	Std. Dev.	[95% Conf. Interval]	
mpg1	12	21	.7881701	2.730301	19.26525	22.73475
mpg2	12	22.75	.9384465	3.250874	20.68449	24.81551
combined	24	21.875	.6264476	3.068954	20.57909	23.17091
diff		-1.75	1.225518		-4.291568	.7915684

```
        diff = mean(mpg1) - mean(mpg2)                          t =  -1.4280
Ho: diff = 0                                   degrees of freedom =       22

    Ha: diff < 0                 Ha: diff != 0                  Ha: diff > 0
 Pr(T < t) = 0.0837      Pr(|T| > |t|) = 0.1673             Pr(T > t) = 0.9163
```

❏

▷ Example 4

ttest can be used to test the equality of a pair of means; see [R] **oneway** for testing the equality of more than two means.

Suppose that we have data on the 50 states. The dataset contains the median age of the population (medage) and the region of the country (region) for each state. Region 1 refers to the Northeast, region 2 to the North Central, region 3 to the South, and region 4 to the West. Using oneway, we can test the equality of all four means.

```
. use http://www.stata-press.com/data/r12/census
(1980 Census data by state)
. oneway medage region
```

	Analysis of Variance				
Source	SS	df	MS	F	Prob > F
Between groups	46.3961903	3	15.4653968	7.56	0.0003
Within groups	94.1237947	46	2.04616945		
Total	140.519985	49	2.8677548		

```
Bartlett's test for equal variances:  chi2(3) =   10.5757  Prob>chi2 = 0.014
```

We find that the means are different, but we are interested only in testing whether the means for the Northeast (region==1) and West (region==4) are different. We could use oneway,

```
. oneway medage region if region==1 | region==4
                         Analysis of Variance
        Source              SS        df      MS          F      Prob > F

Between groups          46.241247      1    46.241247    20.02    0.0002
Within groups           46.1969169    20    2.30984584

        Total           92.4381638    21    4.40181733

Bartlett's test for equal variances:  chi2(1) =    2.4679  Prob>chi2 = 0.116
```

or we could use `ttest`:

```
. ttest medage if region==1 | region==4, by(region)
Two-sample t test with equal variances

   Group |     Obs        Mean     Std. Err.    Std. Dev.   [95% Conf. Interval]

      NE |       9    31.23333     .3411581     1.023474    30.44662    32.02005
    West |      13    28.28462     .4923577     1.775221    27.21186    29.35737

combined |      22    29.49091     .4473059     2.098051    28.56069    30.42113

    diff |            2.948718     .6590372                  1.57399    4.323445

        diff = mean(NE) - mean(West)                              t =     4.4743
    Ho: diff = 0                                  degrees of freedom =         20

     Ha: diff < 0                 Ha: diff != 0                 Ha: diff > 0
  Pr(T < t) = 0.9999      Pr(|T| > |t|) = 0.0002        Pr(T > t) = 0.0001
```

The significance levels of both tests are the same.

◁

Immediate form

▷ Example 5

`ttesti` is like `ttest`, except that we specify summary statistics rather than variables as arguments. For instance, we are reading an article that reports the mean number of sunspots per month as 62.6 with a standard deviation of 15.8. There are 24 months of data. We wish to test whether the mean is 75:

```
. ttesti 24 62.6 15.8 75
One-sample t test

         |     Obs        Mean     Std. Err.    Std. Dev.   [95% Conf. Interval]

       x |      24        62.6     3.225161         15.8    55.92825    69.27175

        mean = mean(x)                                            t =    -3.8448
    Ho: mean = 75                                 degrees of freedom =         23

     Ha: mean < 75               Ha: mean != 75                Ha: mean > 75
  Pr(T < t) = 0.0004      Pr(|T| > |t|) = 0.0008        Pr(T > t) = 0.9996
```

◁

▷ Example 6

There is no immediate form of `ttest` with paired data because the test is also a function of the covariance, a number unlikely to be reported in any published source. For nonpaired data, however, we might type

```
. ttesti 20 20 5  32 15 4
Two-sample t test with equal variances
```

	Obs	Mean	Std. Err.	Std. Dev.	[95% Conf.	Interval]
x	20	20	1.118034	5	17.65993	22.34007
y	32	15	.7071068	4	13.55785	16.44215
combined	52	16.92308	.6943785	5.007235	15.52905	18.3171
diff		5	1.256135		2.476979	7.523021

```
    diff = mean(x) - mean(y)                                    t =    3.9805
Ho: diff = 0                                   degrees of freedom =       50

    Ha: diff < 0                 Ha: diff != 0                  Ha: diff > 0
 Pr(T < t) = 0.9999       Pr(|T| > |t|) = 0.0002        Pr(T > t) = 0.0001
```

If we had typed `ttesti 20 20 5  32 15 4, unequal`, the test would have assumed unequal variances.

◁

Saved results

`ttest` and `ttesti` save the following in `r()`:

Scalars

r(N_1)	sample size n_1	r(sd_1)	standard deviation for first variable
r(N_2)	sample size n_2	r(sd_2)	standard deviation for second variable
r(p_l)	lower one-sided p-value	r(sd)	combined standard deviation
r(p_u)	upper one-sided p-value	r(mu_1)	$\bar{x}_1$ mean for population 1
r(p)	two-sided p-value	r(mu_2)	$\bar{x}_2$ mean for population 2
r(se)	estimate of standard error	r(df_t)	degrees of freedom
r(t)	t statistic		

Methods and formulas

`ttest` and `ttesti` are implemented as ado-files.

See, for instance, Hoel (1984, 140–161) or Dixon and Massey (1983, 121–130) for an introduction and explanation of the calculation of these tests. Acock (2010, 155–166) and Hamilton (2009, 157–162) describe t tests using applications in Stata.

The test for $\mu = \mu_0$ for unknown σ is given by

$$ t = \frac{(\bar{x} - \mu_0)\sqrt{n}}{s} $$

The statistic is distributed as Student's t with $n-1$ degrees of freedom (Gosset [Student, pseud.] 1908).

The test for $\mu_x = \mu_y$ when σ_x and σ_y are unknown but $\sigma_x = \sigma_y$ is given by

$$t = \frac{\overline{x} - \overline{y}}{\left\{\frac{(n_x-1)s_x^2+(n_y-1)s_y^2}{n_x+n_y-2}\right\}^{1/2}\left(\frac{1}{n_x}+\frac{1}{n_y}\right)^{1/2}}$$

The result is distributed as Student's t with $n_x + n_y - 2$ degrees of freedom.

You could perform `ttest` (without the `unequal` option) in a regression setting given that regression assumes a homoskedastic error model. To compare with the `ttest` command, denote the underlying observations on x and y by x_j, $j = 1, \ldots, n_x$, and y_j, $j = 1, \ldots, n_y$. In a regression framework, typing `ttest` without the `unequal` option is equivalent to

1. creating a new variable z_j that represents the stacked observations on x and y (so that $z_j = x_j$ for $j = 1, \ldots, n_x$ and $z_{n_x+j} = y_j$ for $j = 1, \ldots, n_y$)

2. and then estimating the equation $z_j = \beta_0 + \beta_1 d_j + \epsilon_j$, where $d_j = 0$ for $j = 1, \ldots, n_x$ and $d_j = 1$ for $j = n_x + 1, \ldots, n_x + n_y$ (that is, $d_j = 0$ when the z observations represent x, and $d_j = 1$ when the z observations represent y).

The estimated value of β_1, b_1, will equal $\overline{y} - \overline{x}$, and the reported t statistic will be the same t statistic as given by the formula above.

The test for $\mu_x = \mu_y$ when σ_x and σ_y are unknown and $\sigma_x \neq \sigma_y$ is given by

$$t = \frac{\overline{x} - \overline{y}}{\left(s_x^2/n_x + s_y^2/n_y\right)^{1/2}}$$

The result is distributed as Student's t with ν degrees of freedom, where ν is given by (with Satterthwaite's [1946] formula)

$$\frac{\left(s_x^2/n_x + s_y^2/n_y\right)^2}{\frac{\left(s_x^2/n_x\right)^2}{n_x-1} + \frac{\left(s_y^2/n_y\right)^2}{n_y-1}}$$

With Welch's formula (1947), the number of degrees of freedom is given by

$$-2 + \frac{\left(s_x^2/n_x + s_y^2/n_y\right)^2}{\frac{\left(s_x^2/n_x\right)^2}{n_x+1} + \frac{\left(s_y^2/n_y\right)^2}{n_y+1}}$$

The test for $\mu_x = \mu_y$ for matched observations (also known as paired observations, correlated pairs, or permanent components) is given by

$$t = \frac{\overline{d}\sqrt{n}}{s_d}$$

where $\overline{d}$ represents the mean of $x_i - y_i$ and s_d represents the standard deviation. The test statistic t is distributed as Student's t with $n - 1$ degrees of freedom.

You can also use `ttest` without the `unpaired` option in a regression setting because a paired comparison includes the assumption of constant variance. The `ttest` with an unequal variance assumption does not lend itself to an easy representation in regression settings and is not discussed here. $(x_j - y_j) = \beta_0 + \epsilon_j$.

William Sealy Gosset (1876–1937) was born in Canterbury, England. He studied chemistry and mathematics at Oxford and worked as a chemist with the brewers Guinness in Dublin. Gosset became interested in statistical problems, which he discussed with Karl Pearson and later with Fisher and Neyman. He published several important papers under the pseudonym "Student", and he lent that name to the t test he invented.

References

Acock, A. C. 2010. *A Gentle Introduction to Stata*. 3rd ed. College Station, TX: Stata Press.

Boland, P. J. 2000. William Sealy Gosset—alias 'Student' 1876–1937. In *Creators of Mathematics: The Irish Connection*, ed. K. Houston, 105–112. Dublin: University College Dublin Press.

Dixon, W. J., and F. J. Massey, Jr. 1983. *Introduction to Statistical Analysis*. 4th ed. New York: McGraw–Hill.

Gleason, J. R. 1999. sg101: Pairwise comparisons of means, including the Tukey wsd method. *Stata Technical Bulletin* 47: 31–37. Reprinted in *Stata Technical Bulletin Reprints*, vol. 8, pp. 225–233. College Station, TX: Stata Press.

Gosset, W. S. 1943. *"Student's" Collected Papers*. London: Biometrika Office, University College.

Gosset [Student, pseud.], W. S. 1908. The probable error of a mean. *Biometrika* 6: 1–25.

Hamilton, L. C. 2009. *Statistics with Stata (Updated for Version 10)*. Belmont, CA: Brooks/Cole.

Hoel, P. G. 1984. *Introduction to Mathematical Statistics*. 5th ed. New York: Wiley.

Pearson, E. S., R. L. Plackett, and G. A. Barnard. 1990. *'Student': A Statistical Biography of William Sealy Gosset*. Oxford: Oxford University Press.

Preece, D. A. 1982. t is for trouble (and textbooks): A critique of some examples of the paired-samples t-test. *Statistician* 31: 169–195.

Satterthwaite, F. E. 1946. An approximate distribution of estimates of variance components. *Biometrics Bulletin* 2: 110–114.

Senn, S. J., and W. Richardson. 1994. The first t-test. *Statistics in Medicine* 13: 785–803.

Welch, B. L. 1947. The generalization of 'student's' problem when several different population variances are involved. *Biometrika* 34: 28–35.

Zelen, M. 1979. A new design for randomized clinical trials. *New England Journal of Medicine* 300: 1242–1245.

Also see

[R] **bitest** — Binomial probability test

[R] **ci** — Confidence intervals for means, proportions, and counts

[R] **mean** — Estimate means

[R] **oneway** — One-way analysis of variance

[R] **prtest** — One- and two-sample tests of proportions

[R] **sdtest** — Variance-comparison tests

[MV] **hotelling** — Hotelling's T-squared generalized means test

Title

> **update** — Update Stata

Syntax

Report on update level of currently installed Stata

    update

Set update source

    update from *location*

Compare update level of currently installed Stata with that of source

    update query [, from(*location*)]

Perform update if necessary

    update all [, from(*location*) detail force exit]

Set automatic updates (Mac and Windows only)

    <u>se</u>t update_query { on | off }
    <u>se</u>t update_interval #
    <u>se</u>t update_prompt { on | off }

Menu

Help > Check for Updates

Description

The update command reports on the current update level and installs official updates to Stata. Official updates are updates to Stata as it was originally shipped from StataCorp, not the additions to Stata published in, for instance, the *Stata Journal* (SJ). Those additions are installed using the net command and updated using the adoupdate command; see [R] **net** and [R] **adoupdate**.

update without arguments reports on the update level of the currently installed Stata.

update from sets an update source, where *location* is a directory name or URL. If you are on the Internet, type update from http://www.stata.com.

update query compares the update level of the currently installed Stata with that available from the update source and displays a report.

update all updates all necessary files. This is what you should type to check for and install updates.

set update_query determines if update query is to be automatically performed when Stata is launched. Only Mac and Windows platforms can be set for automatic updating.

set update_interval # sets the number of days to elapse before performing the next automatic update query. The default # is 7. The interval starts from the last time an update query was performed (automatically or manually). Only Mac and Windows platforms can be set for automatic updating.

set update_prompt determines whether a dialog is to be displayed before performing an automatic update query. The dialog allows you to perform an update query now, perform one the next time Stata is launched, perform one after the next interval has passed, or disable automatic update query. Only Mac and Windows platforms can be set for automatic updating.

Options

from(*location*) specifies the location of the update source. You can specify the from() option on the individual update commands or use the update from command. Which you do makes no difference. You typically do not need to use this option.

detail specifies to display verbose output during the update process.

force specifies to force downloading of all files even if, based on the date comparison, Stata does not think it is necessary. There is seldom a reason to specify this option.

exit instructs Stata to exit when the update has successfully completed. There is seldom a reason to specify this option.

Remarks

update updates the official components of Stata from the official source: http://www.stata.com. If you are connected to the Internet, the easy thing to do is to type

    . update all

and follow the instructions. If Stata is up to date, update all will do nothing. Otherwise, it will download whatever is necessary and install the files. If you just want to know what updates are available, type

    . update query

update query will check if any updates are available and report that information. If updates are available, it will recommend that you type update all.

If you want to report the current update level, type

    . update

update will report the update level of the Stata installation. update will also show you the date that updates were last checked and if any updates were available at that time.

Saved results

update without a subcommand, update from, and update query save the following in r():

Scalars

r(inst_exe)	date of executable installed (*)
r(avbl_exe)	date of executable available over web (*) (**)
r(inst_ado)	date of ado-files installed (*)
r(avbl_ado)	date of ado-files available over web (*) (**)
r(inst_utilities)	date of utilities installed (*)
r(avbl_utilities)	date of utilities available over web (*) (**)
r(inst_docs)	date of documentation installed (*)
r(avbl_docs)	date of documentation available over web (*) (**)

Macros

r(name_exe)	name of the Stata executable
r(dir_exe)	directory in which executable is stored
r(dir_ado)	directory in which ado-files are stored
r(dir_utilities)	directory in which utilities are stored
r(dir_docs)	directory in which PDF documentation is stored

Notes:

* Dates are stored as integers counting the number of days since January 1, 1960; see [D] **datetime**.

** These dates are not saved by update without a subcommand because update by itself reports information solely about the local computer and does not check what is available on the web.

Also see

[R] **adoupdate** — Update user-written ado-files

[R] **net** — Install and manage user-written additions from the Internet

[R] **ssc** — Install and uninstall packages from SSC

[P] **sysdir** — Query and set system directories

[U] **28 Using the Internet to keep up to date**

[GSM] **19 Updating and extending Stata—Internet functionality**

[GSU] **19 Updating and extending Stata—Internet functionality**

[GSW] **19 Updating and extending Stata—Internet functionality**

Title

> *vce_option* — Variance estimators

Syntax

estimation_cmd ... [, vce(*vcetype*) ...]

vcetype	Description
Likelihood based	
oim	observed information matrix (OIM)
opg	outer product of the gradient (OPG) vectors
Sandwich estimators	
<u>r</u>obust	Huber/White/sandwich estimator
<u>cl</u>uster *clustvar*	clustered sandwich estimator
Replication based	
<u>boot</u>strap [, *bootstrap_options*]	bootstrap estimation
<u>jack</u>knife [, *jackknife_options*]	jackknife estimation

Description

This entry describes the vce() option, which is common to most estimation commands. vce() specifies how to estimate the variance–covariance matrix (VCE) corresponding to the parameter estimates. The standard errors reported in the table of parameter estimates are the square root of the variances (diagonal elements) of the VCE.

Options

⌐ SE/Robust ¬

vce(oim) is usually the default for models fit using maximum likelihood. vce(oim) uses the observed information matrix (OIM); see [R] **ml**.

vce(opg) uses the sum of the outer product of the gradient (OPG) vectors; see [R] **ml**. This is the default VCE when the technique(bhhh) option is specified; see [R] **maximize**.

vce(robust) uses the robust or sandwich estimator of variance. This estimator is robust to some types of misspecification so long as the observations are independent; see [U] **20.20 Obtaining robust variance estimates**.

If the command allows pweights and you specify them, vce(robust) is implied; see [U] **20.22.3 Sampling weights**.

vce(cluster *clustvar*) specifies that the standard errors allow for intragroup correlation, relaxing the usual requirement that the observations be independent. That is, the observations are independent across groups (clusters) but not necessarily within groups. *clustvar* specifies to which group each observation belongs, for example, vce(cluster personid) in data with repeated observations on individuals. vce(cluster *clustvar*) affects the standard errors and variance–covariance matrix of the estimators but not the estimated coefficients; see [U] **20.20 Obtaining robust variance estimates**.

vce(bootstrap [, *bootstrap_options*]) uses a bootstrap; see [R] **bootstrap**. After estimation with vce(bootstrap), see [R] **bootstrap postestimation** to obtain percentile-based or bias-corrected confidence intervals.

vce(jackknife [, *jackknife_options*]) uses the delete-one jackknife; see [R] **jackknife**.

Remarks

Remarks are presented under the following headings:

> Prefix commands
> Passing options in vce()

Prefix commands

Specifying vce(bootstrap) or vce(jackknife) is often equivalent to using the corresponding prefix command. Here is an example using jackknife with regress.

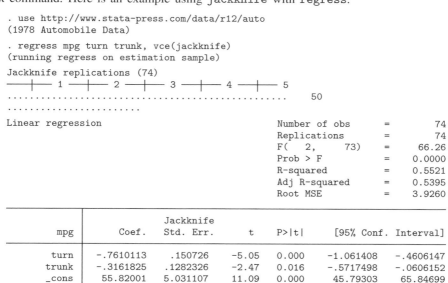

```
. use http://www.stata-press.com/data/r12/auto
(1978 Automobile Data)

. regress mpg turn trunk, vce(jackknife)
(running regress on estimation sample)

Jackknife replications (74)
———+——— 1 ———+——— 2 ———+——— 3 ———+——— 4 ———+——— 5
.................................................. 50
......................
Linear regression                    Number of obs    =         74
                                     Replications     =         74
                                     F(  2,     73)   =      66.26
                                     Prob > F         =     0.0000
                                     R-squared        =     0.5521
                                     Adj R-squared    =     0.5395
                                     Root MSE         =     3.9260
```

		Jackknife				
mpg	Coef.	Std. Err.	t	P>|t|	[95% Conf.	Interval]
turn	-.7610113	.150726	-5.05	0.000	-1.061408	-.4606147
trunk	-.3161825	.1282326	-2.47	0.016	-.5717498	-.0606152
_cons	55.82001	5.031107	11.09	0.000	45.79303	65.84699

```
. jackknife: regress mpg turn trunk
(running regress on estimation sample)
```

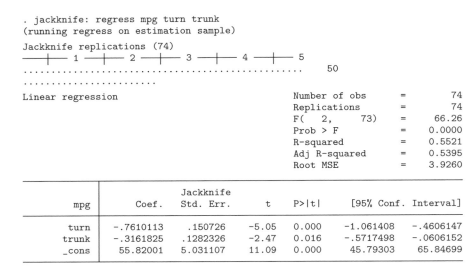

```
Jackknife replications (74)
        1       2       3       4       5
................................................  50
......................
```

```
Linear regression                      Number of obs    =        74
                                       Replications     =        74
                                       F(   2,     73)  =     66.26
                                       Prob > F         =    0.0000
                                       R-squared        =    0.5521
                                       Adj R-squared    =    0.5395
                                       Root MSE         =    3.9260
```

| | | Jackknife | | | | |
mpg	Coef.	Std. Err.	t	P>|t|	[95% Conf.	Interval]
turn	-.7610113	.150726	-5.05	0.000	-1.061408	-.4606147
trunk	-.3161825	.1282326	-2.47	0.016	-.5717498	-.0606152
_cons	55.82001	5.031107	11.09	0.000	45.79303	65.84699

Here it does not matter whether we specify the vce(jackknife) option or instead use the jackknife prefix.

However, vce(jackknife) should be used in place of the jackknife prefix whenever available because they are not always equivalent. For example, to use the jackknife prefix with clogit properly, you must tell jackknife to omit whole groups rather than individual observations. Specifying vce(jackknife) does this automatically.

```
. use http://www.stata-press.com/data/r12/clogitid
. jackknife, cluster(id): clogit y x1 x2, group(id)
  (output omitted )
```

This extra information is automatically communicated to jackknife by clogit when the vce() option is specified.

```
. clogit y x1 x2, group(id) vce(jackknife)
(running clogit on estimation sample)
```

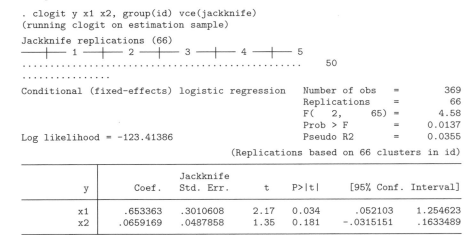

```
Jackknife replications (66)
        1       2       3       4       5
.................................................... 50
..............
```

```
Conditional (fixed-effects) logistic regression   Number of obs   =        369
                                                  Replications    =         66
                                                  F(   2,     65) =       4.58
                                                  Prob > F        =     0.0137
Log likelihood = -123.41386                       Pseudo R2       =     0.0355

                           (Replications based on 66 clusters in id)
```

| | | Jackknife | | | | |
y	Coef.	Std. Err.	t	P>|t|	[95% Conf.	Interval]
x1	.653363	.3010608	2.17	0.034	.052103	1.254623
x2	.0659169	.0487858	1.35	0.181	-.0315151	.1633489

Passing options in vce()

If you wish to specify more options to the bootstrap or jackknife estimation, you can include them within the vce() option. Below we request 300 bootstrap replications and save the replications in bsreg.dta:

```
. use http://www.stata-press.com/data/r12/auto
(1978 Automobile Data)
. regress mpg turn trunk, vce(bootstrap, nodots seed(123) rep(300) saving(bsreg))
```

Linear regression

```
                                    Number of obs   =         74
                                    Replications    =        300
                                    Wald chi2(2)    =     127.28
                                    Prob > chi2     =     0.0000
                                    R-squared       =     0.5521
                                    Adj R-squared   =     0.5395
                                    Root MSE        =     3.9260
```

	Observed Coef.	Bootstrap Std. Err.	z	P>|z|	Normal-based [95% Conf. Interval]	
turn	-.7610113	.1361786	-5.59	0.000	-1.027916	-.4941062
trunk	-.3161825	.1145728	-2.76	0.006	-.540741	-.0916239
_cons	55.82001	4.69971	11.88	0.000	46.60875	65.03127

```
. bstat using bsreg
```

Bootstrap results

```
                                    Number of obs   =         74
                                    Replications    =        300
```

```
          command:  regress mpg turn trunk
```

	Observed Coef.	Bootstrap Std. Err.	z	P>|z|	Normal-based [95% Conf. Interval]	
turn	-.7610113	.1361786	-5.59	0.000	-1.027916	-.4941062
trunk	-.3161825	.1145728	-2.76	0.006	-.540741	-.0916239
_cons	55.82001	4.69971	11.88	0.000	46.60875	65.03127

Methods and formulas

By default, Stata's maximum likelihood estimators display standard errors based on variance estimates given by the inverse of the negative Hessian (second derivative) matrix. If vce(robust), vce(cluster *clustvar*), or pweights is specified, standard errors are based on the robust variance estimator (see [U] **20.20 Obtaining robust variance estimates**); likelihood-ratio tests are not appropriate here (see [SVY] **survey**), and the model χ^2 is from a Wald test. If vce(opg) is specified, the standard errors are based on the outer product of the gradients; this option has no effect on likelihood-ratio tests, though it does affect Wald tests.

If vce(bootstrap) or vce(jackknife) is specified, the standard errors are based on the chosen replication method; here the model χ^2 or F statistic is from a Wald test using the respective replication-based covariance matrix. The t distribution is used in the coefficient table when the vce(jackknife) option is specified. vce(bootstrap) and vce(jackknife) are also available with some commands that are not maximum likelihood estimators.

Also see

[R] **bootstrap** — Bootstrap sampling and estimation

[R] **jackknife** — Jackknife estimation

[XT] *vce_options* — Variance estimators

[U] **20 Estimation and postestimation commands**

Title

> **view** — View files and logs

Syntax

Display file in Viewer

    view [file] ["]*filename*["] [, asis adopath]

Bring up browser pointed to specified URL

    view browse ["]*url*["]

Display help results in Viewer

    view help [*topic_or_command_name*]

Display search results in Viewer

    view search *keywords*

Display news results in Viewer

    view news

Display net results in Viewer

    view net [*netcmd*]

Display ado-results in Viewer

    view ado [*adocmd*]

Display update results in Viewer

    view update [*updatecmd*]

Programmer's analog to view file and view browse

    view view_d

Programmer's analog to view help

    view help_d

Programmer's analog to view search

    view search_d

Programmer's analog to view net

```
view net_d
```

Programmer's analog to view ado

```
view ado_d
```

Programmer's analog to view update

```
view update_d
```

Menu

File > View...

Description

view displays file contents in the Viewer.

view file displays the specified file. file is optional, so if you had a SMCL session log created by typing log using mylog, you could view it by typing view mylog.smcl. view file can properly display .smcl files (logs and the like), .sthlp files, and text files. view file's asis option specifies that the file be displayed as plain text, regardless of the *filename*'s extension.

view browse opens your browser pointed to *url*. Typing view browse http://www.stata.com would bring up your browser pointed to the http://www.stata.com website.

view help does the same as the help command—see [R] **help**—but displays the result in the Viewer. For example, to review the help for Stata's print command, you could type view help print.

view search does the same as the search command—see [R] **search**—but displays the result in the Viewer. For instance, to search the online help for information on robust regression, you could type view search robust regression.

view news does the same as the news command—see [R] **news**—but displays the results in the Viewer. (news displays the latest news from http://www.stata.com.)

view net does the same as the net command—see [R] **net**—but displays the result in the Viewer. For instance, typing view net search hausman test would search the Internet for additions to Stata related to the Hausman test. Typing view net from http://www.stata.com would go to the Stata download site at http://www.stata.com.

view ado does the same as the ado command—see [R] **net**—but displays the result in the Viewer. For instance, typing view ado dir would show a list of files you have installed.

view update does the same as the update command—see [R] **update**—but displays the result in the Viewer. Typing view update would show the dates of what you have installed, and from there you could click to compare those dates with the latest updates available. Typing view update query would skip the first step and show the comparison.

The view *_d commands are more useful in programming contexts than they are interactively.

view view_d displays a dialog box from which you may type the name of a file or a URL to be displayed in the Viewer.

`view help_d` displays a help dialog box from which you may obtain interactive help on any Stata command.

`view search_d` displays a search dialog box from which you may obtain interactive help based on keywords.

`view net_d` displays a search dialog box from which you may search the Internet for additions to Stata (which you could then install).

`view ado_d` displays a dialog box from which you may search the user-written routines you have previously installed.

`view update_d` displays an update dialog box in which you may type the source from which updates are to be obtained.

Options

asis, allowed with `view file`, specifies that the file be displayed as text, regardless of the *filename*'s extension. `view file`'s default action is to display files ending in `.smcl` and `.sthlp` as SMCL; see [P] **smcl**.

adopath, allowed with `view file`, specifies that Stata search the S_ADO path for *filename* and display it, if found.

Remarks

Most users access the Viewer by selecting **File > View...** and proceeding from there. The `view` command allows you to skip that step. Some common interactive uses of `view` are

```
. view mysession.smcl
. view mysession.log

. view help print
. view help regress

. view news
. view browse http://www.stata.com

. view net search hausman test

. view net
. view ado

. view update query
```

Also, programmers find `view` useful for creating special effects.

Also see

[R] **help** — Display online help

[R] **net** — Install and manage user-written additions from the Internet

[R] **news** — Report Stata news

[R] **search** — Search Stata documentation

[R] **update** — Update Stata

[D] **type** — Display contents of a file

[GSM] **3 Using the Viewer**

[GSU] **3 Using the Viewer**

[GSW] **3 Using the Viewer**

Title

vwls — Variance-weighted least squares

Syntax

vwls *depvar indepvars* [*if*] [*in*] [*weight*] [, *options*]

options	Description
Model	
<u>nocon</u>stant	suppress constant term
sd(*varname*)	variable containing estimate of conditional standard deviation
Reporting	
<u>level</u>(#)	set confidence level; default is level(95)
display_options	control column formats, row spacing, line width, and display of omitted variables and base and empty cells
<u>coefl</u>egend	display legend instead of statistics

indepvars may contain factor variables; see [U] **11.4.3 Factor variables**.

bootstrap, by, jackknife, rolling, and statsby are allowed; see [U] **11.1.10 Prefix commands**.

Weights are not allowed with the bootstrap prefix; see [R] **bootstrap**.

fweights are allowed; see [U] **11.1.6 weight**.

coeflegend does not appear in the dialog box.

See [U] **20 Estimation and postestimation commands** for more capabilities of estimation commands.

Menu

Statistics > Linear models and related > Other > Variance-weighted least squares

Description

vwls estimates a linear regression using variance-weighted least squares. It differs from ordinary least-squares (OLS) regression in that it does not assume homogeneity of variance, but requires that the conditional variance of *depvar* be estimated prior to the regression. The estimated variance need not be constant across observations. vwls treats the estimated variance as if it were the true variance when it computes the standard errors of the coefficients.

You must supply an estimate of the conditional standard deviation of *depvar* to vwls by using the sd(*varname*) option, or you must have grouped data with the groups defined by the *indepvars* variables. In the latter case, vwls treats all *indepvars* as categorical variables, computes the mean and standard deviation of *depvar* separately for each subgroup, and computes the regression of the subgroup means on *indepvars*.

regress with analytic weights can be used to produce another kind of "variance-weighted least squares"; see *Remarks* for an explanation of the difference.

Options

noconstant; see [R] **estimation options**.

sd(*varname*) is an estimate of the conditional standard deviation of *depvar* (that is, it can vary observation by observation). All values of *varname* must be > 0. If you specify sd(), you cannot use fweights.

If sd() is not given, the data will be grouped by *indepvars*. Here *indepvars* are treated as categorical variables, and the means and standard deviations of *depvar* for each subgroup are calculated and used for the regression. Any subgroup for which the standard deviation is zero is dropped.

level(*#*); see [R] **estimation options**.

display_options: noomitted, vsquish, noemptycells, baselevels, allbaselevels, cformat(% *fmt*), pformat(% *fmt*), sformat(% *fmt*), and nolstretch; see [R] **estimation options**.

The following option is available with vwls but is not shown in the dialog box:

coeflegend; see [R] **estimation options**.

Remarks

The vwls command is intended for use with two special—and different—types of data. The first contains data that consist of measurements from physical science experiments in which all error is due solely to measurement errors and the sizes of the measurement errors are known.

You can also use variance-weighted least-squares linear regression for certain problems in categorical data analysis, such as when all the independent variables are categorical and the outcome variable is either continuous or a quantity that can sensibly be averaged. If each of the subgroups defined by the categorical variables contains a reasonable number of subjects, then the variance of the outcome variable can be estimated independently within each subgroup. For the purposes of estimation, vwls treats each subgroup as one observation, with the dependent variable being the subgroup mean of the outcome variable.

The vwls command fits the model

$$y_i = \mathbf{x}_i \boldsymbol{\beta} + \varepsilon_i$$

where the errors ε_i are independent normal random variables with the distribution $\varepsilon_i \sim N(0, \nu_i)$. The independent variables $\mathbf{x}_i$ are assumed to be known without error.

As described above, vwls assumes that you already have estimates s_i^2 for the variances ν_i. The error variance is not estimated in the regression. The estimates s_i^2 are used to compute the standard errors of the coefficients; see *Methods and formulas* below.

In contrast, weighted OLS regression assumes that the errors have the distribution $\varepsilon_i \sim N(0, \sigma^2/w_i)$, where the w_i are known weights and σ^2 is an unknown parameter that is estimated in the regression. This is the difference from variance-weighted least squares: in weighted OLS, the magnitude of the error variance is estimated in the regression using all the data.

▷ Example 1

An artificial, but informative, example illustrates the difference between variance-weighted least squares and weighted OLS.

We measure the quantities x_i and y_i and estimate that the standard deviation of y_i is s_i. We enter the data into Stata:

```
. use http://www.stata-press.com/data/r12/vwlsxmpl
. list
```

	x	y	s
1.	1	1.2	.5
2.	2	1.9	.5
3.	3	3.2	1
4.	4	4.3	1
5.	5	4.9	1
6.	6	6.0	2
7.	7	7.2	2
8.	8	7.9	2

Because we want observations with smaller variance to carry larger weight in the regression, we compute an OLS regression with analytic weights proportional to the inverse of the squared standard deviations:

```
. regress y x [aweight=s^(-2)]
(sum of wgt is    1.1750e+01)
```

Source	SS	df	MS
Model	22.6310183	1	22.6310183
Residual	.193355117	6	.032225853
Total	22.8243734	7	3.26062477

```
Number of obs =        8
F(  1,     6) =   702.26
Prob > F      =   0.0000
R-squared     =   0.9915
Adj R-squared =   0.9901
Root MSE      =   .17952
```

y	Coef.	Std. Err.	t	P>|t|	[95% Conf. Interval]	
x	.9824683	.0370739	26.50	0.000	.8917517	1.073185
_cons	.1138554	.1120078	1.02	0.349	-.1602179	.3879288

If we compute a variance-weighted least-squares regression by using vwls, we get the same results for the coefficient estimates but very different standard errors:

```
. vwls y x, sd(s)
Variance-weighted least-squares regression        Number of obs   =        8
Goodness-of-fit chi2(6)     =    0.28              Model chi2(1)   =    33.24
Prob > chi2                 =  0.9996              Prob > chi2     =   0.0000
```

y	Coef.	Std. Err.	z	P>|z|	[95% Conf. Interval]	
x	.9824683	.170409	5.77	0.000	.6484728	1.316464
_cons	.1138554	.51484	0.22	0.825	-.8952124	1.122923

Although the values of y_i were nicely linear with x_i, the `vwls` regression used the large estimates for the standard deviations to compute large standard errors for the coefficients. For weighted OLS regression, however, the scale of the analytic weights has no effect on the standard errors of the coefficients—only the relative proportions of the analytic weights affect the regression.

If we are sure of the sizes of our error estimates for y_i, using `vwls` is valid. However, if we can estimate only the relative proportions of error among the y_i, then `vwls` is not appropriate.

◁

▷ Example 2

Let's now consider an example of the use of `vwls` with categorical data. Suppose that we have blood pressure data for $n = 400$ subjects, categorized by gender and race (black or white). Here is a description of the data:

```
. use http://www.stata-press.com/data/r12/bp
. table gender race, c(mean bp sd bp freq) row col format(%8.1f)
```

		Race	
Gender	White	Black	Total
Female	117.1	118.5	117.8
	10.3	11.6	10.9
	100	100	200
Male	122.1	125.8	124.0
	10.6	15.5	13.3
	100	100	200
Total	119.6	122.2	120.9
	10.7	14.1	12.6
	200	200	400

Performing a variance-weighted regression using `vwls` gives

```
. vwls bp gender race
Variance-weighted least-squares regression        Number of obs   =     400
Goodness-of-fit chi2(1)      =    0.88             Model chi2(2)   =   27.11
Prob > chi2                  =  0.3486             Prob > chi2     =  0.0000
```

bp	Coef.	Std. Err.	z	P>|z|	[95% Conf.	Interval]
gender	5.876522	1.170241	5.02	0.000	3.582892	8.170151
race	2.372818	1.191683	1.99	0.046	.0371631	4.708473
_cons	116.6486	.9296297	125.48	0.000	114.8266	118.4707

By comparison, an OLS regression gives the following result:

```
. regress bp gender race
```

Source	SS	df	MS
Model	4485.66639	2	2242.83319
Residual	58442.7305	397	147.210908
Total	62928.3969	399	157.71528

```
                              Number of obs =     400
                              F(  2,   397) =   15.24
                              Prob > F      =  0.0000
                              R-squared     =  0.0713
                              Adj R-squared =  0.0666
                              Root MSE      =  12.133
```

bp	Coef.	Std. Err.	t	P>|t|	[95% Conf. Interval]
gender	6.1775	1.213305	5.09	0.000	3.792194 8.562806
race	2.5875	1.213305	2.13	0.034	.2021938 4.972806
_cons	116.4862	1.050753	110.86	0.000	114.4205 118.552

Note the larger value for the `race` coefficient (and smaller p-value) in the OLS regression. The assumption of homogeneity of variance in OLS means that the mean for black men pulls the regression line higher than in the `vwls` regression, which takes into account the larger variance for black men and reduces its effect on the regression.

◁

Saved results

`vwls` saves the following in `e()`:

Scalars
e(N)	number of observations
e(df_m)	model degrees of freedom
e(chi2)	model χ^2
e(df_gf)	goodness-of-fit degrees of freedom
e(chi2_gf)	goodness-of-fit χ^2
e(rank)	rank of e(V)

Macros
e(cmd)	vwls
e(cmdline)	command as typed
e(depvar)	name of dependent variable
e(properties)	b V
e(predict)	program used to implement predict
e(asbalanced)	factor variables fvset as asbalanced
e(asobserved)	factor variables fvset as asobserved

Matrices
e(b)	coefficient vector
e(V)	variance–covariance matrix of the estimators

Functions
e(sample)	marks estimation sample

Methods and formulas

`vwls` is implemented as an ado-file.

Let $\mathbf{y} = (y_1, y_2, \ldots, y_n)'$ be the vector of observations of the dependent variable, where n is the number of observations. When `sd()` is specified, let s_1, s_2, $\ldots$, s_n be the standard deviations supplied by `sd()`. For categorical data, when `sd()` is not given, the means and standard deviations of y for each subgroup are computed, and n becomes the number of subgroups, $\mathbf{y}$ is the vector of subgroup means, and s_i are the standard deviations for the subgroups.

Let $\mathbf{V} = \mathrm{diag}(s_1^2, s_2^2, \ldots, s_n^2)$ denote the estimate of the variance of $\mathbf{y}$. Then the estimated regression coefficients are

$$\mathbf{b} = (\mathbf{X}'\mathbf{V}^{-1}\mathbf{X})^{-1}\mathbf{X}'\mathbf{V}^{-1}\mathbf{y}$$

and their estimated covariance matrix is

$$\widehat{\mathrm{Cov}}(\mathbf{b}) = (\mathbf{X}'\mathbf{V}^{-1}\mathbf{X})^{-1}$$

A statistic for the goodness of fit of the model is

$$Q = (\mathbf{y} - \mathbf{X}\mathbf{b})'\,\mathbf{V}^{-1}(\mathbf{y} - \mathbf{X}\mathbf{b})$$

where Q has a χ^2 distribution with $n - k$ degrees of freedom (k is the number of independent variables plus the constant, if any).

References

Gini, R., and J. Pasquini. 2006. Automatic generation of documents. *Stata Journal* 6: 22–39.

Grizzle, J. E., C. F. Starmer, and G. G. Koch. 1969. Analysis of categorical data by linear models. *Biometrics* 25: 489–504.

Press, W. H., S. A. Teukolsky, W. T. Vetterling, and B. P. Flannery. 2007. *Numerical Recipes in C: The Art of Scientific Computing*. 3rd ed. Cambridge: Cambridge University Press.

Also see

[R] **vwls postestimation** — Postestimation tools for vwls

[R] **regress** — Linear regression

[U] **11.1.6 weight**

[U] **20 Estimation and postestimation commands**

Title

vwls postestimation — Postestimation tools for vwls

Description

The following postestimation commands are available after vwls:

Command	Description
contrast	contrasts and ANOVA-style joint tests of estimates
estat	VCE and estimation sample summary
estimates	cataloging estimation results
lincom	point estimates, standard errors, testing, and inference for linear combinations of coefficients
linktest	link test for model specification
margins	marginal means, predictive margins, marginal effects, and average marginal effects
marginsplot	graph the results from margins (profile plots, interaction plots, etc.)
nlcom	point estimates, standard errors, testing, and inference for nonlinear combinations of coefficients
predict	predictions, residuals, influence statistics, and other diagnostic measures
predictnl	point estimates, standard errors, testing, and inference for generalized predictions
pwcompare	pairwise comparisons of estimates
test	Wald tests of simple and composite linear hypotheses
testnl	Wald tests of nonlinear hypotheses

See the corresponding entries in the *Base Reference Manual* for details.

Syntax for predict

predict [*type*] *newvar* [*if*] [*in*] [, xb stdp]

These statistics are available both in and out of sample; type predict ... if e(sample) ... if wanted only for the estimation sample.

Menu

Statistics > Postestimation > Predictions, residuals, etc.

Options for predict

◖ Main ◗

xb, the default, calculates the linear prediction.

stdp calculates the standard error of the linear prediction.

Methods and formulas

All postestimation commands listed above are implemented as ado-files.

Also see

[R] **vwls** — Variance-weighted least squares

[U] **20 Estimation and postestimation commands**

Title

which — Display location and version for an ado-file

Syntax

which *fname*[.*ftype*] [, all]

Description

which looks for *fname*.*ftype* along the S_ADO path. If Stata finds the file, which displays the full path and filename, along with, if the file is text, all lines in the file that begin with "*!" in the first column. If Stata cannot find the file, which issues the message "file not found along ado-path" and sets the return code to 111. *ftype* must be a file type for which Stata usually looks along the ado-path to find. Allowable *ftype*s are

.ado, .class, .dlg, .idlg, .sthlp, .ihlp, .hlp, .key, .maint, .mata, .mlib, .mo, .mnu, .plugin, .scheme, .stbcal, and .style

If *ftype* is omitted, which assumes .ado. When searching for .ado files, if Stata cannot find the file, Stata then checks to see if *fname* is a built-in Stata command, allowing for valid abbreviations. If it is, the message "built-in command" is displayed; if not, the message "command not found as either built-in or ado-file" is displayed and the return code is set to 111.

For information about internal version control, see [P] **version**.

Option

all forces which to report the location of all files matching the *fname*.*ftype* found along the search path. The default is to report just the first one found.

Remarks

If you write programs, you know that you make changes to the programs over time. If you are like us, you also end up with multiple versions of the program stored on your disk, perhaps in different directories. You may even have given copies of your programs to other Stata users, and you may not remember which version of a program you or your friends are using. The which command helps you solve this problem.

▷ Example 1

The which command displays the path for *filename*.ado and any lines in the code that begin with "*!". For example, we might want information about the test command, described in [R] **test**, which is an ado-file written by StataCorp. Here is what happens when we type which test:

```
. which test
C:\Program Files\Stata12\ado\base\t\test.ado
*! version 2.2.1  18feb2011
```

2311

which displays the path for the `test.ado` file and also a line beginning with "`*!`" that indicates the version of the file. This is how we, at StataCorp, do version control—see [U] **18.11.1 Version** for an explanation of our version control numbers.

We do not need to be so formal. `which` will display anything typed after lines that begin with '`*!`'. For instance, we might write `myprog.ado`:

```
. which myprog
.\myprog.ado
*! first written 1/03/2011
*! bug fix on 1/05/2011 (no variance case)
*! updated 1/24/2011 to include noconstant option
*! still suspicious if variable takes on only two values
```

It does not matter where in the program the lines beginning with `*!` are—`which` will list them (in particular, our "still suspicious" comment was buried about 50 lines down in the code). All that is important is that the `*!` marker appear in the first two columns of a line.

◁

▷ Example 2

If we type `which` *command*, where *command* is a built-in command rather than an ado-file, Stata responds with

```
. which summarize
built-in command:  summarize
```

If *command* was neither a built-in command nor an ado-file, Stata would respond with

```
. which junk
command junk not found as either built-in or ado-file
r(111);
```

◁

Also see

[P] **findfile** — Find file in path

[P] **version** — Version control

[U] **17 Ado-files**

[U] **18.11.1 Version**

Title

xi — Interaction expansion

Syntax

xi $\left[\right.$, $\underline{pre}fix(string)$ noomit $\left.\right]$ $term(s)$

xi $\left[\right.$, $\underline{pre}fix(string)$ noomit $\left.\right]$: $any\_stata\_command$ $varlist\_with\_terms$ ...

where a *term* has the form

	or	
i. *varname*		I. *varname*
i. *varname*$_1$ *i. *varname*$_2$		I. *varname*$_1$ *I. *varname*$_2$
i. *varname*$_1$ *varname*$_3$		I. *varname*$_1$ *varname*$_3$
i. *varname*$_1$ | *varname*$_3$		I. *varname*$_1$ | *varname*$_3$

varname, *varname*$_1$, and *varname*$_2$ denote numeric or string categorical variables. *varname*$_3$ denotes a continuous, numeric variable.

Menu

Data > Create or change data > Other variable-creation commands > Interaction expansion

---

Most commands in Stata now allow factor variables; see [U] **11.4.3 Factor variables**. To determine if a command allows factor variables, see the information printed below the options table for the command. If the command allows factor variables, it will say something like "*indepvars* may contain factor variables".

We recommend that you use factor variables instead of xi if a command allows factor variables.

We include [R] **xi** in our documentation so that readers can consult it when using a Stata command that does not allow factor variables.

Description

xi expands terms containing categorical variables into indicator (also called dummy) variable sets by creating new variables and, in the second syntax (xi: *any_stata_command*), executes the specified command with the expanded terms. The dummy variables created are

i. *varname*	creates dummies for categorical variable *varname*
i. *varname*$_1$ *i. *varname*$_2$	creates dummies for categorical variables *varname*$_1$ and *varname*$_2$: all interactions and main effects
i. *varname*$_1$ *varname*$_3$	creates dummies for categorical variable *varname*$_1$ and continuous variable *varname*$_3$: all interactions and main effects
i. *varname*$_1$ | *varname*$_3$	creates dummies for categorical variable *varname*$_1$ and continuous variable *varname*$_3$: all interactions and main effect of *varname*$_3$, but no main effect of *varname*$_1$

Options

prefix(*string*) allows you to choose a prefix other than _I for the newly created interaction variables. The prefix cannot be longer than four characters. By default, xi will create interaction variables starting with _I. When you use xi, it drops all previously created interaction variables starting with the prefix specified in the prefix(*string*) option or with _I by default. Therefore, if you want to keep the variables with a certain prefix, specify a different prefix in the prefix(*string*) option.

noomit prevents xi from omitting groups. This option provides a way to generate an indicator variable for every category having one or more variables, which is useful when combined with the noconstant option of an estimation command.

Remarks

Remarks are presented under the following headings:

> *Background*
> *Indicator variables for simple effects*
> *Controlling the omitted dummy*
> *Categorical variable interactions*
> *Interactions with continuous variables*
> *Using xi: Interpreting output*
> *How xi names variables*
> *xi as a command rather than a command prefix*
> *Warnings*

xi provides a convenient way to include dummy or indicator variables when fitting a model (say, with regress or logistic). For instance, assume that the categorical variable agegrp contains 1 for ages 20–24, 2 for ages 25–39, 3 for ages 40–44, etc. Typing

```
. xi: logistic outcome weight i.agegrp bp
```

estimates a logistic regression of outcome on weight, dummies for each agegrp category, and bp. That is, xi searches out and expands terms starting with "i." or "I." but ignores the other variables. xi will expand both numeric and string categorical variables, so if you had a string variable race containing "white", "black", and "other", typing

```
. xi: logistic outcome weight bp i.agegrp i.race
```

would include indicator variables for the race group as well.

The i. indicator variables xi expands may appear anywhere in the *varlist*, so

```
. xi: logistic outcome i.agegrp weight i.race bp
```

would fit the same model.

You can also create interactions of categorical variables; typing

```
xi: logistic outcome weight bp i.agegrp*i.race
```

fits a model with indicator variables for all agegrp and race combinations, including the agegrp and race main-effect terms (that is, the terms that are created when you just type i.agegrp i.race).

You can interact dummy variables with continuous variables; typing

```
xi: logistic outcome bp i.agegrp*weight i.race
```

fits a model with indicator variables for all `agegrp` categories interacted with `weight`, plus the main-effect terms `weight` and `i.agegrp`.

You can get the interaction terms without the `agegrp` main effect (but with the `weight` main effect) by typing

```
xi: logistic outcome bp i.agegrp|weight i.race
```

You can also include multiple interactions:

```
xi: logistic outcome bp i.agegrp*weight i.agegrp*i.race
```

We will now back up and describe the construction of dummy variables in more detail.

Background

The terms *continuous*, *categorical*, and *indicator* or *dummy* variables are used below. Continuous variables measure something—such as height or weight—and at least conceptually can take on any real number over some range. Categorical variables, on the other hand, take on a finite number of values, each denoting membership in a subclass—for example, excellent, good, and poor, which might be coded 0, 1, 2, or 1, 2, 3, or even "Excellent", "Good", and "Poor". An indicator or dummy variable—the terms are used interchangeably—is a special type of two-valued categorical variable that contains values 0, denoting false, and 1, denoting true. The information contained in any k-valued categorical variable can be equally well represented by k indicator variables. Instead of one variable recording values representing excellent, good, and poor, you can have three indicator variables, indicating the truth or falseness of "result is excellent", "result is good", and "result is poor".

`xi` provides a convenient way to convert categorical variables to dummy or indicator variables when you fit a model (say, with `regress` or `logistic`).

▷ Example 1

For instance, assume that the categorical variable `agegrp` contains 1 for ages 20–24, 2 for ages 25–39, and 3 for ages 40–44. (There is no one over 44 in our data.) As it stands, `agegrp` would be a poor candidate for inclusion in a model even if we thought age affected the outcome. The reason is that the coding would restrict the effect of being in the second age group to be twice the effect of being in the first, and, similarly, the effect of being in the third to be three times the first. That is, if we fit the model,

$$y = \beta_0 + \beta_1 \, \texttt{agegrp} + X\beta_2$$

the effect of being in the first age group is β_1, the second $2\beta_1$, and the third $3\beta_1$. If the coding 1, 2, and 3 is arbitrary, we could just as well have coded the age groups 1, 4, and 9, making the effects β_1, $4\beta_1$, and $9\beta_1$.

The solution is to convert the categorical variable `agegrp` to a set of indicator variables, a_1, a_2, and a_3, where a_i is 1 if the individual is a member of the ith age group and 0 otherwise. We can then fit the model

$$y = \beta_0 + \beta_{11}a_1 + \beta_{12}a_2 + \beta_{13}a_3 + X\beta_2$$

The effect of being in age group 1 is now β_{11}; 2, β_{12}; and 3, β_{13}; and these results are independent of our (arbitrary) coding. The only difficulty at this point is that the model is unidentified in the sense that there are an infinite number of $(\beta_0, \beta_{11}, \beta_{12}, \beta_{13})$ that fit the data equally well.

To see this, pretend that $(\beta_0, \beta_{11}, \beta_{12}, \beta_{13}) = (1, 1, 3, 4)$. The predicted values of y for the various age groups are

$$y = \begin{cases} 1 + 1 + X\beta_2 = 2 + X\beta_2 & \text{(age group 1)} \\ 1 + 3 + X\beta_2 = 4 + X\beta_2 & \text{(age group 2)} \\ 1 + 4 + X\beta_2 = 5 + X\beta_2 & \text{(age group 3)} \end{cases}$$

Now pretend that $(\beta_0, \beta_{11}, \beta_{12}, \beta_{13}) = (2, 0, 2, 3)$. The predicted values of y are

$$y = \begin{cases} 2 + 0 + X\beta_2 = 2 + X\beta_2 & \text{(age group 1)} \\ 2 + 2 + X\beta_2 = 4 + X\beta_2 & \text{(age group 2)} \\ 2 + 3 + X\beta_2 = 5 + X\beta_2 & \text{(age group 3)} \end{cases}$$

These two sets of predictions are indistinguishable: for age group 1, $y = 2 + X\beta_2$ regardless of the coefficient vector used, and similarly for age groups 2 and 3. This arises because we have three equations and four unknowns. Any solution is as good as any other, and, for our purposes, we merely need to choose one of them. The popular selection method is to set the coefficient on the first indicator variable to 0 (as we have done in our second coefficient vector). This is equivalent to fitting the model

$$y = \beta_0 + \beta_{12}a_2 + \beta_{13}a_3 + X\beta_2$$

How we select a particular coefficient vector (identifies the model) does not matter. It does, however, affect the *interpretation* of the coefficients.

For instance, we could just as well choose to omit the second group. In our artificial example, this would yield $(\beta_0, \beta_{11}, \beta_{12}, \beta_{13}) = (4, -2, 0, 1)$ instead of $(2, 0, 2, 3)$. These coefficient vectors are the same in the sense that

$$y = \begin{cases} 2 + 0 + X\beta_2 = 2 + X\beta_2 = 4 - 2 + X\beta_2 & \text{(age group 1)} \\ 2 + 2 + X\beta_2 = 4 + X\beta_2 = 4 + 0 + X\beta_2 & \text{(age group 2)} \\ 2 + 3 + X\beta_2 = 5 + X\beta_2 = 4 + 1 + X\beta_2 & \text{(age group 3)} \end{cases}$$

But what does it mean that β_{13} can just as well be 3 or 1? We obtain $\beta_{13} = 3$ when we set $\beta_{11} = 0$, so $\beta_{13} = \beta_{13} - \beta_{11}$ and β_{13} measures the difference between age groups 3 and 1.

In the second case, we obtain $\beta_{13} = 1$ when we set $\beta_{12} = 0$, so $\beta_{13} - \beta_{12} = 1$ and β_{13} measures the difference between age groups 3 and 2. There is no inconsistency. According to our $\beta_{12} = 0$ model, the difference between age groups 3 and 1 is $\beta_{13} - \beta_{11} = 1 - (-2) = 3$, the same result we got in the $\beta_{11} = 0$ model.

◁

▷ Example 2

The issue of interpretation is important because it can affect the way we discuss results. Imagine that we are studying recovery after a coronary bypass operation. Assume that the age groups are children under 13 (we have two of them), young adults under 25 (we have a handful of them), adults under 46 (of which we have even more), mature adults under 56, older adults under 65, and elderly adults. We follow the prescription of omitting the first group, so all our results are reported relative to children under 13. While there is nothing statistically wrong with this, readers will be suspicious when we make statements like "compared with young children, older and elder adults . . . ". Moreover, we will probably have to end each statement with "although results are not statistically significant" because we have only two children in our comparison group. Of course, even with results reported in this way, we can do reasonable comparisons (say, with mature adults), but we will have to do extra work to perform the appropriate linear hypothesis test using Stata's `test` command.

Here it would be better to force the omitted group to be more reasonable, such as mature adults. There is, however, a generic rule for automatic comparison group selection that, although less popular, tends to work better than the omit-the-first-group rule. That rule is to omit the most prevalent group. The most prevalent is usually a reasonable baseline.

◁

In any case, the prescription for categorical variables is

1. Convert each k-valued categorical variable to k indicator variables.

2. Drop one of the k indicator variables; any one will do, but dropping the first is popular, dropping the most prevalent is probably better in terms of having the computer guess at a reasonable interpretation, and dropping a specified one often eases interpretation the most.

3. Fit the model on the remaining $k - 1$ indicator variables.

xi automates this procedure.

We will now consider each of xi's features in detail.

Indicator variables for simple effects

When you type i.*varname*, xi internally tabulates *varname* (which may be a string or a numeric variable) and creates indicator (dummy) variables for each observed value, omitting the indicator for the smallest value. For instance, say that agegrp takes on the values 1, 2, 3, and 4. Typing

```
xi: logistic outcome i.agegrp
```

creates indicator variables named _Iagegrp_2, _Iagegrp_3, and _Iagegrp_4. (xi chooses the names and tries to make them readable; xi guarantees that the names are unique.) The expanded logistic model is

```
. logistic outcome _Iagegrp_2 _Iagegrp_3 _Iagegrp_4
```

Afterward, you can drop the new variables xi leaves behind by typing 'drop _I*' (note the capitalization).

xi provides the following features when you type i.*varname*:

- *varname* may be string or numeric.

- Dummy variables are created automatically.

- By default, the dummy-variable set is identified by dropping the dummy corresponding to the smallest value of the variable (how to specify otherwise is discussed below).

- The new dummy variables are left in your dataset. By default, the names of the new dummy variables start with _I; therefore, you can drop them by typing 'drop _I*'. You do not have to do this; each time you use xi, any automatically generated dummies with the same prefix as the one specified in the prefix(*string*) option, or _I by default, are dropped and new ones are created.

- The new dummy variables have variable labels so that you can determine what they correspond to by typing 'describe'.

- xi may be used with any Stata command (not just logistic).

Controlling the omitted dummy

By default, i.*varname* omits the dummy corresponding to the smallest value of *varname*; for a string variable, this is interpreted as dropping the first in an alphabetical, case-sensitive sort. xi provides two alternatives to dropping the first: xi will drop the dummy corresponding to the most prevalent value of *varname*, or xi will let you choose the particular dummy to be dropped.

To change xi's behavior to dropping the most prevalent dummy, type

```
. char _dta[omit] prevalent
```

although whether you type "prevalent" or "yes" or anything else does not matter. Setting this characteristic affects the expansion of all categorical variables in the dataset. If you resave your dataset, the prevalent preference will be remembered. If you want to change the behavior back to the default drop-the-first rule, type

```
. char _dta[omit]
```

to clear the characteristic.

Once you set _dta[omit], i.*varname* omits the dummy corresponding to the most prevalent value of *varname*. Thus the coefficients on the dummies have the interpretation of change from the most prevalent group. For example,

```
. char _dta[omit] prevalent
. xi: regress y i.agegrp
```

might create _Iagegrp_1 through _Iagegrp_4, resulting in _Iagegrp_2 being omitted if agegrp $=$ 2 is most common (as opposed to the default dropping of _Iagegrp_1). The model is then

$$y = b_0 + b_1 \text{ _Iagegrp_1} + b_3 \text{ _Iagegrp_3} + b_4 \text{ _Iagegrp_4} + u$$

Then

Predicted y for agegrp $1 = b_0 + b_1$	Predicted y for agegrp $3 = b_0 + b_3$
Predicted y for agegrp $2 = b_0$	Predicted y for agegrp $4 = b_0 + b_4$

Thus the model's reported t or Z statistics are for a test of whether each group is different from the most prevalent group.

Perhaps you wish to omit the dummy for agegrp 3 instead. You do this by setting the variable's omit characteristic:

```
. char agegrp[omit] 3
```

This overrides _dta[omit] if you have set it. Now when you type

```
. xi: regress y i.agegrp
```

_Iagegrp_3 will be omitted, and you will fit the model

$$y = b'_0 + b'_1 \text{ _Iagegrp_1} + b'_2 \text{ _Iagegrp_2} + b'_4 \text{ _Iagegrp_4} + u$$

Later if you want to return to the default omission, type

```
. char agegrp[omit]
```

to clear the characteristic.

In summary, i.*varname* omits the first group by default, but if you define

```
. char _dta[omit] prevalent
```

the default behavior changes to dropping the most prevalent group. Either way, if you define a characteristic of the form

    . char *varname*[omit] #

or, if *varname* is a string,

    . char *varname*[omit] *string-literal*

the specified value will be omitted.

Examples: . char agegrp[omit] 1
                   . char race[omit] White   (for race, a string variable)
                   . char agegrp[omit]   (to restore default for agegrp)

Categorical variable interactions

i.*varname*$_1$*i.*varname*$_2$ creates the dummy variables associated with the interaction of the categorical variables *varname*$_1$ and *varname*$_2$. The identification rules—which categories are omitted—are the same as those for i.*varname*. For instance, assume that agegrp takes on four values and race takes on three values. Typing

    . xi: regress y i.agegrp*i.race

results in

model: dummies for:

$$y = a + b_2 \_\text{Iagegrp\_2} + b_3 \_\text{Iagegrp\_3} + b_4 \_\text{Iagegrp\_4} \quad \text{(agegrp)}$$
$$+ c_2 \_\text{Irace\_2} + c_3 \_\text{Irace\_3} \quad \text{(race)}$$
$$+ d_{22} \_\text{IageXrac\_2\_2} + d_{23} \_\text{IageXrac\_2\_3}$$
$$+ d_{32} \_\text{IageXrac\_3\_2} + d_{33} \_\text{IageXrac\_3\_3} \quad \text{(agegrp*race)}$$
$$+ d_{42} \_\text{IageXrac\_4\_2} + d_{43} \_\text{IageXrac\_4\_3}$$
$$+ u$$

That is, typing

    . xi: regress y i.agegrp*i.race

is the same as typing

    . xi: regress y i.agegrp i.race i.agegrp*i.race

Although there are many other ways the interaction could have been parameterized, this method has the advantage that you can test the joint significance of the interactions by typing

    . testparm _IageXrac*

When you perform the estimation step, whether you specify i.agegrp*i.race or i.race*i.agegrp makes no difference (other than in the names given to the interaction terms; in the first case, the names will begin with _IageXrac; in the second, _IracXage). Thus

    . xi: regress y i.race*i.agegrp

fits the same model.

You may also include multiple interactions simultaneously:

    . xi: regress y i.agegrp*i.race i.agegrp*i.sex

The model fit is

model :   dummies for:

$$y = a + b_2 \text{ \_Iagegrp\_2} + b_3 \text{ \_Iagegrp\_3} + b_4 \text{ \_Iagegrp\_4}$$ (agegrp)

$$+ c_2 \text{ \_Irace\_2} + c_3 \text{ \_Irace\_3}$$ (race)

$$+ d_{22} \text{ \_IageXrac\_2\_2} + d_{23} \text{ \_IageXrac\_2\_3}$$

$$+ d_{32} \text{ \_IageXrac\_3\_2} + d_{33} \text{ \_IageXrac\_3\_3}$$ (agegrp*race)

$$+ d_{42} \text{ \_IageXrac\_4\_2} + d_{43} \text{ \_IageXrac\_4\_3}$$

$$+ e_2 \text{ \_Isex\_2}$$ (sex)

$$+ f_{22} \text{ \_IageXsex\_2\_2} + f_{23} \text{ \_IageXsex\_2\_3} + f_{24} \text{ \_IageXsex\_2\_4}$$ (agegrp*sex)

$$+ u$$

The agegrp dummies are (correctly) included only once.

Interactions with continuous variables

i.*varname*$_1$*varname*$_2$ (as distinguished from i.*varname*$_1$*i.*varname*$_2$—note the second i.) specifies an interaction of a categorical variable with a continuous variable. For instance,

    . xi: regress y i.agegr*wgt

results in the model

$$y = a + b_2 \text{ \_Iagegrp\_2} + b_3 \text{ \_Iagegrp\_3} + b_4 \text{ \_Iagegrp\_4}$$ (agegrp dummies)

$$+ c \, \text{wgt}$$ (continuous wgt effect)

$$+ d_2 \text{ \_IageXwgt\_2} + d_3 \text{ \_IageXwgt\_3} + d_4 \text{ \_IageXwgt\_4}$$ (agegrp*wgt interactions)

$$+ u$$

A variation on this notation, using | rather than *, omits the agegrp dummies. Typing

    . xi: regress y i.agegrp|wgt

fits the model

$$y = a' + c' \, \text{wgt}$$ (continuous wgt effect)

$$+ d'_2 \text{ \_IageXwgt\_2} + d'_3 \text{ \_IageXwgt\_3} + d'_4 \text{ \_IageXwgt\_4}$$ (agegrp*wgt interactions)

$$+ u'$$

The predicted values of y are

agegrp*wgt model	agegrp|wgt model	
$y = a + c \, \text{wgt}$	$a' + c' \, \text{wgt}$	if agegrp $= 1$
$a + c \, \text{wgt} + b_2 + d_2 \, \text{wgt}$	$a' + c' \text{wgt} + d'_2 \, \text{wgt}$	if agegrp $= 2$
$a + c \, \text{wgt} + b_3 + d_3 \, \text{wgt}$	$a' + c' \text{wgt} + d'_3 \, \text{wgt}$	if agegrp $= 3$
$a + c \, \text{wgt} + b_4 + d_4 \, \text{wgt}$	$a' + c' \text{wgt} + d'_4 \, \text{wgt}$	if agegrp $= 4$

That is, typing

    . xi: regress y i.agegrp*wgt

is equivalent to typing

```
. xi: regress y i.agegrp i.agegrp|wgt
```

In either case, you do not need to specify separately the continuous variable wgt; it is included automatically.

Using xi: Interpreting output

```
. xi: regress mpg i.rep78
i.rep78              _Irep78_1-5    (naturally coded; _Irep78_1 omitted)
   (output from regress appears )
```

Interpretation: i.rep78 expanded to the dummies _Irep78_1, _Irep78_2, ..., _Irep78_5. The numbers on the end are "natural" in the sense that _Irep78_1 corresponds to rep78 = 1, _Irep78_2 to rep78 = 2, and so on. Finally, the dummy for rep78 = 1 was omitted.

```
. xi: regress mpg i.make
i.make               _Imake_1-74    (_Imake_1 for make==AMC Concord omitted)
   (output from regress appears )
```

Interpretation: i.make expanded to _Imake_1, _Imake_2, ..., _Imake_74. The coding is not natural because make is a string variable. _Imake_1 corresponds to one make, _Imake_2 to another, and so on. You can find out the coding by typing describe. _Imake_1 for the AMC Concord was omitted.

How xi names variables

By default, xi assigns to the dummy variables it creates names having the form

$$\text{\_I}stub\text{\_}groupid$$

You may subsequently refer to the entire set of variables by typing '_I*stub**'. For example,

name	= _I +	stub	+ _ +	groupid	Entire set
_Iagegrp_1	_I	agegrp	_	1	_Iagegrp*
_Iagegrp_2	_I	agegrp	_	2	_Iagegrp*
_IageXwgt_1	_I	ageXwgt	_	1	_IageXwgt*
_IageXrac_1_2	_I	ageXrac	_	1_2	_IageXrac*
_IageXrac_2_1	_I	ageXrac	_	2_1	_IageXrac*

If you specify a prefix in the prefix(*string*) option, say, _S, then xi will name the variables starting with the prefix

$$\text{\_S}stub\text{\_}groupid$$

xi as a command rather than a command prefix

xi can be used as a command prefix or as a command by itself. In the latter form, xi merely creates the indicator and interaction variables. Typing

```
. xi: regress y i.agegrp*wgt
i.agegrp              _Iagegrp_1-4    (naturally coded; _Iagegrp_1 omitted)
i.agegrp*wgt          _IageXwgt_1-4   (coded as above)
  (output from regress appears )
```

is equivalent to typing

```
. xi i.agegrp*wgt
i.agegrp              _Iagegrp_1-4    (naturally coded; _Iagegrp_1 omitted)
i.agegrp*wgt          _IageXwgt_1-4   (coded as above)

. regress y _Iagegrp* _IageXwgt*
  (output from regress appears )
```

Warnings

1. xi creates new variables in your dataset; most are bytes, but interactions with continuous variables will have the storage type of the underlying continuous variable. You may get the message "insufficient memory". If so, you will need to increase the amount of memory allocated to Stata's data areas; see [U] **6 Managing memory**.

2. When using xi with an estimation command, you may get the message "matsize too small". If so, see [R] **matsize**.

Saved results

xi saves the following characteristics:

```
_dta[__xi__Vars__Prefix__]      prefix names
_dta[__xi__Vars__To__Drop__]    variables created
```

Methods and formulas

xi is implemented as an ado-file.

References

Hendrickx, J. 1999. dm73: Using categorical variables in Stata. *Stata Technical Bulletin* 52: 2–8. Reprinted in *Stata Technical Bulletin Reprints*, vol. 9, pp. 51–59. College Station, TX: Stata Press.

——. 2000. dm73.1: Contrasts for categorical variables: Update. *Stata Technical Bulletin* 54: 7. Reprinted in *Stata Technical Bulletin Reprints*, vol. 9, pp. 60–61. College Station, TX: Stata Press.

——. 2001a. dm73.2: Contrasts for categorical variables: Update. *Stata Technical Bulletin* 59: 2–5. Reprinted in *Stata Technical Bulletin Reprints*, vol. 10, pp. 9–14. College Station, TX: Stata Press.

——. 2001b. dm73.3: Contrasts for categorical variables: Update. *Stata Technical Bulletin* 61: 5. Reprinted in *Stata Technical Bulletin Reprints*, vol. 10, pp. 14–15. College Station, TX: Stata Press.

Also see

[U] **11.1.10 Prefix commands**

[U] **20 Estimation and postestimation commands**

Title

> **zinb** — Zero-inflated negative binomial regression

Syntax

> zinb *depvar* [*indepvars*] [*if*] [*in*] [*weight*] ,
>
>   <u>inf</u>late(*varlist*[, <u>off</u>set(*varname*)] | _cons) [*options*]

options	Description
Model	
*<u>inf</u>late( )**	equation that determines whether the count is zero
<u>nocon</u>stant	suppress constant term
<u>exp</u>osure(*varname$_e$*)	include ln(*varname$_e$*) in model with coefficient constrained to 1
<u>off</u>set(*varname$_o$*)	include *varname$_o$* in model with coefficient constrained to 1
<u>constr</u>aints(*constraints*)	apply specified linear constraints
<u>colli</u>near	keep collinear variables
probit	use probit model to characterize excess zeros; default is logit
SE/Robust	
vce(*vcetype*)	*vcetype* may be oim, <u>r</u>obust, <u>c</u>luster *clustvar*, opg, <u>boot</u>strap, or <u>jack</u>knife
Reporting	
<u>level</u>(#)	set confidence level; default is level(95)
irr	report incidence-rate ratios
vuong	perform Vuong test
zip	perform ZIP likelihood-ratio test
<u>nocnsr</u>eport	do not display constraints
display_options	control column formats, row spacing, line width, and display of omitted variables and base and empty cells
Maximization	
maximize_options	control the maximization process; seldom used
<u>coef</u>legend	display legend instead of statistics

*<u>inf</u>late(*varlist*[, <u>off</u>set(*varname*)] | _cons) is required.

indepvars and *varlist* may contain factor variables; see [U] **11.4.3 Factor variables**.

bootstrap, by, jackknife, rolling, statsby, and svy are allowed; see [U] **11.1.10 Prefix commands**.

Weights are not allowed with the bootstrap prefix; see [R] **bootstrap**.

vce(), vuong, zip, and weights are not allowed with the svy prefix; see [SVY] **svy**.

fweights, iweights, and pweights are allowed; see [U] **11.1.6 weight**.

coeflegend does not appear in the dialog box.

See [U] **20 Estimation and postestimation commands** for more capabilities of estimation commands.

Menu

Statistics > Count outcomes > Zero-inflated negative binomial regression

Description

zinb estimates a zero-inflated negative binomial (ZINB) regression of *depvar* on *indepvars*, where *depvar* is a nonnegative count variable.

Options

╭─── Model ───

inflate(*varlist*[, offset(*varname*)] | _cons) specifies the equation that determines whether the observed count is zero. Conceptually, omitting inflate() would be equivalent to fitting the model with nbreg.

  inflate(*varlist*[, offset(*varname*)]) specifies the variables in the equation. You may optionally include an offset for this *varlist*.

  inflate(_cons) specifies that the equation determining whether the count is zero contains only an intercept. To run a zero-inflated model of *depvar* with only an intercept in both equations, type zinb *depvar*, inflate(_cons).

noconstant, exposure(*varname_e*), offset(*varname_o*), constraints(*constraints*), collinear; see [R] **estimation options**.

probit requests that a probit, instead of logit, model be used to characterize the excess zeros in the data.

╭─── SE/Robust ───

vce(*vcetype*) specifies the type of standard error reported, which includes types that are derived from asymptotic theory, that are robust to some kinds of misspecification, that allow for intragroup correlation, and that use bootstrap or jackknife methods; see [R] **vce_option**.

╭─── Reporting ───

level(*#*); see [R] **estimation options**.

irr reports estimated coefficients transformed to incidence-rate ratios, that is, e^{β_i} rather than β_i. Standard errors and confidence intervals are similarly transformed. This option affects how results are displayed, not how they are estimated or stored. irr may be specified at estimation or when replaying previously estimated results.

vuong specifies that the Vuong (1989) test of ZINB versus negative binomial be reported. This test statistic has a standard normal distribution with large positive values favoring the ZINB model and large negative values favoring the negative binomial model.

zip requests that a likelihood-ratio test comparing the ZINB model with the zero-inflated Poisson model be included in the output.

nocnsreport; see [R] **estimation options**.

display_options: noomitted, vsquish, noemptycells, baselevels, allbaselevels, cformat(%*fmt*), pformat(%*fmt*), sformat(%*fmt*), and nolstretch; see [R] **estimation options**.

⌐ Maximization ⌐

maximize_options: <u>diff</u>icult, <u>tech</u>nique(*algorithm_spec*), <u>iter</u>ate(*#*), [<u>no</u>]<u>log</u>, <u>tra</u>ce,
gradient, showstep, <u>hess</u>ian, <u>showtol</u>erance, <u>tol</u>erance(*#*), <u>ltol</u>erance(*#*),
<u>nrtol</u>erance(*#*), <u>nonrtol</u>erance, and from(*init_specs*); see [R] **maximize**. These options are
seldom used.

Setting the optimization type to technique(bhhh) resets the default *vcetype* to vce(opg).

The following option is available with zinb but is not shown in the dialog box:

coeflegend; see [R] **estimation options**.

Remarks

See Long (1997, 242–247) and Greene (2012, 821–826) for a discussion of zero-modified count
models. For information about the test developed by Vuong (1989), see Greene (2012, 823–824) and
Long (1997). Greene (1994) applied the test to zero-inflated Poisson and negative binomial models,
and there is a description of that work in Greene (2012).

Negative binomial regression fits models of the number of occurrences (counts) of an event. You
could use nbreg for this (see [R] **nbreg**), but in some count-data models, you might want to account
for the prevalence of zero counts in the data.

For instance, you could count how many fish each visitor to a park catches. Many visitors may
catch zero, because they do not fish (as opposed to being unsuccessful). You may be able to model
whether a person fishes depending on several covariates related to fishing activity and model how
many fish a person catches depending on several covariates having to do with the success of catching
fish (type of lure/bait, time of day, temperature, season, etc.). This is the type of data for which the
zinb command is useful.

The zero-inflated (or zero-altered) negative binomial model allows overdispersion through the
splitting process that models the outcomes as zero or nonzero.

▷ Example 1

We have data on the number of fish caught by visitors to a national park. Some of the visitors do
not fish, but we do not have the data on whether a person fished; we have data merely on how many
fish were caught, together with several covariates. Because our data have a preponderance of zeros
(142 of 250), we use the zinb command to model the outcome.

```
. use http://www.stata-press.com/data/r12/fish
. zinb count persons livebait, inf(child camper) vuong
Fitting constant-only model:
Iteration 0:   log likelihood = -519.33992
  (output omitted)
Iteration 8:   log likelihood = -442.66299
Fitting full model:
Iteration 0:   log likelihood = -442.66299  (not concave)
  (output omitted)
Iteration 8:   log likelihood = -401.54776
```

```
Zero-inflated negative binomial regression          Number of obs    =         250
                                                    Nonzero obs      =         108
                                                    Zero obs         =         142

Inflation model = logit                             LR chi2(2)       =       82.23
Log likelihood  = -401.5478                         Prob > chi2      =      0.0000
```

	Coef.	Std. Err.	z	P>|z|	[95% Conf. Interval]	
count						
persons	.9742984	.1034938	9.41	0.000	.7714543	1.177142
livebait	1.557523	.4124424	3.78	0.000	.7491503	2.365895
_cons	-2.730064	.476953	-5.72	0.000	-3.664874	-1.795253
inflate						
child	3.185999	.7468551	4.27	0.000	1.72219	4.649808
camper	-2.020951	.872054	-2.32	0.020	-3.730146	-.3117567
_cons	-2.695385	.8929071	-3.02	0.003	-4.44545	-.9453189
/lnalpha	.5110429	.1816816	2.81	0.005	.1549535	.8671323
alpha	1.667029	.3028685			1.167604	2.380076

```
Vuong test of zinb vs. standard negative binomial: z =      5.59  Pr>z = 0.0000
```

In general, Vuong test statistics that are significantly positive favor the zero-inflated models, whereas those that are significantly negative favor the non–zero-inflated models. Thus, in the above model, the zero inflation is significant.   ◁

Saved results

zinb saves the following in e():

Scalars

e(N)	number of observations
e(N_zero)	number of zero observations
e(k)	number of parameters
e(k_eq)	number of equations in e(b)
e(k_eq_model)	number of equations in overall model test
e(k_aux)	number of auxiliary parameters
e(k_dv)	number of dependent variables
e(df_m)	model degrees of freedom
e(ll)	log likelihood
e(ll_0)	log likelihood, constant-only model
e(df_c)	degrees of freedom for comparison test
e(N_clust)	number of clusters
e(chi2)	χ^2
e(p)	significance of model test
e(chi2_cp)	χ^2 for test of $\alpha = 0$
e(vuong)	Vuong test statistic
e(rank)	rank of e(V)
e(ic)	number of iterations
e(rc)	return code
e(converged)	1 if converged, 0 otherwise

Macros
    e(cmd)   zinb
    e(cmdline)   command as typed
    e(depvar)   name of dependent variable
    e(inflate)   logit or probit
    e(wtype)   weight type
    e(wexp)   weight expression
    e(title)   title in estimation output
    e(clustvar)   name of cluster variable
    e(offset1)   offset
    e(offset2)   offset for inflate()
    e(chi2type)   Wald or LR; type of model χ^2 test
    e(chi2_cpt)   Wald or LR; type of model χ^2 test corresponding to e(chi2_cp)
    e(vce)   *vcetype* specified in vce()
    e(vcetype)   title used to label Std. Err.
    e(opt)   type of optimization
    e(which)   max or min; whether optimizer is to perform maximization or minimization
    e(ml_method)   type of ml method
    e(user)   name of likelihood-evaluator program
    e(technique)   maximization technique
    e(properties)   b V
    e(predict)   program used to implement predict
    e(asbalanced)   factor variables fvset as asbalanced
    e(asobserved)   factor variables fvset as asobserved

Matrices
    e(b)   coefficient vector
    e(Cns)   constraints matrix
    e(ilog)   iteration log (up to 20 iterations)
    e(gradient)   gradient vector
    e(V)   variance–covariance matrix of the estimators
    e(V_modelbased)   model-based variance

Functions
    e(sample)   marks estimation sample

Methods and formulas

zinb is implemented as an ado-file.

Several models in the literature are (correctly) described as zero inflated. The zinb command maximizes the log likelihood $\ln L$, defined by

$$m = 1/\alpha$$

$$p_j = 1/(1 + \alpha\mu_j)$$

$$\xi_j^\beta = \mathbf{x}_j\boldsymbol{\beta} + \text{offset}_j^\beta$$

$$\xi_j^\gamma = \mathbf{z}_j\boldsymbol{\gamma} + \text{offset}_j^\gamma$$

$$\mu_j = \exp(\xi_j^\beta)$$

$$\ln L = \sum_{j \in S} w_j \ln\left[F(\xi_j^\gamma) + \left\{1 - F(\xi_j^\gamma)\right\}p_j^m\right]$$

$$+ \sum_{j \notin S} w_j\left[\ln\left\{1 - F(\xi_j^\gamma)\right\} + \ln\Gamma(m + y_j) - \ln\Gamma(y_j + 1)\right.$$

$$\left. - \ln\Gamma(m) + m\ln p_j + y_j\ln(1 - p_j)\right]$$

where w_j are the weights, F is the logit link (or probit link if `probit` was specified), and S is the set of observations for which the outcome $y_j = 0$.

This command supports the Huber/White/sandwich estimator of the variance and its clustered version using `vce(robust)` and `vce(cluster clustvar)`, respectively. See [P] **_robust**, particularly *Maximum likelihood estimators* and *Methods and formulas*.

`zinb` also supports estimation with survey data. For details on VCEs with survey data, see [SVY] **variance estimation**.

References

Greene, W. H. 1994. Accounting for excess zeros and sample selection in Poisson and negative binomial regression models. Working paper EC-94-10, Department of Economics, Stern School of Business, New York University. http://ideas.repec.org/p/ste/nystbu/94-10.html.

——. 2012. *Econometric Analysis*. 7th ed. Upper Saddle River, NJ: Prentice Hall.

Long, J. S. 1997. *Regression Models for Categorical and Limited Dependent Variables*. Thousand Oaks, CA: Sage.

Long, J. S., and J. Freese. 2001. Predicted probabilities for count models. *Stata Journal* 1: 51–57.

——. 2006. *Regression Models for Categorical Dependent Variables Using Stata*. 2nd ed. College Station, TX: Stata Press.

Mullahy, J. 1986. Specification and testing of some modified count data models. *Journal of Econometrics* 33: 341–365.

Vuong, Q. H. 1989. Likelihood ratio tests for model selection and non-nested hypotheses. *Econometrica* 57: 307–333.

Also see

[R] **zinb postestimation** — Postestimation tools for zinb

[R] **zip** — Zero-inflated Poisson regression

[R] **nbreg** — Negative binomial regression

[R] **poisson** — Poisson regression

[R] **tnbreg** — Truncated negative binomial regression

[R] **tpoisson** — Truncated Poisson regression

[SVY] **svy estimation** — Estimation commands for survey data

[XT] **xtnbreg** — Fixed-effects, random-effects, & population-averaged negative binomial models

[U] **20 Estimation and postestimation commands**

Title

zinb postestimation — Postestimation tools for zinb

Description

The following postestimation commands are available after `zinb`:

Command	Description
contrast	contrasts and ANOVA-style joint tests of estimates
estat	AIC, BIC, VCE, and estimation sample summary
estat (svy)	postestimation statistics for survey data
estimates	cataloging estimation results
lincom	point estimates, standard errors, testing, and inference for linear combinations of coefficients
lrtest[1]	likelihood-ratio test
margins	marginal means, predictive margins, marginal effects, and average marginal effects
marginsplot	graph the results from margins (profile plots, interaction plots, etc.)
nlcom	point estimates, standard errors, testing, and inference for nonlinear combinations of coefficients
predict	predictions, residuals, influence statistics, and other diagnostic measures
predictnl	point estimates, standard errors, testing, and inference for generalized predictions
pwcompare	pairwise comparisons of estimates
suest	seemingly unrelated estimation
test	Wald tests of simple and composite linear hypotheses
testnl	Wald tests of nonlinear hypotheses

[1] `lrtest` is not appropriate with svy estimation results.

See the corresponding entries in the *Base Reference Manual* for details, but see [SVY] **estat** for details about `estat` (svy).

Syntax for predict

> predict [*type*] *newvar* [*if*] [*in*] [, *statistic* <u>nooff</u>set]

> predict [*type*] { *stub** | *newvar*$_{reg}$ *newvar*$_{inflate}$ *newvar*$_{lnalpha}$ } [*if*] [*in*] , <u>sc</u>ores

statistic	Description
Main	
n	number of events; the default
ir	incidence rate
<u>pr</u>	probability of a degenerate zero
pr(*n*)	probability $\Pr(y_j = n)$
pr(*a*,*b*)	probability $\Pr(a \leq y_j \leq b)$
xb	linear prediction
stdp	standard error of the linear prediction

These statistics are available both in and out of sample; type `predict ... if e(sample) ...` if wanted only for the estimation sample.

Menu

Statistics > Postestimation > Predictions, residuals, etc.

Options for predict

    &boxopen; Main &boxopen;

n, the default, calculates the predicted number of events, which is $(1 - p_j) \exp(\mathbf{x}_j \boldsymbol{\beta})$ if neither offset() nor exposure() was specified when the model was fit, where p_j is the predicted probability of a zero outcome; $(1 - p_j) \exp\{(\mathbf{x}_j \boldsymbol{\beta}) + \text{offset}_j\}$ if offset() was specified; or $(1 - p_j)\{\exp(\mathbf{x}_j \boldsymbol{\beta}) \times \text{exposure}_j\}$ if exposure() was specified.

ir calculates the incidence rate $\exp(\mathbf{x}_j \boldsymbol{\beta})$, which is the predicted number of events when exposure is 1. This is equivalent to specifying both the n and the nooffset options.

pr calculates the probability $\Pr(y_j = 0)$, where this zero was obtained from the degenerate distribution $F(\mathbf{z}_j \boldsymbol{\gamma})$. If offset() was specified within the inflate() option, then $F(\mathbf{z}_j \boldsymbol{\gamma} + \text{offset}_j^\gamma)$ is calculated.

pr(n) calculates the probability $\Pr(y_j = n)$, where n is a nonnegative integer that may be specified as a number or a variable. Note that pr is not equivalent to pr(0).

pr(a,b) calculates the probability $\Pr(a \leq y_j \leq b)$, where a and b are nonnegative integers that may be specified as numbers or variables;

    b missing ($b \geq .$) means $+\infty$;
    pr(20,.) calculates $\Pr(y_j \geq 20)$;
    pr(20,b) calculates $\Pr(y_j \geq 20)$ in observations for which $b \geq .$ and calculates $\Pr(20 \leq y_j \leq b)$ elsewhere.

    pr(.,b) produces a syntax error. A missing value in an observation of the variable a causes a missing value in that observation for pr(a,b).

xb calculates the linear prediction, which is $\mathbf{x}_j \boldsymbol{\beta}$ if neither offset() nor exposure() was specified; $\mathbf{x}_j \boldsymbol{\beta} + \text{offset}_j$ if offset() was specified; or $\mathbf{x}_j \boldsymbol{\beta} + \ln(\text{exposure}_j)$ if exposure() was specified; see nooffset below.

stdp calculates the standard error of the linear prediction.

nooffset is relevant only if you specified offset() or exposure() when you fit the model. It modifies the calculations made by predict so that they ignore the offset or exposure variable; the linear prediction is treated as $\mathbf{x}_j \boldsymbol{\beta}$ rather than as $\mathbf{x}_j \boldsymbol{\beta} + \text{offset}_j$ or $\mathbf{x}_j \boldsymbol{\beta} + \ln(\text{exposure}_j)$. Specifying predict ... , nooffset is equivalent to specifying predict ... , ir.

scores calculates equation-level score variables.

    The first new variable will contain $\partial \ln L / \partial(\mathbf{x}_j \boldsymbol{\beta})$.

    The second new variable will contain $\partial \ln L / \partial(\mathbf{z}_j \boldsymbol{\gamma})$.

    The third new variable will contain $\partial \ln L / \partial \ln \alpha$.

Methods and formulas

All postestimation commands listed above are implemented as ado-files.

The probabilities calculated using the `pr(n)` option are the probability $\Pr(y_i = n)$. These are calculated using

$$\Pr(0|\mathbf{x}_i) = \omega_i + (1 - \omega_i)\,p_2(0|\mathbf{x}_i)$$

$$\Pr(n|\mathbf{x}_i) = (1 - \omega_i)\,p_2(n|\mathbf{x}_i) \qquad \text{for } n = 1, 2, \ldots$$

where ω_i is the probability of obtaining an observation from the degenerate distribution whose mass is concentrated at zero, and $p_2(n|\mathbf{x}_i)$ is the probability of $y_i = n$ from the nondegenerate, negative binomial distribution. ω_i can be obtained from the `pr` option.

See Cameron and Trivedi (1998, sec. 4.7) for further details.

Reference

Cameron, A. C., and P. K. Trivedi. 1998. *Regression Analysis of Count Data*. Cambridge: Cambridge University Press.

Also see

[R] **zinb** — Zero-inflated negative binomial regression

[U] **20 Estimation and postestimation commands**

Title

> **zip** — Zero-inflated Poisson regression

Syntax

> zip *depvar* [*indepvars*] [*if*] [*in*] [*weight*] ,
>
>   <u>inf</u>late(*varlist*[, <u>off</u>set(*varname*)] | _cons) [*options*]

options	Description
Model	
* <u>inf</u>late( )	equation that determines whether the count is zero
<u>nocon</u>stant	suppress constant term
<u>exp</u>osure(*varname*_e)	include $\ln(varname_e)$ in model with coefficient constrained to 1
<u>off</u>set(*varname*_o)	include $varname_o$ in model with coefficient constrained to 1
<u>constraints</u>(*constraints*)	apply specified linear constraints
<u>col</u>linear	keep collinear variables
<u>probit</u>	use probit model to characterize excess zeros; default is logit
SE/Robust	
vce(*vcetype*)	*vcetype* may be oim, <u>r</u>obust, <u>c</u>luster *clustvar*, opg, <u>boot</u>strap, or <u>jack</u>knife
Reporting	
<u>level</u>(#)	set confidence level; default is level(95)
irr	report incidence-rate ratios
vuong	perform Vuong test
<u>nocnsre</u>port	do not display constraints
display_options	control column formats, row spacing, line width, and display of omitted variables and base and empty cells
Maximization	
maximize_options	control the maximization process; seldom used
<u>coefl</u>egend	display legend instead of statistics

* <u>inf</u>late(*varlist*[, <u>off</u>set(*varname*)] | _cons) is required.

indepvars and *varlist* may contain factor variables; see [U] **11.4.3 Factor variables**.

bootstrap, by, jackknife, rolling, statsby, and svy are allowed; see [U] **11.1.10 Prefix commands**.

Weights are not allowed with the bootstrap prefix; see [R] **bootstrap**.

vce(), vuong, and weights are not allowed with the svy prefix; see [SVY] **svy**.

fweights, iweights, and pweights are allowed; see [U] **11.1.6 weight**.

coeflegend does not appear in the dialog box.

See [U] **20 Estimation and postestimation commands** for more capabilities of estimation commands.

Menu

Statistics > Count outcomes > Zero-inflated Poisson regression

Description

zip estimates a zero-inflated Poisson (ZIP) regression of *depvar* on *indepvars*, where *depvar* is a nonnegative count variable.

Options

    ☐ Model ☐

inflate(*varlist*[, offset(*varname*)] | _cons) specifies the equation that determines whether the observed count is zero. Conceptually, omitting inflate() would be equivalent to fitting the model with poisson; see [R] **poisson**.

    inflate(*varlist*[, offset(*varname*)]) specifies the variables in the equation. You may optionally include an offset for this *varlist*.

    inflate(_cons) specifies that the equation determining whether the count is zero contains only an intercept. To run a zero-inflated model of *depvar* with only an intercept in both equations, type zip *depvar*, inflate(_cons).

noconstant, exposure(*varname$_e$*), offset(*varname$_o$*), constraints(*constraints*), collinear; see [R] **estimation options**.

probit requests that a probit, instead of logit, model be used to characterize the excess zeros in the data.

    ☐ SE/Robust ☐

vce(*vcetype*) specifies the type of standard error reported, which includes types that are derived from asymptotic theory, that are robust to some kinds of misspecification, that allow for intragroup correlation, and that use bootstrap or jackknife methods; see [R] *vce_option*.

    ☐ Reporting ☐

level(*#*); see [R] **estimation options**.

irr reports estimated coefficients transformed to incidence-rate ratios, that is, e^b rather than b. Standard errors and confidence intervals are similarly transformed. This option affects how results are displayed, not how they are estimated or stored. irr may be specified at estimation or when replaying previously estimated results.

vuong specifies that the Vuong (1989) test of ZIP versus Poisson be reported. This test statistic has a standard normal distribution with large positive values favoring the ZIP model and large negative values favoring the Poisson model.

nocnsreport; see [R] **estimation options**.

display_options: noomitted, vsquish, noemptycells, baselevels, allbaselevels, cformat(*%fmt*), pformat(*%fmt*), sformat(*%fmt*), and nolstretch; see [R] **estimation options**.

```
                    Maximization
```

maximize_options: <u>diff</u>icult, <u>tech</u>nique(*algorithm_spec*), <u>iter</u>ate(*#*), [<u>no</u>]<u>log</u>, <u>tr</u>ace,
     <u>grad</u>ient, showstep, <u>hess</u>ian, <u>showtol</u>erance, <u>tol</u>erance(*#*), <u>ltol</u>erance(*#*),
     <u>nrtol</u>erance(*#*), <u>nonrtol</u>erance, and from(*init_specs*); see [R] **maximize**. These options are
     seldom used.

Setting the optimization type to technique(bhhh) resets the default *vcetype* to vce(opg).

The following option is available with zip but is not shown in the dialog box:

coeflegend; see [R] **estimation options**.

Remarks

See Long (1997, 242–247) and Greene (2012, 821–826) for a discussion of zero-modified count
models. For information about the test developed by Vuong (1989), see Greene (2012, 823–824) and
Long (1997). Greene (1994) applied the test to ZIP and ZINB models, as described in Greene (2012,
824).

Poisson regression fits models of the number of occurrences (counts) of an event. You could use
poisson for this (see [R] **poisson**), but in some count-data models, you might want to account for
the prevalence of zero counts in the data.

For instance, you might count how many fish each visitor to a park catches. Many visitors may
catch zero, because they do not fish (as opposed to being unsuccessful). You may be able to model
whether a person fishes depending on several covariates related to fishing activity and model how
many fish a person catches depending on several covariates having to do with the success of catching
fish (type of lure/bait, time of day, temperature, season, etc.). This is the type of data for which the
zip command is useful.

The zero-inflated (or zero-altered) Poisson model allows overdispersion through the splitting process
that models the outcomes as zero or nonzero.

▷ Example 1

We have data on the number of fish caught by visitors to a national park. Some of the visitors do
not fish, but we do not have the data on whether a person fished; we merely have data on how many
fish were caught together with several covariates. Because our data have a preponderance of zeros
(142 of 250), we use the zip command to model the outcome.

```
    . use http://www.stata-press.com/data/r12/fish
    . zip count persons livebait, inf(child camper) vuong
    Fitting constant-only model:
    Iteration 0:    log likelihood =  -1347.807
    Iteration 1:    log likelihood = -1305.3245
      (output omitted)
    Iteration 4:    log likelihood = -1103.9425
    Fitting full model:
    Iteration 0:    log likelihood = -1103.9425
    Iteration 1:    log likelihood =  -896.2346
      (output omitted)
    Iteration 5:    log likelihood = -850.70142
```

```
Zero-inflated Poisson regression                 Number of obs   =        250
                                                 Nonzero obs     =        108
                                                 Zero obs        =        142
Inflation model = logit                          LR chi2(2)      =     506.48
Log likelihood  = -850.7014                      Prob > chi2     =     0.0000
```

	Coef.	Std. Err.	z	P>|z|	[95% Conf. Interval]	
count						
persons	.8068853	.0453288	17.80	0.000	.7180424	.8957281
livebait	1.757289	.2446082	7.18	0.000	1.277866	2.236713
_cons	-2.178472	.2860289	-7.62	0.000	-2.739078	-1.617865
inflate						
child	1.602571	.2797719	5.73	0.000	1.054228	2.150913
camper	-1.015698	.365259	-2.78	0.005	-1.731593	-.2998038
_cons	-.4922872	.3114562	-1.58	0.114	-1.10273	.1181558

```
Vuong test of zip vs. standard Poisson:          z =       3.95  Pr>z = 0.0000
```

In general, Vuong test statistics that are significantly positive favor the zero-inflated models, while those that are significantly negative favor the non–zero-inflated models. Thus, in the above model, the zero inflation is significant.

◁

Saved results

zip saves the following in e():

Scalars

e(N)	number of observations
e(N_zero)	number of zero observations
e(k)	number of parameters
e(k_eq)	number of equations in e(b)
e(k_eq_model)	number of equations in overall model test
e(k_dv)	number of dependent variables
e(df_m)	model degrees of freedom
e(ll)	log likelihood
e(ll_0)	log likelihood, constant-only model
e(ll_c)	log likelihood, comparison model
e(df_c)	degrees of freedom for comparison test
e(N_clust)	number of clusters
e(chi2)	χ^2
e(p)	significance of model test
e(vuong)	Vuong test statistic
e(rank)	rank of e(V)
e(ic)	number of iterations
e(rc)	return code
e(converged)	1 if converged, 0 otherwise

Macros
    e(cmd)   zip
    e(cmdline)   command as typed
    e(depvar)   name of dependent variable
    e(inflate)   logit or probit
    e(wtype)   weight type
    e(wexp)   weight expression
    e(title)   title in estimation output
    e(clustvar)   name of cluster variable
    e(offset1)   offset
    e(offset2)   offset for inflate()
    e(chi2type)   Wald or LR; type of model χ^2 test
    e(vce)   *vcetype* specified in vce()
    e(vcetype)   title used to label Std. Err.
    e(opt)   type of optimization
    e(which)   max or min; whether optimizer is to perform maximization or minimization
    e(ml_method)   type of ml method
    e(user)   name of likelihood-evaluator program
    e(technique)   maximization technique
    e(properties)   b V
    e(predict)   program used to implement predict
    e(asbalanced)   factor variables fvset as asbalanced
    e(asobserved)   factor variables fvset as asobserved

Matrices
    e(b)   coefficient vector
    e(Cns)   constraints matrix
    e(ilog)   iteration log (up to 20 iterations)
    e(gradient)   gradient vector
    e(V)   variance–covariance matrix of the estimators
    e(V_modelbased)   model-based variance

Functions
    e(sample)   marks estimation sample

Methods and formulas

zip is implemented as an ado-file.

Several models in the literature are (correctly) described as zero inflated. The zip command maximizes the log-likelihood $\ln L$, defined by

$$\xi_j^\beta = \mathbf{x}_j\boldsymbol{\beta} + \text{offset}_j^\beta$$

$$\xi_j^\gamma = \mathbf{z}_j\boldsymbol{\gamma} + \text{offset}_j^\gamma$$

$$\ln L = \sum_{i \in S} w_j \ln\left[F(\xi_j^\gamma) + \left\{1 - F(\xi_j^\gamma)\right\}\exp(-\lambda_j)\right] +$$

$$\sum_{i \notin S} w_j \left[\ln\left\{1 - F(\xi_j^\gamma)\right\} - \lambda_j + \xi_j^\beta y_j - \ln(y_j!)\right]$$

where w_j are the weights, F is the logit link (or probit link if `probit` was specified), and S is the set of observations for which the outcome $y_j = 0$.

This command supports the Huber/White/sandwich estimator of the variance and its clustered version using `vce(robust)` and `vce(cluster` *clustvar*`)`, respectively. See [P] **_robust**, particularly *Maximum likelihood estimators* and *Methods and formulas*.

`zip` also supports estimation with survey data. For details on VCEs with survey data, see [SVY] **variance estimation**.

References

Greene, W. H. 1994. Accounting for excess zeros and sample selection in Poisson and negative binomial regression models. Working paper EC-94-10, Department of Economics, Stern School of Business, New York University. http://ideas.repec.org/p/ste/nystbu/94-10.html.

———. 2012. *Econometric Analysis*. 7th ed. Upper Saddle River, NJ: Prentice Hall.

Lambert, D. 1992. Zero-inflated Poisson regression, with an application to defects in manufacturing. *Technometrics* 34: 1–14.

Long, J. S. 1997. *Regression Models for Categorical and Limited Dependent Variables*. Thousand Oaks, CA: Sage.

Long, J. S., and J. Freese. 2001. Predicted probabilities for count models. *Stata Journal* 1: 51–57.

———. 2006. *Regression Models for Categorical Dependent Variables Using Stata*. 2nd ed. College Station, TX: Stata Press.

Mullahy, J. 1986. Specification and testing of some modified count data models. *Journal of Econometrics* 33: 341–365.

Vuong, Q. H. 1989. Likelihood ratio tests for model selection and non-nested hypotheses. *Econometrica* 57: 307–333.

Also see

[R] **zip postestimation** — Postestimation tools for zip

[R] **zinb** — Zero-inflated negative binomial regression

[R] **nbreg** — Negative binomial regression

[R] **poisson** — Poisson regression

[R] **tnbreg** — Truncated negative binomial regression

[R] **tpoisson** — Truncated Poisson regression

[SVY] **svy estimation** — Estimation commands for survey data

[XT] **xtpoisson** — Fixed-effects, random-effects, and population-averaged Poisson models

[U] **20 Estimation and postestimation commands**

Title

> **zip postestimation** — Postestimation tools for zip

Description

The following postestimation commands are available after `zip`:

Command	Description
contrast	contrasts and ANOVA-style joint tests of estimates
estat	AIC, BIC, VCE, and estimation sample summary
estat (svy)	postestimation statistics for survey data
estimates	cataloging estimation results
lincom	point estimates, standard errors, testing, and inference for linear combinations of coefficients
lrtest[1]	likelihood-ratio test
margins	marginal means, predictive margins, marginal effects, and average marginal effects
marginsplot	graph the results from margins (profile plots, interaction plots, etc.)
nlcom	point estimates, standard errors, testing, and inference for nonlinear combinations of coefficients
predict	predictions, residuals, influence statistics, and other diagnostic measures
predictnl	point estimates, standard errors, testing, and inference for generalized predictions
pwcompare	pairwise comparisons of estimates
suest	seemingly unrelated estimation
test	Wald tests of simple and composite linear hypotheses
testnl	Wald tests of nonlinear hypotheses

[1] `lrtest` is not appropriate with svy estimation results.

See the corresponding entries in the *Base Reference Manual* for details, but see [SVY] **estat** for details about `estat` (svy).

Syntax for predict

> predict [*type*] *newvar* [*if*] [*in*] [, *statistic* <u>nooff</u>set]

> predict [*type*] { *stub** | *newvar*$_{\text{reg}}$ *newvar*$_{\text{inflate}}$ } [*if*] [*in*] , <u>sc</u>ores

statistic	Description
Main	
n	number of events; the default
ir	incidence rate
<u>pr</u>	probability of a degenerate zero
pr(*n*)	probability $\Pr(y_j = n)$
pr(*a*,*b*)	probability $\Pr(a \le y_j \le b)$
xb	linear prediction
stdp	standard error of the linear prediction

These statistics are available both in and out of sample; type `predict ... if e(sample) ...` if wanted only for the estimation sample.

Menu

Statistics > Postestimation > Predictions, residuals, etc.

Options for predict

⌐ Main ⌐

n, the default, calculates the predicted number of events, which is $(1 - p_j) \exp(\mathbf{x}_j\boldsymbol{\beta})$ if neither offset() nor exposure() was specified when the model was fit, where p_j is the predicted probability of a zero outcome; $(1 - p_j) \exp\{(\mathbf{x}_j\boldsymbol{\beta}) + \text{offset}_j\}$ if offset() was specified; or $(1 - p_j)\{\exp(\mathbf{x}_j\boldsymbol{\beta}) \times \text{exposure}_j\}$ if exposure() was specified.

ir calculates the incidence rate $\exp(\mathbf{x}_j\boldsymbol{\beta})$, which is the predicted number of events when exposure is 1. This is equivalent to specifying both the n and the nooffset options.

pr calculates the probability $\Pr(y_j = 0)$, where this zero was obtained from the degenerate distribution $F(\mathbf{z}_j\boldsymbol{\gamma})$. If offset() was specified within the inflate() option, then $F(\mathbf{z}_j\boldsymbol{\gamma} + \text{offset}_j^\gamma)$ is calculated.

pr(n) calculates the probability $\Pr(y_j = n)$, where n is a nonnegative integer that may be specified as a number or a variable. Note that pr is not equivalent to pr(0).

pr(a,b) calculates the probability $\Pr(a \le y_j \le b)$, where a and b are nonnegative integers that may be specified as numbers or variables;

b missing ($b \ge .$) means $+\infty$;
pr(20,.) calculates $\Pr(y_j \ge 20)$;
pr(20,b) calculates $\Pr(y_j \ge 20)$ in observations for which $b \ge .$ and calculates $\Pr(20 \le y_j \le b)$ elsewhere.

pr(.,b) produces a syntax error. A missing value in an observation of the variable a causes a missing value in that observation for pr(a,b).

xb calculates the linear prediction, which is $\mathbf{x}_j\boldsymbol{\beta}$ if neither offset() nor exposure() was specified; $\mathbf{x}_j\boldsymbol{\beta} + \text{offset}_j$ if offset() was specified; or $\mathbf{x}_j\boldsymbol{\beta} + \ln(\text{exposure}_j)$ if exposure() was specified; see nooffset below.

stdp calculates the standard error of the linear prediction.

nooffset is relevant only if you specified offset() or exposure() when you fit the model. It modifies the calculations made by predict so that they ignore the offset or exposure variable; the linear prediction is treated as $\mathbf{x}_j\boldsymbol{\beta}$ rather than as $\mathbf{x}_j\boldsymbol{\beta} + \text{offset}_j$ or $\mathbf{x}_j\boldsymbol{\beta} + \ln(\text{exposure}_j)$. Specifying predict ..., nooffset is equivalent to specifying predict ..., ir.

scores calculates equation-level score variables.

The first new variable will contain $\partial \ln L / \partial(\mathbf{x}_j\boldsymbol{\beta})$.

The second new variable will contain $\partial \ln L / \partial(\mathbf{z}_j\boldsymbol{\gamma})$.

Methods and formulas

All postestimation commands listed above are implemented as ado-files.

The probabilities calculated using the **pr**(*n*) option are the probability $\Pr(y_i = n)$. These are calculated using

$$\Pr(0|\mathbf{x}_i) = \omega_i + (1 - \omega_i) \exp(-\lambda_i)$$

$$\Pr(n|\mathbf{x}_i) = (1 - \omega_i) \frac{\lambda_i^n \exp(-\lambda_i)}{n!} \qquad \text{for } n = 1, 2, \ldots$$

where ω_i is the probability of obtaining an observation from the degenerate distribution whose mass is concentrated at zero. ω_i can be obtained from the **pr** option.

See Cameron and Trivedi (1998, sec. 4.7) for further details.

Reference

Cameron, A. C., and P. K. Trivedi. 1998. *Regression Analysis of Count Data.* Cambridge: Cambridge University Press.

Also see

[R] **zip** — Zero-inflated Poisson regression

[U] **20 Estimation and postestimation commands**

Author index

This is the author index for the 4-volume *Stata Base Reference Manual.*

Subject index

This is the subject index for the 4-volume *Base Reference Manual*. Readers may also want to consult the combined subject index (and the combined author index) in the *Quick Reference and Index*.

Semicolons set off the most important entries from the rest. Sometimes no entry will be set off with semicolons, meaning that all entries are equally important.

M

N

V

varabbrev, set subcommand, [R] **set**
variables,
    categorical, *see* categorical data
    dummy, *see* indicator variables
    factor, *see* factor variables
    in model, maximum number, [R] **matsize**
    orthogonalize, [R] **orthog**
variance
    estimators, [R] *vce_option*
    inflation factors, [R] **regress postestimation**
    stabilizing transformations, [R] **boxcox**
variance,
    analysis of, [R] **anova**, [R] **loneway**, [R] **oneway**
    displaying, [R] **summarize**, [R] **table**, [R] **tabstat**,
        [R] **tabulate, summarize()**; [R] **lv**
    Huber/White/sandwich estimator, *see* robust,
        Huber/White/sandwich estimator of variance
    nonconstant, *see* robust, Huber/White/sandwich
        estimator of variance
    testing equality of, [R] **sdtest**
variance-comparison test, [R] **sdtest**
variance–covariance matrix of estimators, [R] **correlate**,
    [R] **estat**
variance-weighted least squares, [R] **vwls**
varkeyboard, set subcommand, [R] **set**
vce, estat subcommand, [R] **estat**
vce() option, [R] *vce_option*
version of ado-file, [R] **which**
version of Stata, [R] **about**
view
    ado command, [R] **view**
    ado_d command, [R] **view**
    browse command, [R] **view**
    command, [R] **view**
    help command, [R] **view**
    help_d command, [R] **view**
    net command, [R] **view**
    net_d command, [R] **view**
    news command, [R] **view**
    search command, [R] **view**
    search_d command, [R] **view**
    update command, [R] **view**
    update_d command, [R] **view**
    view_d command, [R] **view**
view_d, view subcommand, [R] **view**
viewing previously typed lines, [R] **#review**
vif, estat subcommand, [R] **regress postestimation**
vwls command, [R] **vwls**, [R] **vwls postestimation**

W

Wald tests, [R] **contrast**, [R] **predictnl**, [R] **test**,
    [R] **testnl**
weak instrument test, [R] **ivregress postestimation**

weighted least squares, [R] **regress**
    for grouped data, [R] **glogit**
    generalized linear models, [R] **glm**
    generalized method of moments estimation,
        [R] **gmm**
    instrumental-variables regression, [R] **gmm**,
        [R] **ivregress**
    nonlinear least-squares estimation, [R] **nl**
    nonlinear systems of equations, [R] **nlsur**
    variance, [R] **vwls**
Welsch distance, [R] **regress postestimation**
which command, [R] **which**
White/Huber/sandwich estimator of variance, *see* robust,
    Huber/White/sandwich estimator of variance
White's test for heteroskedasticity, [R] **regress**
    postestimation
Wilcoxon
    rank-sum test, [R] **ranksum**
    signed-ranks test, [R] **signrank**

X

xchart command, [R] **qc**
xi prefix command, [R] **xi**

Z

Zellner's seemingly unrelated regression, [R] **sureg**;
    [R] **reg3**, [R] **suest**
zero-altered, *see* zero-inflated
zero-inflated
    negative binomial regression, [R] **zinb**
    Poisson regression, [R] **zip**
zero-skewness transform, [R] **lnskew0**
zinb command, [R] **zinb**, [R] **zinb postestimation**
zip command, [R] **zip**, [R] **zip postestimation**